Techniques and Concepts
of High-Energy Physics VII

NATO ASI Series

Advanced Science Institutes Series

A series presenting the results of activities sponsored by the NATO Science Committee, which aims at the dissemination of advanced scientific and technological knowledge, with a view to strengthening links between scientific communities.

The series is published by an international board of publishers in conjunction with the NATO Scientific Affairs Division

A	Life Sciences	Plenum Publishing Corporation
B	Physics	New York and London
C	Mathematical and Physical Sciences	Kluwer Academic Publishers
D	Behavioral and Social Sciences	Dordrecht, Boston, and London
E	Applied Sciences	
F	Computer and Systems Sciences	Springer-Verlag
G	Ecological Sciences	Berlin, Heidelberg, New York, London,
H	Cell Biology	Paris, Tokyo, Hong Kong, and Barcelona
I	Global Environmental Change	

Recent Volumes in this Series

Series B: Physics

Techniques and Concepts of High-Energy Physics VII

Edited by

Thomas Ferbel

University of Rochester
Rochester, New York

Springer Science+Business Media, LLC

Proceedings of a NATO Advanced Study Institute on
Techniques and Concepts of High-Energy Physics,
held July 15–26, 1992,
at St. Croix, Virgin Islands

NATO-PCO-DATA BASE

The electronic index to the NATO ASI Series provides full bibliographical references (with keywords and/or abstracts) to more than 30,000 contributions from international scientists published in all sections of the NATO ASI Series. Access to the NATO-PCO-DATA BASE is possible in two ways:

—via online FILE 128 (NATO-PCO-DATA BASE) hosted by ESRIN, Via Galileo Galilei, I-00044 Frascati, Italy

—via CD-ROM "NATO Science and Technology Disk" with user-friendly retrieval software in English, French, and German (©WTV GmbH and DATAWARE Technologies, Inc. 1989). The CD-ROM also contains the AGARD Aerospace Database.

The CD-ROM can be ordered through any member of the Board of Publishers or through NATO-PCO, Overijse, Belgium.

Additional material to this book can be downloaded from http://extra.springer.com.

Library of Congress Cataloging-in-Publication Data

Techniques and concepts of high-energy physics VII / edited by Thomas
   Ferbel.
       p.   cm. -- (NATO ASI series. Series B, Physics ; v. 322)
     "Proceedings of a NATO Advanced Study Institute on Techniques and
   Concepts of High-Energy Physics, held July 15-26, 1992, at St.
   Croix, Virgin Islands"--T.p. verso.
     Includes bibliographical references and index.
     ISBN 978-1-4613-6026-1   ISBN 978-1-4615-2419-9 (eBook)
     DOI 10.1007/978-1-4615-2419-9
     1. Particles (Nuclear physics)--Technique--Congresses.
   I. Ferbel, Thomas. II. North Atlantic Treaty Organization.
   Scientific Affairs Division. III. NATO Advanced Study Institute on
   Techniques and Concepts of High-Energy Physics (7th : 1992 : St.
   Croix, V.I.) IV. Series.
   QC793.46.T43   1994
   539.7'2--dc20   94-4101
                                                                CIP

ISBN 978-1-4613-6026-1

© 1994 Springer Science+Business Media New York
Originally published by Plenum Press, New York in 1994
Softcover reprint of the hardcover 1st edition 1994

LECTURERS

M. Danilov	ITEP, Moscow, Russia
F. Dydak	CERN, Geneva, Switzerland
D. Edwards	Fermilab, Batavia, Illinois
C. Jarlskog	Stockholm University, Stockholm, Sweden
S. Rudaz	University of Minnesota, Minneapolis, Minnesota
D. Saxon	Glasgow University, Glasgow, United Kingdom
J. Siegrist	SSC Lab, Dallas, Texas
M. Spiro	CEN Saclay, Gif-sur-Yvette, France
P. Tipton	University of Rochester, Rochester, New York

ADVISORY COMMITTEE

B. Barish	Caltech, Pasadena, California
L. DiLella	CERN, Geneva, Switzerland
C. Fabjan	CERN, Geneva, Switzerland
J. Iliopoulos	Ecole Normale Superieure, Paris, France
M. Jacob	CERN, Geneva, Switzerland
C. Quigg	Fermilab, Batavia, Illinois
A. Sessler	University California at Berkeley, California
P. Soding	DESY, Hamburg, Federal Republic of Germany

DIRECTOR

T. Ferbel	University of Rochester, Rochester, New York

PREFACE

The seventh Advanced Study Institute (ASI) on Techniques and Concepts of High Energy Physics was held for the second time at the Club St. Croix, in St. Croix, U.S. Virgin Islands. The ASI brought together a total of 75 participants, from 19 countries. The primary support for the meeting was again provided by the Scientific Affairs Division of NATO. The ASI was cosponsored by the U.S. Department of Energy, by Fermilab, by the National Science Foundation, and by the University of Rochester. A special contribution from the Oliver S. and Jennie R. Donaldson Charitable Trust provided an important degree of flexibility, as well as support for worthy students from developing countries.

As in the case of the previous ASIs, the scientific program was designed for advanced graduate students and recent PhD recipients in experimental particle physics. The present volume of lectures should complement the material published in the first six ASIs, and prove to be of value to a wider audience of physicists.

It is a pleasure to acknowledge the encouragement and support that I have continued to receive from colleagues and friends in organizing this meeting. I am indebted to the members of my Advisory Committee for their infinite patience and excellent advice. I am grateful to the distinguished lecturers for their enthusiastic participation in the ASI, and, of course for their hard work in preparing the lectures and providing the superb manuscripts for the proceedings. I thank Dane Skow for organizing the student presentations, and Betsey and Bill Gladselter for their fascinating description of the geology and marine life of St. Croix (and Ann and Albert Lang for pointing me to them).

I thank P.K. Williams for support from the Department of Energy and David Berley for assistance from the National Science Foundation. I am grateful to John Peoples for providing the talents of Angela Gonzales for designing the poster for the School. At Rochester, I am indebted to Ovide Corriveau, Judy Mack and especially to Connie Jones for exceptional organizational assistance and typing.

I owe thanks to Priscilla and Jack Dodds of the Club St. Croix for their and their staff's hospitality. I wish to acknowledge the generosity of Chris Lirakis and Mrs. Marjorie Atwood of the Donaldson Trust, and support from George Blanar and Bert Yost of the LeCroy Research Systems Corporation. Finally, I thank Luis da Cunha of NATO for his cooperation and confidence.

T. Ferbel

Rochester, New York

CONTENTS

SCALAR FIELDS IN PARTICLE PHYSICS AND COSMOLOGY: SELECTED INTRODUCTORY TOPICS

Serge Rudaz

School of Physics and Astronomy
University of Minnesota
Minneapolis, MN 55455

These lectures were presented to an audience of students of experimental high energy physics, and are informal in tone and introductory in nature. It was felt that the audience had been familiarized with the Standard Model of Particle Physics, at least in outline, through countless previous lectures: in fact, all aspects of the Standard Model have been admirably covered in the six previous NATO ASI's of this series, the proceedings of which provide an invaluable source of information.

On the other hand, it is likely that a discussion of basic principles will be very useful in providing a stronger background for understanding some of the theoretical issues surrounding the Standard Model. This is what motivated the choice of topics covered in these lectures. The first four lectures lead the student to an understanding of the quantum corrections to scalar field theories, up to and including a calculation of the renormalized effective potential.

The reader will find no references in the main body of the text, in keeping with the informal nature of these notes. Indeed, while the presentation may be at times slightly unconventional, the material of the first four lectures is ably covered in any modern field theory textbook. These lectures are meant to serve as an introduction to further reading: A good review of radiative corrections to scalar potentials in the standard model will be found in an article by M. Sher (Physics Reports, volume 179, numbers 5 and 6, August 1989).

The fifth lecture, on scalar fields, cosmology and inflation makes use of some of the material developed earlier. The student wishing to further acquaint him or herself with these topics is referred to the review of inflation by K. Olive which appeared in Physics Reports (volume 190, number 6, June 1990), as well as to a recent survey of issues surrounding inflationary cosmology, published as part of the NATO ASI series: Observational Tests of Cosmological Inflation, edited by T. Shanks et al., Kluwer Academic Publishers (1991).

I. RELATIVISTIC INVARIANCE: NOTATIONS AND CONVENTIONS

Modern physics is founded on the assumption of invariance under Lorentz transformations. This section will serve to establish notations and conventions. The speed of light is always set to unity: $c = 1$.

The space-time coordinates are written as the components of <u>contravariant</u> (four-) vectors x^μ, y^μ, ... with an upper greek index μ, ν, ..., running from 0 to 3; $x^\mu = (x^0, x^1, x^2, x^3) \equiv (t, x, y, z)$. Latin indices i, j, ... will label space (-like) directions and run from 1 to 3. A Lorentz transformation from one reference frame S with coordinates x^μ to another frame S', x'^μ is a linear, invertible transformation that leaves the space-time interval between two events invariant, i.e.

$$
\begin{aligned}
(x' - y')^2 = (x - y)^2 &= x^2 + y^2 - 2x \cdot y \\
&\equiv (x^0 - y^0)^2 - (\vec{x} - \vec{y})^2
\end{aligned}
\tag{1}
$$

where the scalar product of two vectors is defined to be

$$
x \cdot y = x^0 y^0 - \vec{x} \cdot \vec{y}
\tag{2}
$$

The scalar product itself is invariant under <u>homogeneous</u> Lorentz transformations, i.e.

$$
x' \cdot y' = x \cdot y
\tag{3}
$$

where

$$
\begin{aligned}
x'^\mu &= \Lambda^\mu_{\ \nu} x^\nu \\
y'^\mu &= \Lambda^\mu_{\ \sigma} x^\sigma
\end{aligned}
\tag{4}
$$

Here $\Lambda^\mu_{\ \nu}$ are a set of constants, and we have introduced the summation convention: repeated superscripts and subscripts are to be summed over. Thus,

$$
x'^\mu = \Lambda^\mu_{\ \nu} x^\nu = \sum_{\nu=0}^{3} \Lambda^\mu_{\ \nu} x^\nu
$$

The finite space-time interval (1) is invariant under the more general inhomogeneous Lorentz transformation

$$
x'^\mu = \Lambda^\mu_{\ \nu} x^\nu + a^\mu
\tag{5}
$$

which also includes a rigid translation of coordinates by a constant four-vector a^μ.

More generally, any four-component object which transforms according to (4) is known as a contravariant vector: for example, consider the four-velocity u^μ, defined to be

$$u^\mu = \frac{dx^\mu}{ds} \tag{6}$$

where

$$ds^2 = dx \cdot dx = dt^2 - d\vec{x} \cdot d\vec{x} \tag{7}$$

is the infinitesimal invariant interval. From (6) and (7), one has

$$u^0 = \frac{dt}{ds} = (1 - \vec{v} \cdot \vec{v})^{-1/2} \equiv \gamma$$
$$\vec{u} = \vec{v}(1 - \vec{v} \cdot \vec{v})^{-1/2} \equiv \gamma\vec{v} \tag{8}$$

so obviously $u^2 = 1$.

To every contravariant vector A^μ one associates a corresponding <u>covariant</u> vector A_μ (with <u>lower</u> indices) according to the rule

$$A_0 = A^0 \quad , \quad A_k = -A^k \tag{9}$$

i.e. $A^\mu = (A^0, \vec{A})$, $A_\mu = (A^0, -\vec{A})$. Then the scalar product can be written, using the summation convention,

$$A \cdot B = A_\mu B^\mu = A^\mu B_\mu$$
$$= A^0 B^0 - \vec{A} \cdot \vec{B} \tag{10}$$

One can introduce a useful matrix notation: let the components of x^μ be gathered in a column vector $x = (x^0, \vec{x})^T$, and those of x_μ into a column vector $\widetilde{x} = (x^0, -\vec{x})^T$. Then the scalar product $x \cdot y$ can be written

$$x \cdot y = x^T \widetilde{y} = \widetilde{y}^T x$$
$$= \widetilde{x}^T y = y^T \widetilde{x} \tag{11}$$

The transformation law (4) for contravariant vectors can be written by considering $\Lambda^\mu{}_\nu$, for fixed μ and ν, to be the $(\mu\nu)$ component of a matrix Λ, i.e. $\Lambda^\mu{}_\nu = (\Lambda)_{\mu\nu}$ whence (4) can be written

$$x' = \Lambda x \tag{12}$$

Similarly, introduce the transformation law

$$\widetilde{x}' = \widetilde{\Lambda}\, \widetilde{x} \tag{13}$$

for covariant vectors. The invariance of the scalar product then implies

$$\widetilde{\Lambda}^T = \Lambda^{-1} \tag{14}$$

3

Thus, in components

$$(\widetilde{\Lambda})_{\mu\nu} = (\Lambda^{-1})_{\nu\mu} \tag{15}$$

and since (because the transformation is linear)

$$(\Lambda)_{\mu\nu} = \Lambda^{\mu}{}_{\nu} \equiv \frac{\partial x'^{\mu}}{\partial x^{\nu}} \tag{16}$$

it follows that

$$(\Lambda^{-1})_{\nu\mu} = \frac{\partial x^{\nu}}{\partial x'^{\mu}} \tag{17}$$

and so one can compare the transformation laws:

$$x'^{\mu} = \frac{\partial x'^{\mu}}{\partial x^{\nu}} x^{\nu} \equiv \Lambda^{\mu}{}_{\nu} x^{\nu} \tag{18a}$$

$$x'_{\mu} = \frac{\partial x^{\nu}}{\partial x'^{\mu}} x_{\nu} \equiv \Lambda_{\mu}{}^{\nu} x_{\nu} \tag{18b}$$

where we have introduced the notation $(\widetilde{\Lambda})_{\mu\nu} \equiv \Lambda_{\mu}{}^{\nu}$ to be justified soon. Note that the four-gradient, the derivative with respect to a contravariant vector, actually transforms like a covariant vector, as can be seen quickly:

$$\frac{\partial}{\partial x'^{\mu}} \equiv \frac{\partial x^{\nu}}{\partial x'^{\mu}} \frac{\partial}{\partial x^{\nu}} \tag{19}$$

from basic rules of calculus and comparison with (18b). Thus, one uses the notation

$$\partial_{\mu} \equiv \frac{\partial}{\partial x^{\mu}} = \left(\frac{\partial}{\partial t} , \vec{\nabla} \right) \tag{20}$$

to remind us of this transformation property. Similarly, $\partial^{\mu} = \partial/\partial x_{\mu} = (\partial/\partial t, -\vec{\nabla})$. The d' Alembertian operator

$$\Box \equiv \partial_{\mu}\partial^{\mu} = \frac{\partial^2}{\partial t^2} - \vec{\nabla}^2 \tag{21}$$

is clearly Lorentz invariant.

One can now introduce tensors. Tensors of rank "n" are 4^n-component objects defined by means of their transformation properties under Lorentz transformations, which is that of a product of n four-vectors. For example, a contravariant tensor of rank-two $T^{\mu\nu}$ represents 16 quantities which transform one into another like the product of coordinate infinitesimals $dx^{\mu}dx^{\nu}$, i.e.

$$T'^{\mu\nu} = \Lambda^{\mu}{}_{\sigma}\Lambda^{\nu}{}_{\rho}T^{\sigma\rho} \tag{22}$$

One can also consider covariant tensors of rank two, with two lower indices, and mixed tensors with both upper and lower indices.

One can still use matrix notation to describe the transformation of rank two tensors, e.g.

$$T' = \Lambda \, T \, \Lambda^T \tag{23}$$

is the obvious counterpart of (22). Note that a rank-two tensor can always be written as a sum of symmetric and antisymmetric parts, e.g.

$$T^{\mu\nu} = \frac{1}{2} \left(T^{\mu\nu} + T^{\nu\mu} \right) + \frac{1}{2} \left(T^{\mu\nu} - T^{\nu\mu} \right)$$

$$\equiv S^{\mu\nu} + A^{\mu\nu}$$

with $S^{\mu\nu} = S^{\nu\mu}$ and $A^{\mu\nu} = -A^{\nu\mu}$. This property of symmetry or antisymmetry is clearly preserved by Lorentz transformations.

Tensor manipulations follow some very simple rules. First, one introduces the very important metric tensor, symmetric and of second rank,

$$\eta^{\mu\nu} = \eta_{\mu\nu} = \begin{pmatrix} 1 & 0 & 0 & 0 \\ 0 & -1 & 0 & 0 \\ 0 & 0 & -1 & 0 \\ 0 & 0 & 0 & -1 \end{pmatrix} \tag{24}$$

which evidently serves to raise and lower indices, e.g.

$$\eta_{\mu\nu} A^\nu = A_\mu \quad , \quad \eta^{\mu\nu} A_\nu = A^\mu \tag{25}$$

The scalar product $x \cdot y$ can now neatly be written as

$$x \cdot y = \eta_{\mu\nu} x^\mu y^\nu = \eta^{\mu\nu} x_\mu y_\nu \tag{26}$$

or, in matrix form $(\eta = \eta^T)$,

$$x \cdot y = x^T \eta y = y^T \eta x$$
$$= \tilde{x}^T \eta \tilde{y} = \tilde{y}^T \eta \tilde{x} \tag{27}$$

The invariance requirement $x' \cdot y' = x \cdot y$ is

$$\Lambda^T \eta \Lambda = \eta = \Lambda \, \eta \, \Lambda^T \tag{28}$$

or

$$\tilde{\Lambda}^T \, \eta \, \tilde{\Lambda} = \eta \tag{29}$$

Comparing (28) with (23) justifies calling $\eta^{\mu\nu}$ a tensor: furthermore, $\eta^{\mu\nu}$ is a Lorentz-invariant tensor. From (28) or (29) and (14) we have (recall $\eta = \eta^{-1}$)

$$\tilde{\Lambda} = \eta \, \Lambda \, \eta \tag{30}$$

or in component form

$$(\widetilde{\Lambda})_{\mu\nu} = (\eta)_{\mu\sigma}(\Lambda)_{\sigma\rho}(\eta)_{\rho\nu}$$
$$= \eta_{\mu\sigma}\Lambda^{\sigma}_{\ \rho}\eta^{\rho\nu} \qquad (31)$$

where the placement of the indices is unambiguous and follows from the definition of Λ and the summation convention. Now we can justify our earlier notation for the elements of $\widetilde{\Lambda}$, namely

$$(\widetilde{\Lambda})_{\mu\nu} = \frac{\partial x^{\nu}}{\partial x'^{\mu}} \equiv \Lambda_{\mu}^{\ \nu}$$

by recalling the role of η in raising and lowering tensor indices: in (31), the upper index σ is lowered and becomes μ, and the lower index ρ is raised and becomes ν,

$$\eta_{\mu\sigma}\Lambda^{\sigma}_{\ \rho}\eta^{\rho\nu} \equiv \Lambda_{\mu}^{\ \nu} \qquad (32)$$

If, in any tensor, a pair of upper and lower indices are set equal and summed over, the resulting object is a tensor of rank reduced by two: this process is known as index contraction. For example, starting with $T^{\mu\nu}$, one can lower an index, i.e.

$$T^{\mu}_{\ \sigma} = T^{\mu\nu}\eta_{\nu\sigma} \qquad (33)$$

and one can further set $\mu = \sigma$ to write

$$T^{\mu}_{\ \mu} = T^{\mu\nu}\eta_{\nu\mu} = T^{\mu\nu}\eta_{\mu\nu}$$
$$= T^{0}_{\ 0} + \sum_{j=1}^{3} T^{j}_{\ j} \qquad (34)$$
$$= T^{00} - T^{11} - T^{22} - T^{33}$$

The object in Eq. (34) is known as the <u>trace of</u> $T^{\mu\nu}$, and transforms like a tensor of rank-zero, namely, a Lorentz <u>scalar</u>. In matrix notation,

$$T^{\mu}_{\ \mu} \equiv \text{Tr}\,(T\eta) = \text{Tr}\,(T\,\Lambda^{T}\,\eta\,\Lambda)$$
$$= \text{Tr}\,(T'\eta) = \text{Tr}(T'\,\eta') \qquad (35)$$

where we have used the definition (23), as well as (28). We have already used the fact that $\eta = \eta^{-1}$ earlier: thus,

$$(\eta)_{\mu\lambda}(\eta)_{\lambda\nu} = (1)_{\mu\nu}$$

which can be written, according to the summation convention, as

$$(1)_{\mu\nu} = \eta_{\mu\lambda}\eta^{\lambda\nu} = \eta^{\mu\lambda}\eta_{\lambda\nu} = \eta_{\mu}^{\ \nu} = \eta^{\mu}_{\ \nu}$$

so the "mixed" metric tensor is actually a representation of the unit matrix.

We now establish some properties of Lorentz transformations that will be needed subsequently. From (28), we note that taking the determinant of both sides, one has

$$(\det \Lambda)^2 = 1 \tag{36}$$

i.e.

$$\det \Lambda = \pm 1 \tag{37}$$

Further, writing the same equation in component form, one has

$$(\eta)_{\mu\nu} = (\Lambda^T)_{\mu\rho}(\eta)_{\rho\sigma}(\Lambda)_{\sigma\nu}$$

i.e.

$$\eta_{\mu\nu} = \Lambda^\rho_{\ \mu}\eta_{\rho\sigma}\Lambda^\sigma_{\ \nu} \tag{38}$$

so, setting $\mu = \nu = 0$ this becomes

$$1 = (\Lambda^0_{\ 0})^2 - \sum_{k=1}^{3}(\Lambda^k_{\ 0})^2 \tag{39}$$

i.e.

$$\left(\Lambda^0_{\ 0}\right)^2 \geq 1 \tag{40}$$

so either $\Lambda^0_{\ 0} \geq 1$ or $\Lambda^0_{\ 0} \leq -1$. Equations (36) and (40) divide Lorentz transformations into four disjoint sectors, according to the sign of the determinant of Λ and of its "00" component. Only those transformations with $\det \Lambda = +1$ and $\Lambda^0_{\ 0} \geq 1$ can be generated by infinitesimal, continuous steps starting from the identity: these are the proper, orthochronous Lorentz transformations, and one requires invariance only under these particular transformations. The other kinds represent discrete space-time transformations such as parity, time-reversal and the combination of the two, which are known not to be invariances of the weak interactions.

The form of the quantities $\Lambda^\mu_{\ 0}$ is easily determined. Recall that the four-velocity vector u^μ for a particle moving with velocity $\vec{v}$ is

$$u^\mu = (\gamma, \gamma\vec{v}) \tag{41}$$

where $\gamma = (1 - \vec{v} \cdot \vec{v})^{-1/2}$. This can be obtained by a Lorentz transformation from the restframe in which $u^\mu_{\text{rest}} = (1, \vec{0})$ through

$$u^\mu = \Lambda^\mu_{\ \nu} u^\nu_{\text{rest}} \tag{42}$$

so in fact, the components $\Lambda^\mu_{\ 0}$ corresponding to a Lorentz "boost" from a frame S to another frame S' with relative velocity $\vec{v}$ are then given by

$$\Lambda^0_{\ 0} = \gamma = u^0 \text{ and } \Lambda^k_{\ 0} = \gamma v^k = u^k$$

which of course satisfies Eq. (39). One can go back to Eq. (38), written in the form

$$\eta^{\mu\nu} = \eta^{\rho\sigma}\Lambda^{\mu}{}_{\rho}\Lambda^{\nu}{}_{\sigma} \equiv \Lambda^{\mu}{}_{0}\Lambda^{\nu}{}_{0} + \eta^{ij}\Lambda^{\mu}{}_{i}\Lambda^{\nu}{}_{j} \tag{43}$$

to conclude that

$$\begin{aligned} \eta^{ij}\Lambda^{\mu}{}_{i}\Lambda^{\nu}{}_{j} &= \eta^{\mu\nu} - \Lambda^{\mu}{}_{0}\Lambda^{\nu}{}_{0} \\ &= \eta^{\mu\nu} - u^{\mu}u^{\nu} \end{aligned} \tag{44}$$

One could go on to use these expressions to write down expressions for the remaining components $\Lambda^{\mu}{}_{k}$ of a boost: but now, these are not uniquely determined, because spatial rotations are a subgroup of Lorentz transformations (i.e. those with $\Lambda^{0}{}_{0} = 1$, $\Lambda^{k}{}_{0} = \Lambda^{0}{}_{k} = 0$ and $\Lambda^{i}{}_{j} = (R)_{ij}$ so that $R^{T} = R^{-1}$ by Eq. (28)), and the successive application of a pure rotation followed by a boost is also a boost.

We note that <u>six</u> parameters characterize the general homogeneous Lorentz transformation, as can be seen from Eq. (28) which is an equality of two symmetric matrices, i.e. provides ten constraints on the sixteen possible entries $\Lambda^{\mu}{}_{\nu}$. These six parameters correspond to $\vec{v}$, the boost velocity, and three angles specifying spatial rotations. The general inhomogeneous Lorentz transformation, including rigid space-time translations, i.e.

$$x' = \Lambda x + a \tag{45}$$

is then a 10-parameter group known as the <u>Poincaré group</u>: denoting the element of this group corresponding to the transformation (45) by (Λ, a), we have the group multiplication law:

$$(\Lambda_{1}, a_{1}) \cdot (\Lambda_{2}, a_{2}) = (\Lambda_{1}\Lambda_{2}, \Lambda_{1}a_{2} + a_{1}) \tag{46}$$

so while two translations $(1, a_{1})$ and $(1, a_{2})$ commute, translations do not commute with homogeneous Lorentz transformations, i.e. boosts and rotations. The Poincaré group is thus not just the direct product of the Lorentz and translation groups, but has a more complicated structure.

To conclude this section, we want to emphasize that the whole point behind the definition of tensors and all the accompanying cumbersome apparatus is that equations relating tensor components retain the same form upon transformation to a different Lorentz frame. This is known as relativistic covariance, i.e. the statement that physical laws take the <u>same</u> form to different observers in frames related by Lorentz transformations. For example, an equation relating different tensor components such as, say,

$$A^{\mu}B_{\mu}C^{\nu} = D^{\sigma}U_{\sigma\rho}T^{\rho\nu}$$

becomes upon transformation to another frame,

$$A'^{\mu}B'_{\mu}C'^{\nu} = D'^{\sigma}U'_{\sigma\rho}T'^{\rho\nu}$$

Verify this! Note that a covariant equation is easy to spot, as the number and placement (up, or down) of uncontracted greek indices necessarily matches on both sides of the equation.

II. CLASSICAL FIELDS: EQUATIONS OF MOTION, LAGRANGIANS AND CONSERVATION LAWS

A classical scalar field $\phi(x)$ is a <u>real</u> function of space-time coordinates with the property that under Lorentz transformations,

$$\phi'(x') = \phi(x) \tag{1}$$

Here $\phi'(x')$ refers to field in a primed frame, at the point $x' = \Lambda x + a$ referred to that frame. Note that x' and x refer to the same space-time point. The simplest Lorentz covariant equation of motion for $\phi(x)$ is the Klein-Gordon equation (somewhat improperly named in this context),

$$\left(\partial_\mu \partial^\mu + m^2\right) \phi(x) = 0 \tag{2}$$

where m^2 is a real constant, and one must require $m^2 \geq 0$ as discussed below. To find the most general solution of this equation, abandoning the reality constraint on $\phi(x)$ for now, write the Fourier transform

$$\phi(x) = \int \frac{d^4 k}{(2\pi)^4}\, e^{-ik \cdot x} \phi(k) \tag{3}$$

and apply Eq. (2) to get

$$\int \frac{d^4 k}{(2\pi)^4}\, \left(-k^2 + m^2\right) \phi(k) e^{-ik \cdot x} = 0 \tag{4}$$

Here $k^2 = k^\mu k_\mu$, and k^μ is a wave-number four-vector, such that $k \cdot x$ and the volume element $d^4 k$ are invariant under Lorentz transformations. The solution to (4) requires that $(k^2 - m^2)\phi(k) = 0$ for all k, so necessarily one has ($f(k)$ is some function of k),

$$\phi(k) = f(k)\delta(k^2 - m^2) \tag{5}$$

To get the most general Lorentz-invariant solution of the Klein-Gordon equation, it is necessary that $f(k)$ be a function only of Lorentz-invariant arguments: here, these can be k^2, and since $k^2 > 0$, one also has the <u>sign</u> of the time-component k^0 that is left invariant by Lorentz-transformations. Thus one can write, in this case

$$f(k) = 2\pi(a(k^2) + \epsilon(k^0)b(k^2)) \tag{6}$$

where $\epsilon(k^0)$ is the so-called sign function: $\epsilon(k^0) = +1$ for $k^0 > 0$ and $\epsilon(k^0) = -1$ for $k^0 < 0$. It can be expressed in a variety of ways, for example in terms of the step function $\theta(k^0)$, namely

$$\epsilon(k^0) = \theta(k^0) - \theta(-k^0) \tag{7}$$

Since $f(k)$ in (5) is multiplied by $\delta(k^2 - m^2)$, the arguments k^2 in (6) can in fact be replaced by m^2, and so a and b reduce to constants, $a = a(m^2)$, $b = b(m^2)$. Hence, the most general Lorentz-invariant solution to the Klein-Gordon equation is, up to a constant factor

$$\phi(x) = \int \frac{d^4 k}{(2\pi)^3} \, \delta(k^2 - m^2) e^{-ik\cdot x}(a + b\epsilon(k^0)) \tag{8}$$

Defining the frequency $\omega_k = (\vec{k}^2 + m^2)^{1/2}$, the k^0- integral can be done using the identity

$$\delta(k^2 - m^2) = \delta(k^{02} - \omega_k{}^2)$$

$$= \frac{1}{2\omega_k} \left[\delta(k^0 - \omega_k) + \delta(k^0 + \omega_k) \right]$$

with the result,

$$\phi(x) = \int \frac{d^3 k}{(2\pi)^3 2\omega_k} \left[(a + b)e^{-ik\cdot x} + (a - b)e^{ik\cdot x} \right] \tag{9}$$

where now $k^\mu = (\omega_k, \vec{k})$ and in the second term we have used the fact that the integral is invariant under a change $\vec{k} \to -\vec{k}$ of the sign of the integration variable. A real field $\phi(x)$ will now be more properly understood as a superposition of the real and imaginary parts of Eq. (9), namely, with A and B real,

$$\phi(x) = \int \frac{d^3 k}{(2\pi)^3 2\omega_k} \left[A \cos k \cdot x + B \sin k \cdot x \right] \tag{10}$$

subject to boundary conditions which must of course be specified. We will need the result (9) in the next section.

All of this has been rather dry, so let's quickly discuss the simplest concrete example of a classical solution to Eq. (2): suppose that in some frame one looks for a static, spherically symmetric solution $\phi(t, \vec{x}) \equiv \phi(r)$. Then the KG equation simplifies to

$$\frac{1}{r} \frac{d^2}{dr^2} \, r\phi(r) = m^2 \phi(r) \tag{11}$$

i.e.

$$\frac{d^2}{dr^2} \, (r\phi(r)) = m^2 (r\phi(r)) \tag{12}$$

from which one trivially finds the solution

$$\phi(r) = \frac{C}{r} \, e^{-mr} \tag{13}$$

that is bounded at infinity, but singular at the origin. This is of course the famous short-range Yukawa potential, interpreted as follows: (13) is the solution of (11) for all $r \neq 0$, excluding the origin where (11) itself cannot be valid. Rather, there must be a point source at $r = 0$, whose strength is related to the constant C, producing the spherically symmetric field (13) whose short-range is characterized by the length scale $1/m$. The interpretation of this length parameter in terms of some particle mass must await the consideration of quantum fields in the next section.

LAGRANGIANS AND VARIATIONAL PRINCIPLES

Following closely the approach to classical point particle mechanics, we postulate the existence of a function of fields and their first derivatives known as the Lagrangian density,

$$\mathcal{L} = \mathcal{L}(\phi(x), \partial_\mu \phi(x)) \tag{14}$$

and we define the action integral depending on an arbitrary volume Ω of space-time

$$S(\Omega) = \int_\Omega d^4x \; \mathcal{L}(\phi, \partial_\mu \phi) \tag{15}$$

The field equations are then postulated to follow from a variational principle, formulated as follows. Let $\phi_0(x)$ be a solution to the desired equation of motion, with corresponding value $S_0(\Omega)$ for the action, and consider changing $\phi_0(x)$ by some arbitrary small function $\delta\phi(x) = \epsilon\eta(x)$ over the space-time volume Ω, subject only to the condition that $\eta(x)$ vanishes at the boundary of Ω, and leading to a change away from S_0 by a corresponding amount $\delta S(\Omega)$. The variational principle is then that $S_0(\Omega)$ is such that $\delta S(\Omega) = 0$.

To have a concrete picture of space-time volumes and their boundaries, you can think of Ω as consisting of the space-time volume consisting of some spatial volume V, fixed, taken between some initial and final times t_i and t_f which define the boundary.

To see how the equations of motion emerge, let $\epsilon \ll 1$ and calculate $\delta S(\Omega)$ away from $S_0(\Omega)$:

$$
\begin{aligned}
\delta S(\Omega) &= \int_\Omega d^4x \left[\epsilon\eta(x) \frac{\partial \mathcal{L}}{\partial \phi} \bigg|_{\phi=\phi_0} + \epsilon\partial_\mu\eta(x) \frac{\partial \mathcal{L}}{\partial \partial_\mu \phi} \bigg|_{\phi=\phi_0} \right] \\
&= \int_\Omega d^4x \left[\epsilon\partial_\mu \left(\eta(x) \frac{\partial \mathcal{L}}{\partial \partial_\mu \phi} \bigg|_{\phi=\phi_0} \right) + \epsilon\eta(x) \frac{\partial \mathcal{L}}{\partial \phi} \bigg|_{\phi=\phi_0} \right. \\
&\quad \left. - \epsilon\eta(x)\partial_\mu \frac{\partial \mathcal{L}}{\partial \partial_\mu \phi} \bigg|_{\phi=\phi_0} \right]
\end{aligned}
\tag{16}
$$

First note that the first term in the second equation (16) will vanish by using the 4-dimensional analog of the divergence theorem, since $\eta(x)$ is vanishing on the boundary of Ω. Next requiring $\delta S(\Omega) = 0$ for otherwise arbitrary $\eta(x)$ leads to the Euler-Lagrange equations of motion

$$\frac{\partial \mathcal{L}}{\partial \phi} = \partial_\mu \frac{\partial \mathcal{L}}{\partial \partial_\mu \phi} \tag{17}$$

Note that if $\mathcal{L}$ is a Lorentz-scalar, then so will the action, and the equations of motion will be covariant: also, there is some arbitrariness in the choice of a Lagrangian, since the same field equations follow if an arbitrary four-divergence $\partial_\mu K^\mu$ is added, and, more trivially, if $\mathcal{L}$ is multiplied by some constant.

For example, the following Lagrangian for a real scalar field $\phi(x)$ will be seen to lead to the Klein-Gordon equation (2),

$$\mathcal{L}_{KG} = \frac{1}{2}\, \partial_\mu \phi(x) \partial^\mu \phi(x) - \frac{1}{2}\, m^2 \phi^2(x) \tag{18}$$

More generally, the following scalar field Lagrangian

$$\mathcal{L} = \frac{1}{2}\, \partial_\mu \phi \partial^\mu \phi - V(\phi) \tag{19}$$

where $V(\phi)$ is a function of $\phi(x)$ only known as the potential term, leads to the equtaion of motion

$$\begin{aligned} \partial_\mu \partial^\mu \phi(x) &= -\frac{dV(\phi)}{d\phi} \\ &\equiv -V'(\phi) \end{aligned} \tag{20}$$

The Lagrangian formalism allows for a systematic exploration of the consequences of the covariance of the equations of motion which follow from the invariance of $\mathcal{L}$ under space-time transformations discussed in Section I, as well as other types of continuous transformations (for example, "internal symmetries"): there is a general result known as Noether's theorem that states that there are as many <u>conservation laws</u> as there are parameters that characterize the continuous symmetries of the Lagrangian. A discussion of this very important result will be found in any good field theory textbook: we will not need the full machinery of Noether's theorem to explore the consequences of invariance under that subset of the Poincaré group consisting only of rigid translations, considered to be infinitesimal $x'^\mu = x^\mu + \epsilon^\mu$, $\epsilon \ll 1$, since any finite translation can be seen as the result of compounding many infinitesimal ones.

To have translational invariance, it is necessary that the Lagrangian density $\mathcal{L}(\phi, \partial_\mu \phi)$ not depend explicitly on space-time coordinates, but only implicitly, through $\phi(x)$ and $\partial_\mu \phi(x)$: thus, a Lorentz-invariant term like $x^2 \phi(x)$ would violate translation invariance. This is the crucial observation that allows the four-gradient $\partial^\mu \mathcal{L}$ of the Lagrangian density to be written as

$$\partial^\mu \mathcal{L} = \frac{\partial \mathcal{L}}{\partial \phi}\, \partial^\mu \phi + \frac{\partial \mathcal{L}}{\partial \partial_\nu \phi}\, \partial^\mu \partial_\nu \phi \tag{21}$$

with no term on the RHS of (21) corresponding to a derivative of $\mathcal{L}$ at fixed ϕ and $\partial_\mu \phi$. Using the equation of motion (17), the fact that partial derivatives can be taken in any order, and writing $\partial^\mu \mathcal{L} = \eta^{\mu\nu} \partial_\nu \mathcal{L}$, (21) can be rearranged to read

$$\partial_\nu \left(\partial^\mu \phi\, \frac{\partial \mathcal{L}}{\partial \partial_\nu \phi} - \eta^{\mu\nu} \mathcal{L} \right) = 0 \tag{22}$$

that is,

$$\partial_\nu T^{\mu\nu} = 0 \qquad (23)$$

which serves to define the canonical energy-momentum tensor $T^{\mu\nu}$,

$$T^{\mu\nu} = \partial^\mu \phi \, \frac{\partial \mathcal{L}}{\partial \partial_\nu \phi} - \eta^{\mu\nu} \mathcal{L} \qquad (24)$$

with the four conservation laws (23) holding by virtue of the equations of motion. For example, from the scalar field Lagrangian (19) one derives the expression,

$$T^{\mu\nu} = \partial^\mu \phi \partial^\nu \phi - \eta^{\mu\nu} \left[\frac{1}{2} \, \partial_\alpha \phi \partial^\alpha \phi - V(\phi) \right] \qquad (25)$$

a form that we will need later.

To motivate and explain the designation energy-momentum tensor, return to (23) and integrate over all space, i.e. obtain

$$\int d^3x \, \frac{\partial}{\partial t} \, T^{\mu 0}(x) = - \int d^3x \, \frac{\partial}{\partial x^k} \, T^{\mu k}(x) \qquad (26)$$

For fields vanishing at infinity, the right hand side of this equation vanishes by the divergence theorem, and one has simply

$$\frac{d}{dt} \int d^3x \, T^{\mu 0}(t, \vec{x}) = 0 \qquad (27)$$

Although it may not be immediately obvious, this can be written as a conservation law for the components of a contravariant four-vector, namely

$$\frac{d}{dt} \, P^\mu(t) = 0 \qquad (28)$$

where

$$P^\mu(t) = \int d^3x \, T^{\mu 0}(t, \vec{x}) \qquad (29)$$

Thus the quantities $P^\mu(t)$ are in fact time-independent, and are interpreted as the components of the conserved field energy-momentum four-vector. To see that (29) is indeed a four-vector, note that we can write

$$P^\mu(t) = P^\mu(0) = \int d^4x \, \delta(x^0) T^{\mu 0}(x) \qquad (30)$$

On the other hand, this expression can be understood as

$$P^\mu = \int d^4x \, \delta(n_\mu x^\mu) \, n_\nu T^{\mu\nu}(x) \qquad (31)$$

where n^μ is a time-like unit four-vector $n^2 = 1$, evaluated in a particular frame where $n^\mu = (1, 0, 0, 0)$. In (31), the space-time volume element as well as the argument of the delta function are Lorentz invariant, and $n_\nu T^{\mu\nu}$ transforms as a four-vector, and so does P^μ. Given all this, we see that $T^{00}(x)$ is the field energy density, or Hamiltonian density and T^{k0} is density of momentum in the direction $\hat{k}$.

To understand the significance of the other components of $T^{\mu\nu}$, return to (26) but now integrate over a small spatial volume. Then, setting $\mu = 0$ for example, one has an equation relating the change in energy in the small volume to a net energy-flow per unit time and unit area, corresponding to T^{0k} in the direction $\hat{k}$. Similarly, set $\mu = 1$ for example, and the change in momentum in the "1" direction is related to a sum of effects: a flow of momentum per unit area per unit time in the same "1" direction (i.e. a pressure, T^{11}) as well as a contribution from momentum flows coming from the "2" and "3" directions, i.e. shear stresses, T^{12} and T^{13}. Given this information, it is easy to write down the energy-momentum tensor of a perfect fluid, i.e. a gas of non-interacting particles with some distribution of speeds, in the rest-frame of the fluid. Such a perfect fluid is free of heat flows (so $T^{0k} = 0$) and of shear stresses (so $T^{ij} = 0$ for $i \neq j$), as well as of net momentum so $T^{k0} = 0$. The pressure p in the fluid, corresponding to the diagonal elements T^{ii} is isotropic, and there is an energy density (including masses) $T^{00} = \rho$: Thus,

$$T^{\mu\nu}_{\text{rest}} = \begin{pmatrix} \rho & 0 & 0 & 0 \\ 0 & p & 0 & 0 \\ 0 & 0 & p & 0 \\ 0 & 0 & 0 & p \end{pmatrix} \tag{32}$$

This can be transformed trivially to a more general form in which the fluid is not globally at rest but moves with four-velocity u^μ.

Recall from the work in Section I that the transformed tensor should be of the form

$$T^{\mu\nu} = \Lambda^\mu_{\sigma} \Lambda^\nu_{\rho} T^{\sigma\rho}_{\text{rest}} \tag{33}$$

With

$$T^{\mu\nu}_{\text{rest}} = T^{00}_{\text{rest}} - \eta^{ij} p \tag{34}$$

one can write (33) as

$$T^{\mu\nu} = \Lambda^\mu_{0} \Lambda^\nu_{0} \rho - p\eta^{ij} \Lambda^\mu_{i} \Lambda^\nu_{j} \tag{35}$$

Now recalling some results from the preceding section, we have $\Lambda^\mu_{0} = u^\mu$ for a transformation effecting a boost from rest to four-velocity u^μ (cf. Eqn. (I.42)), and using (I.44) we arrive at the standard form for the relativistic perfect fluid energy-momentum tensor

$$T^{\mu\nu} = (p + \rho)u^\mu u^\nu - p\eta^{\mu\nu} \tag{36}$$

where ρ and p are Lorentz invariant characteristics of rest-frame properties. We will need this result later as well.

Returning to Eq. (25) for a moment, in the special case $V(\phi) = \frac{1}{2} m^2 \phi^2$, we see that the energy density of a classical field satisfying the Klein-Gordon equation is given by

$$T_{\text{KG}}^{00} = \dot{\phi}^2 - \left[\frac{1}{2} \dot{\phi}^2 - \frac{1}{2} (\vec{\nabla}\phi)^2 - \frac{1}{2} m^2 \phi^2 \right]$$

i.e.

$$T_{\text{KG}}^{00} = \frac{1}{2} \dot{\phi}^2 + \frac{1}{2} (\vec{\nabla}\phi)^2 + \frac{1}{2} m^2 \phi^2 \tag{37}$$

This energy density must be positive definite for any field configuration, and it is this requirement that fixes $m^2 \geq 0$.

In the rest of this section, we shall outline the treatment of the classical vector field $A^\mu(x)$ in the Lagrangian framework. Although these lectures are devoted to topics in scalar fields, this is a useful exercise in the application of the concepts described up to this point.

DIGRESSION: THE CLASSICAL VECTOR FIELD

By a vector field $A^\mu(x)$ one means a set of four functions of space-time which transform like the components of a four-vector under Lorentz-Poincaré transformations, that is

$$A'^\mu(x') = \Lambda^\mu_{\ \nu} A^\nu(x) \tag{38}$$

where $x' = \Lambda x + a$. To obtain the analog of the Klein-Gordon equation for the vector field, we should write down the most general Lagrangian density function that is quadratic in A^μ and $\partial^\mu A^\nu$. Write this in the form

$$\mathcal{L} = -\frac{1}{2} (\partial_\mu \partial_\nu)(\partial^\mu A^\nu) + \frac{\xi}{2} (\partial_\mu A_\nu)(\partial^\nu A^\nu) + \frac{1}{2} \mu^2 A_\mu A^\mu \tag{39}$$

where ξ and μ^2 are unspecified real constants. One might object that another possible Lorentz invariant term of the form $(\partial_\mu A^\mu)(\partial_\nu A^\nu)$ is absent from the Lagrangian (39). However, one can easily show that this is actually equivalent to a term already present in (39), up to a total divergence which does not affect the equations of motion. Indeed,

$$
\begin{aligned}
(\partial_\mu A^\mu)(\partial_\nu A^\nu) &= \partial_\mu (A^\mu \partial_\nu A^\nu) - A^\mu \partial_\mu \partial_\nu A^\nu \\
&= \partial_\mu (A^\mu \partial_\nu A^\nu) - \partial_\nu (A^\mu \partial_\mu A^\nu) + (\partial_\nu A^\mu)(\partial_\mu A^\nu) \\
&\equiv (\partial_\nu A^\mu)(\partial_\mu A^\nu) + \partial_\mu (A^\mu \partial_\nu A^\nu - A^\nu \partial_\nu A^\mu)
\end{aligned}
$$

after a relabeling of dummy indices. One can get the equations of motion from (17), simply generalized to the form

$$\frac{\partial \mathcal{L}}{\partial A_\rho} = \partial_\sigma \frac{\partial \mathcal{L}}{\partial \partial_\sigma A_\rho} \tag{40}$$

a compact notation indicating that one has four equations, one for each component field of $A^\mu(x)$. Now one can use simple relations like

$$\frac{\partial A^\mu}{\partial A_\rho} = \eta^{\mu\rho} \quad , \quad \frac{\partial \partial_\mu A_\nu}{\partial \partial_\sigma A_\rho} = \eta^\sigma{}_\mu \eta^\rho{}_\nu$$

etc. (check these: be careful of the positioning of the indices! These relations simply express the independence of the different components of A^μ and $\partial^\mu A^\nu$, with due account taken of sign differences in contravariant and covariant components), to derive that it follows from the form (39) that

$$\begin{aligned}\frac{\partial \mathcal{L}}{\partial A_\rho} &= \mu^2 A^\rho \\ \frac{\partial \mathcal{L}}{\partial \partial_\sigma A_\rho} &= -\partial^\sigma A^\rho + \xi \partial^\rho A^\sigma\end{aligned} \tag{41}$$

whence one has the equation of motion

$$\Box A^\rho - \xi \partial^\rho \partial_\sigma A^\sigma + \mu^2 A^\rho = 0 \tag{42}$$

which differs only in the presence of the ξ-dependent term from a set of four Klein-Gordon equations, one for each component of A^μ. One can also simply calculate the energy-momentum tensor, using the obvious analog of formula (24),

$$T^{\mu\nu} = \partial^\mu A_\sigma \frac{\partial \mathcal{L}}{\partial \partial_\nu A_\sigma} - \eta^{\mu\nu} \mathcal{L} \tag{43}$$

i.e.

$$\begin{aligned}T^{\mu\nu} = &- \partial^\mu A_\sigma \partial^\nu A^\sigma + \xi \partial^\mu A_\sigma \partial^\sigma A^\nu \\ &+ \eta^{\mu\nu} \frac{1}{2} \left[\partial_\lambda A_\sigma \partial^\lambda A^\sigma - \xi \partial_\lambda A_\sigma \partial^\sigma A^\lambda - \mu^2 A_\lambda A^\lambda \right]\end{aligned} \tag{44}$$

To see that vector fields have some perplexing aspects, consider, for simplicity, only time-dependent solutions $A^\mu(t)$ of the equations of motion, which take the form

$$\ddot{A}^0(1 - \xi) + \mu^2 A^0 = 0 \tag{45}$$

$$\ddot{A}^k + \mu^2 A^k = 0 \tag{46}$$

while for such field configurations, the energy density takes the form

$$T^{00} = \frac{1}{2}(\xi - 1)(\dot{A}^0)^2 + \frac{1}{2}\dot{\vec{A}} \cdot \dot{\vec{A}} - \frac{\mu^2}{2}(A^0)^2 + \frac{\mu^2}{2}\vec{A} \cdot \vec{A} \tag{47}$$

which can be compared to the analogous result for the scalar field (with $\vec{\nabla}\phi = 0$)

$$T^{00}_{\mathrm{KG}} = \frac{1}{2}\dot{\phi}^2 + \frac{1}{2}m^2\phi^2 \tag{48}$$

16

For general ξ and μ^2, T^{00} is not always positive definite for the vector field. In fact, it will suffice to take $\mu^2 > 0$ so that any possible problem will be with terms involving A^0: the limiting case $\mu^2 \to 0$ will be considered separately. Thus, one can look at : 1) if $\xi < 1$, the equations of motion are of standard Klein-Gordon-like form for all components of A^μ, but the terms involving A^0 in the energy density are always negative definite, which is not acceptable. 2) if $\xi > 1$, the "kinetic" part of T^{00} involving $\dot{A}^0$ is now positive, but on the contrary the equation of motion satisfied by A^0 is of a different character entirely as compared to that for the spatial components $\vec{A}$, the one leading to exponential, the other to oscillating solutions. This is also not acceptable, and the simplest, most reasonable choice is to take $\xi = 1$. With this choice, the kinetic terms of A^0 vanish in T^{00}, and by (45) so do the terms in $(A^0)^2$: actually, in general one would look at fields with some spatial dependence, so A^0 does not always vanish by the equation of motion. However, for $\xi = 1$ A^0 has no time dependence and (45) is just an equation of constraint: A^0 does not go anywhere and is irrelevant.

Thus according to this argument (often made by J. S. Bell) one should choose $\xi = 1$ for the massive vector field, whence the Lagrangian will take the form

$$\mathcal{L} = -\frac{1}{2}\,\partial_\mu A_\nu \partial^\mu A^\nu + \frac{1}{2}\,\partial_\mu A_\nu \partial^\nu A^\mu + \frac{\mu^2}{2}\,A^\mu A_\mu \qquad (49)$$

With some index gymnastics, one can write

$$\partial_\mu A_\nu (\partial^\mu A^\nu - \partial^\nu A^\mu) \equiv \frac{1}{2}\,(\partial_\mu A_\nu - \partial_\nu A_\mu)(\partial^\mu A^\nu - \partial^\nu A^\mu) \qquad (50)$$

so that defining the antisymmetric field strength tensor

$$F^{\mu\nu} = \partial^\mu A^\nu - \partial^\nu A^\mu \qquad (51)$$

the Lagrangian density becomes

$$\mathcal{L} = -\frac{1}{4}\,F^{\mu\nu} F_{\mu\nu} + \frac{1}{2}\,\mu^2 A^\mu A_\mu \qquad (52)$$

In the limit $\mu^2 \to 0$, this Lagrangian is invariant under so-called local gauge transformations,

$$A^{\mu\prime}(x) = A^\mu(x) + \partial^\mu \alpha(x) \qquad (53)$$

where $\alpha(x)$ is an arbitrary scalar function. The equation of motion following from (52) is the so-called Proca equation

$$\partial_\mu F^{\mu\nu} + \mu^2 A^\nu = 0 \qquad (54)$$

and the equation $\partial_\nu A^\mu = 0$ follows as well. Note that if there are interactions, i.e. (54) becomes the inhomogeneous equation

$$\partial_\mu F^{\mu\nu} + \mu^2 A^\nu = g j^\nu \qquad (55)$$

where g is some coupling constant and j^ν a four-vector current, the $\mu^2 \to 0$ limit is not necessarily smooth. For example, one now has $\mu^2 \partial_\nu A^\nu = g \partial_\nu j^\nu$ (by differentiating (55), which can be put back into (55) so that one has

$$\Box A^\nu - \frac{g}{\mu^2}\, \partial^\nu \partial_\mu j^\mu + \mu^2 A^\nu = g j^\nu \tag{56}$$

or

$$(\Box + \mu^2) A^\nu = g \left(\eta^{\mu\nu} + \partial^\mu \partial^\nu / \mu^2 \right) j_\mu \tag{57}$$

A smooth $\mu^2 \to 0$ limit can only follow if the current j^μ is conserved, i.e. $\partial_\mu j^\mu = 0$: one can also see directly from (55) that for $\mu^2 = 0$, the same follows from the antisymmetry of $F^{\mu\nu}$. Finally, note that the way the Lorentz condition $\partial_\mu A^\mu = 0$ emerges in the massless case is not as a consequence of the equation of motion, but rather takes advantage of the gauge invariance of the equations, whereby solutions $A_\mu(x)$ and $A'_\mu(x)$ differing only by a total derivative are physically equivalent.

This concludes our short excursion into classical fields: we next turn to the quantum theory of the hermitian scalar field.

III. RELATIVISTIC QUANTUM THEORY OF THE FREE SCALAR FIELD

In remarks made at a 1965 UNESCO Colloquium marking the tenth anniversary of the passing of Einstein and Teilhard de Chardin, Werner Heisenberg gave the following characterization of quantum field theory:

"In quantum theory, the field distinguishes, as in classical physics, between something and nothing: but its essential function is to change the state of the world, which is characterized by a probability amplitude, by a statement concerning potentialities. In this way, experimental situations in elementary particle physics can be described by applying operators constructed from products of field operators on the ground state "world"."

In this lecture we shall try to clarify the content of these remarks. We assume that the scalar field $\phi(x)$ is now a hermitian <u>operator</u>, $\phi(x) = \phi^\dagger(x)$, which satisfies the Klein-Gordon equation

$$\left(\Box + m^2\right) \phi(x) = 0 \tag{1}$$

We need to specify the analog of Eq. (II.1) regarding the transformation properties of $\phi(x)$ under transformations of the Poincaré group. Consider the behavior of the matrix element $\langle \beta | \phi(x) | \alpha \rangle$ between two arbitrary states of the Hilbert space in which $\phi(x)$ is assumed to act: this matrix element is the probability amplitude referred to by Heisenberg. The states $|\alpha\rangle$ and $|\beta\rangle$ are defined with respect to some observer S: we assume that the corresponding states $|\alpha'\rangle$ and $|\beta'\rangle$, referring to a different observer S', for which $x' = \Lambda x + a$, are related to the original ones by means of a unitary transformation $U(\Lambda, a)$, i.e.

$$
\begin{aligned}
|\alpha'\rangle &= U(\Lambda, a)|\alpha\rangle \\
|\beta'\rangle &= U(\Lambda, a)|\beta\rangle
\end{aligned} \tag{2}
$$

The analog of the invariance statement $\phi'(x') = \phi(x)$ for a classical field will then be a statement about matrix elements: by definition, for a scalar field operator,

$$
\langle\beta'|\phi(x')|\alpha'\rangle = \langle\beta|\phi(x)|\alpha\rangle \tag{3}
$$

For a vector field operator, the quantum analog of the classical equation (II.36) will be

$$
\langle\beta'|A^\mu(x')|\alpha'\rangle = \Lambda^\mu{}_\nu \langle\beta|A^\nu(x)|\alpha\rangle \tag{4}
$$

Since (3) and (4) must hold for arbitrary states, one has the operator relation

$$
\phi(x') = U(\Lambda, a)\phi(x)U^{-1}(\Lambda, a) \tag{5}
$$

for the scalar field, and an analogous result for the vector field.

Consider space-time translations $U(1, a)$: they can be written as follows

$$
U(1, a) = \exp(iP^\mu a_\mu/\hbar) \tag{6}
$$

for some Hermitian four-vector operator P^μ. Thus,

$$
\phi(x + a) = e^{iP\cdot a/\hbar}\phi(x)e^{-iP\cdot a/\hbar} \tag{7}
$$

Put $x^\mu + a^\mu = 0$ in (7), and take the gradient ∂^μ on both sides: remembering that P^μ and ϕ do not in general commute, one gets

$$
[P^\mu, \phi(x)] = -i\hbar\partial^\mu\phi(x) \tag{8}
$$

where $[A, B] \equiv AB - BA$ is the commutator bracket. Eq. (8) is the covariant form of the Heisenberg equation motion familiar from quantum mechanics, provided one identifies $P^0 = H$, the quantum Hamiltonian operator. A good first guess for P^μ would be that it would, by appropriate substitution of operators for classical fields, be given by (II.29), given the Lagrangian (II.18): the leading candidate for $P^0 = H$ in the quantum theory would then be

$$
H = \int d^3x \left(\frac{1}{2}\,\dot\phi^2 + \frac{1}{2}\left(\vec\nabla\phi\right)^2 + \frac{1}{2}\,m^2\phi^2 \right) \tag{9}
$$

and correspondingly for $\vec P$. Note that the obvious requirement on translations, namely

$$
\begin{aligned}
U(1, a)U(1, b) &= U(1, b)U(1, a) \\
&= U(1, a + b)
\end{aligned} \tag{10}
$$

translates into

$$[P^\mu, P^\nu] = 0 \tag{11}$$

The generators of translations in quantum theory are then to be identified with the field four-momentum operator.

To further specify the quantum theory, one requires a rule on the properties of $\phi(x)$ under commutation: these should be such that for the appropriate choice of H (e.g. (9), perhaps), the Heisenberg equation of motion (8) reduces to the Klein-Gordon equation (1). Note that H is bilinear in the field and its derivatives: the commutator on the left hand side of (8) is of the generic type $[AB, C]$, where A, B and C are in general non-commuting. This can be written as

$$
\begin{aligned}
[AB, C] &= ABC - CAB \\
&= ABC + ACB - ACB - CAB
\end{aligned} \tag{12}
$$

where in the second line we have added zero to the first. Note that the terms in (12) can be grouped in two ways:

$$
\begin{aligned}
[AB, C] &= ABC \pm ACB \mp ACB - CAB \\
&\equiv A[B, C]_\pm \mp [A, C]_\pm B
\end{aligned} \tag{13}
$$

involving <u>either</u> the commutator or anticommutator bracket

$$[A, B]_\pm = AB \pm BA \tag{14}$$

When no subscript is given, the square bracket always refers to the commutator ("$-$"). Thus, what is needed is a rule for $[\phi(x), \phi(y)]_\pm$. Here, relativistic invariance will be of crucial importance.

We first make the crucial assumption that $[\phi(x), \phi(y)]_\pm$ is <u>not</u> an operator at all, but just an ordinary function: this only holds for free fields, satisfying <u>linear</u> field equations, such as Eq. (1). Thus, we postulate

$$[\phi(x), \phi(y)]_\pm = \hbar F(x, y) \tag{15}$$

By applying the operator $U(1, -y)$ on the left and $U(1, y) = U^{-1}(1, -y)$ on the right, one has

$$
\begin{aligned}
U(1, &-y)\phi(x)U^{-1}(1, -y)U(1, -y)\phi(y)U^{-1}(1, -y) \\
&\pm U(1, -y)\phi(y)U^{-1}(1, -y)U(1, -y)\phi(x)U^{-1}(1, -y) \\
&= \phi(x - y)\phi(0) \pm \phi(0)\phi(x - y)
\end{aligned} \tag{16}
$$

whence $F(x, y) = F(x - y)$ is only a function of the difference $x - y$. The requirement of invariance under Lorentz transformations $U(\Lambda, 0)$, i.e.

$$U(\Lambda, 0)\phi(x)U^{-1}(\Lambda, 0) = \phi(\Lambda x) \tag{17}$$

also requires that $F(x-y) = F(\Lambda(x-y))$, i.e. that $F(x-y)$ be a Lorentz invariant function of $x-y$. Finally, since $\phi(x)$ satisfies the Klein-Gordon equation, so does $F(x-y)$, i.e.

$$\left(\Box_x + m^2\right) F(x-y) = 0 \tag{18}$$

Thus, at this stage, $F(x-y)$ is the most general Lorentz invariant solution of the Klein-Gordon equation (1), which we have already derived in the second lecture (Eq. II.9).

The final crucial requirement is known as microcausality: reasoning that relativistic invariance implies that two space-like separated events can have no causal effect on one another, one finally requires that, for $(x-y)^2 < 0$,

$$[\phi(x), \phi(y)]_\pm = 0 \tag{19}$$

One easily checks that the only possibility is now to set $a = 0$ in Eq. (II. 9). The resulting Lorentz invariant solution of the Klein-Gordon equation, vanishing for space-like values of its argument, is now unique up to a constant of proportionality: one defines

$$i\Delta(x-y) = \int \frac{d^3k}{(2\pi)^3 2\omega_k} \left(e^{-ik\cdot(x-y)} - e^{ik\cdot(x-y)}\right) \tag{20}$$

and finally writes

$$[\phi(x), \phi(y)]_\pm = \hbar\eta i\Delta(x-y) \tag{21}$$

where η is a constant. Here and elsewhere, factors of $\hbar$ are present for dimensional reasons. Now we can show that in fact one cannot choose the anticommutator ("+") alternative in (21). Note that $i\Delta(x-y) = -i\Delta(y-x)$, while, by definition

$$[\phi(x), \phi(y)]_\pm = \pm[\phi(y), \phi(x)]_\pm \tag{22}$$

The only choice consistent with Poincaré invariance and microcausality is to quantize the free scalar field using commutators. Thus, one has the result

$$[\phi(x), \phi(y)] = \hbar\eta i\Delta(x-y) \tag{23}$$

and by taking the Hermitian conjugate one easily sees that η is real, i.e. $\eta = \eta^*$. This has profound consequences, as we shall see.

One can now use (23) to check whether (1) follows from (8) with $\mu = 0$, given (9), i.e.

$$H = \int d^3y \, \mathcal{H}(y)$$
$$= \int d^3y \, \frac{1}{2} \left(\dot{\phi}^2(y) + \left(\vec{\nabla}\phi(y)\right)^2 + m^2\phi^2(y)\right)$$

and (23). Recall that H is time independent, and write $[H, \phi(x)] = -i\hbar\dot{\phi}(x)$: now take the time-derivative $\partial/\partial x^0$ of this equation, to get

$$[H, \dot{\phi}(x)] = -i\hbar\ddot{\phi}(x)$$

i.e.

$$\int d^3y \left[\mathcal{H}(\vec{y}, x^0), \dot{\phi}(x) \right] = -i\hbar \ddot{\phi}(x) \tag{24}$$

This should reduce to the Klein-Gordon equation: note that since H is, again, time-independent, we can choose to evaluate the commutators at underline-times, which is all that is really necessary. Since it is readily apparent that

$$[\dot{\phi}(\vec{y}, x^0), \dot{\phi}(\vec{x}, x^0)] = 0 \tag{25}$$

we need only know

$$[\phi(\vec{y}, x^0), \dot{\phi}(\vec{x}, x^0)] = i\hbar\eta\delta^3(\vec{x} - \vec{y}) \tag{26}$$

as easily follows by differentiation of (23) given (20). Using Eq. 13, one calculates in turn

$$\begin{aligned}
\int d^3y \, \frac{1}{2} \, & \left[\left(\vec{\nabla}\phi(\vec{y}) \right)^2, \dot{\phi}(\vec{x}) \right] \\
&= \int d^3y \, \vec{\nabla}\phi(\vec{y}) i\hbar\eta \vec{\nabla}\delta^3(\vec{y} - \vec{x}) \\
&= -i\hbar\eta \int d^3y \, \delta^3(\vec{y} - \vec{x}) \vec{\nabla}^2\phi(\vec{y}) \\
&= -i\hbar\eta \vec{\nabla}^2\phi(x)
\end{aligned} \tag{27}$$

and

$$\begin{aligned}
\int d^3y \, \frac{1}{2} \, m^2 \, & \left[\phi^2(\vec{y}), \dot{\phi}(\vec{x}) \right] \\
&= m^2 \int d^3y \, \phi(\vec{y}) i\hbar\eta\delta^3(\vec{y} - \vec{x}) \\
&= i\hbar\eta m^2 \phi(x)
\end{aligned} \tag{28}$$

where we have suppressed the common time arguments x^0 until the end and discarded a surface term after integration by parts. Putting (24), (27) and (28) together one gets

$$-i\hbar\eta \vec{\nabla}^2\phi(x) + i\hbar\eta m^2 \phi(x) = -i\hbar\ddot{\phi}(x) \tag{29}$$

so one must set $\eta = 1$ to indeed see that (29) reduces to the Klein-Gordon equation.

Note that whereas the commutation relation (23) cannot be generalized to the case of fields satisfying non-linear equations, the commutation rule (26) at equal-times can be so generalized and is at the basis of the so-called canonical quantization program.

Next, we assume that there are free particles of mass M, and that as simultaneous (cf. 11) eigenstates of energy and momentum they form a complete set of states that can be used as a basis for the Hilbert space on which $\phi(x)$ acts. We will also assume that they are eigenstates of particle number (given that such an operator can be defined). There will be a ground state, or vacuum, $|0\rangle$, required to be

Lorentz invariant: this means that the expectation value of the energy-momentum tensor can only be proportional to the Lorentz invariant tensor $\eta^{\mu\nu}$, i.e.

$$\langle 0|T^{\mu\nu}(x)|0\rangle = \langle 0|T^{\mu\nu}(0)|0\rangle$$
$$= \eta^{\mu\nu}\epsilon_{\text{vac}} \tag{30}$$

The constant of proportionality ϵ_{vac} is the vacuum energy density, as can be seen by setting $\mu = \nu = 0$. Note that the momentum density T^{k0} must vanish in the ground state. It will be convenient to <u>define</u> the quantum Hamiltonian density as follows

$$\mathcal{H} = T^{00}(x) - \langle 0|T^{00}(x)|0\rangle$$

so that

$$\langle 0|\mathcal{H}|0\rangle = 0 \tag{31}$$

This does not change the equation of motion since all that is involved is the commutator of $\mathcal{H}(x)$ with $\phi(x)$. The vacuum is normalized to unity, $\langle 0|0\rangle = 1$ and given (31), $P^\mu|0\rangle = 0$. Single-particle states $|\vec{p}\rangle$ are eigenstates of P^μ:

$$P^\mu|\vec{p}\rangle = p^\mu|\vec{p}\rangle \tag{32}$$

with $p^\mu = (E_p, \vec{p})$ and $E_p^2 - \vec{p}^{\,2} = M^2$, with M the particle mass. Multiparticle states are just tensor products of single particle states, i.e.

$$|\vec{p}_1, \vec{p}_2\rangle = |\vec{p}_1\rangle|\vec{p}_2\rangle$$
$$P^\mu|\vec{p}_1, \vec{p}_2\rangle = (p_1^\mu + p_2^\mu)|\vec{p}_1, \vec{p}_2\rangle \tag{33}$$

etc. We shall assume for now that these particles satisfy Bose-Einstein statistics, i.e. that states are symmetric under interchange, $|\vec{p}_1, \vec{p}_2\rangle = +|\vec{p}_2, \vec{p}_1\rangle$: this will actually be shown to follow from the formalism later, at the end of this section.

We shall work in a box of volume $V = L^3$, whence the momenta will take discrete values $\vec{p} = \frac{2\pi\hbar\vec{n}}{L}$: as a result, the orthogonality relations are

$$\langle \vec{p}_2|\vec{p}_1\rangle = \delta(p_1, p_2)$$
$$\langle \vec{p}_3, \vec{p}_4|\vec{p}_1, \vec{p}_2\rangle = \delta(p_1, p_4)\delta(p_2, p_3) + \delta(p_1, p_3)\delta(p_2, p_4) \tag{34}$$
$$\equiv \langle \vec{p}_4|\vec{p}_1\rangle\langle \vec{p}_3|\vec{p}_2\rangle + \langle \vec{p}_3|\vec{p}_1\rangle\langle \vec{p}_4|\vec{p}_2\rangle$$

etc. where $\delta(p_1, p_2)$ is a Kronecker delta, equal to 1 if $\vec{p}_1 = \vec{p}_2$, zero otherwise. Finally, the completeness relation is

$$\mathbf{1} = |0\rangle\langle 0| + \sum_{\vec{p}} |\vec{p}\rangle\langle \vec{p}|$$
$$+ \frac{1}{2!} \sum_{\vec{p}_1, \vec{p}_2} |\vec{p}_1, \vec{p}_2\rangle\langle \vec{p}_1, \vec{p}_2|$$
$$+ \frac{1}{3!} \sum_{\vec{p}_1, \vec{p}_2, \vec{p}_3} |\vec{p}_1, \vec{p}_2, \vec{p}_3\rangle\langle \vec{p}_1, \vec{p}_2, \vec{p}_3| \tag{35}$$
$$+ \cdots$$

Note that states with different numbers of particles are orthogonal, i.e. if $k \neq n$, then

$$\langle \vec{p}_1{}', \vec{p}_2{}', \dots \vec{p}_n{}' | \vec{p}_1, \vec{p}_2, \dots, \vec{p}_k \rangle = 0 \tag{36}$$

In general, for four-momentum eigenstates $|\alpha\rangle$ and $|\beta\rangle$ with $P^\mu |\alpha\rangle = p_\alpha^\mu |\alpha\rangle$, etc. it follows from translation invariance that

$$\begin{aligned}
\langle \beta | \phi(x) | \alpha \rangle &= \langle \beta | e^{iP \cdot x/\hbar} \phi(0) e^{-iP \cdot x/\hbar} | \alpha \rangle \\
&= e^{i(p_\beta - p_\alpha) \cdot x/\hbar} \langle \beta | \phi(0) | \alpha \rangle
\end{aligned} \tag{37}$$

Thus, applying the Klein-Gordon equation,

$$\begin{aligned}
\Box \langle \beta | \phi(x) | \alpha \rangle &= -m^2 \langle \beta | \phi(x) | \alpha \rangle \\
&= -\hbar^{-2} \left(p_\beta - p_\alpha \right)^2 \langle \beta | \phi(x) | \alpha \rangle
\end{aligned} \tag{38}$$

In particular, $\langle \beta | \phi(x) | \alpha \rangle$ will vanish unless the equation

$$(p_\beta - p_\alpha)^2 = \hbar^2 m^2 \tag{39}$$

can have a solution. From this one can conclude a number of things:

1. The vacuum expectation value of the field must vanish if $m^2 \neq 0$:

$$\langle 0 | \phi(0) | 0 \rangle = 0 \tag{40}$$

2. Choosing $|\alpha\rangle = |\vec{p}\rangle$ and $|\beta\rangle = |0\rangle$ one finds

$$p^2 \equiv M^2 = \hbar^2 m^2 \tag{41}$$

showing that for $m^2 > 0$ the particle interpretation is consistent, and that $\langle 0 | \phi(x) | \vec{p} \rangle$ satisfies the Klein-Gordon equation

$$\left(-\hbar^2 \Box + M^2 \right) \langle 0 | \phi(x) | \vec{p} \rangle = 0 \tag{42}$$

so that

$$\langle 0 | \phi(x) | \vec{p} \rangle \equiv f_{\vec{p}}(x) \equiv e^{-ip \cdot x/\hbar} f_{\vec{p}}(0) \tag{43}$$

can be identified as the single particle wave function of the relativistic quantum mechanics based on the Klein-Gordon equation. We will come back to this.

3. Choosing $|\alpha\rangle = |\vec{p}_1, \vec{p}_2, \dots, \vec{p}_n\rangle$ and $|\beta\rangle = |0\rangle$ one sees that because

$$(p_1 + p_2 + \dots + p_n)^2 \geq (nM)^2 \tag{44}$$

from simple kinematics, Eq. (39) has no solution except for $n = 1$ only.

4. Choosing $|\alpha\rangle = |\vec{p}_1, \vec{p}_2, \dots, \vec{p}_n\rangle$ and $|\beta\rangle = |\vec{p}_1{}', \vec{p}_2{}', \dots, \vec{p}_k{}'\rangle$, one sees that (39) takes the form

$$(p_1 + p_2 + \dots + p_n - p_1' - p_2' - \dots - p_k')^2 = M^2 \tag{45}$$

given (41), and that this can only have a solution if $|n-k| = 1$, in which case the only possibilities are, say with $n = k+1$: p_1 is arbitrary while $p_2 + p_3 + \ldots + p_n = p'_1 + \vec{p}\,'_2 + \ldots + \vec{p}\,'_k$, and similarly running through $p_2, p_3, \ldots$ to p_n. Thus, we know that up to a proportionality constant

$$
\begin{aligned}
&\langle p'_1 p'_2 \cdots p'_{n-1} | \phi(x) | p_1 p_2 \cdots p_n \rangle \\
&\propto \sum_{i=1}^{n} \langle p'_1 p'_2 \cdots p'_{n-1} | p_1 p_2 \cdots p_{i-1} p_{i+1} \cdots p_n \rangle e^{-i p_i \cdot x / \hbar}
\end{aligned}
\tag{46}
$$

The proportionality constant can be shown (see the end of this lecture) to be $f_{\vec{p}_i}(0)$ (cf. Eq. 43) so that all matrix elements of $\phi(x)$ between free particle states can be expressed in terms of only one basic quantity, $\langle 0 | \phi(x) | \vec{p} \rangle$. This follows only because of the linearity of the equation of motion: in the next lecture we will see explicitly how interactions (non-linearities) modify the picture.

The expression for $f_{\vec{p}}(0)$ can now be determined by considering the vacuum expectation value of (23) with $\eta = 1$, i.e. given $\langle 0 | 0 \rangle = 1$,

$$
\langle 0 | [\phi(x), \phi(y)] | 0 \rangle = \hbar i \Delta(x - y)
\tag{47}
$$

Inserting a complete set of states, and noting that only single particle states contribute, one has

$$
\begin{aligned}
&\sum_{\vec{p}} |\langle 0 | \phi(0) | \vec{p} \rangle|^2 \left(e^{-i p \cdot (x-y)/\hbar} - e^{i p \cdot (x-y)/\hbar} \right) \\
&= \hbar i \Delta(x - y)
\end{aligned}
\tag{48}
$$

Recognizing that one can go to the infinite volume limit through the replacement

$$
\sum_{\vec{p}} \equiv V \int \frac{d^3 p}{(2\pi\hbar)^3}
\tag{49}
$$

one finds easily, given (20) and with the definition $\vec{p} = \hbar \vec{k}$, that

$$
\langle 0 | \phi(0) | \vec{p} \rangle = \sqrt{\frac{\hbar}{2\omega_k V}}
\tag{50}
$$

with $\omega_k^2 = \vec{k}^2 + m^2 = E_p^2 / \hbar^2$. Thus, the relativistic single-particle wave-function is necessarily of the form

$$
\langle 0 | \phi(x) | \vec{p} \rangle = \sqrt{\frac{\hbar}{2\omega_{\vec{k}} V}} \; e^{-i k \cdot x}
\tag{51}
$$

with $k^\mu = p^\mu / \hbar$. Note the inevitable presence of the factor $(2\omega_k)^{-1/2}$: this has a profound consequence. In the spirit of quantum theory, one may wish to interpret

$\phi(\vec{x}, t)|0\rangle$ as an eigenstate of position (at some time t), so that the plane-wave solution of the free particle Klein-Gordon equation (51) would be understood as

$$\begin{aligned}
\langle 0|\phi(\vec{x}, t)|\vec{p}\rangle &= e^{-iE_p t/\hbar}\langle 0|\phi(\vec{x}, 0)|\vec{p}\rangle \\
&\equiv e^{-iE_p t/\hbar}\langle \vec{x}|\vec{p}\rangle
\end{aligned} \tag{52}$$

i.e. as the projection onto the space of position eigenstates $|\vec{x}\rangle$ of the momentum eigenstates $|\vec{p}\rangle$. One can determine whether this is a tenable position by calculating the overlap $\langle \vec{x}\,'|\vec{x}\rangle$ of two such "position eigenstates":

$$\langle 0|\phi(\vec{x}\,', 0)\phi(\vec{x}, 0)|0\rangle = \langle \vec{x}\,'|\vec{x}\rangle \tag{53}$$

But now, by going through the same steps that led from (47) through (51), one easily gets

$$\langle 0|\phi(\vec{x}\,', 0)\phi(\vec{x}, 0)|0\rangle = \int \frac{d^3 k}{(2\pi)^3 2\omega_k}\, \hbar e^{i\vec{k}\cdot(\vec{x}'-\vec{x})} \tag{54}$$

Because of the factor $1/\omega_k$ in the integrand, this is not a delta-function, unlike what is assumed in non-relativistic quantum mechanics, where one has, by assumption $\langle \vec{x}\,'|\vec{x}\rangle = \delta^3(\vec{x} - \vec{x}\,')$. In fact, for $r \equiv |\vec{x} - \vec{x}\,'| \gg m^{-1}$, (54) has the behavior

$$\langle 0|\phi(\vec{x}\,', 0)\phi(\vec{x}, 0)|0\rangle \sim \hbar m^2 (mr)^{-3/2} e^{-mr} \tag{55}$$

showing an exponential decrease, the onset of which is the typical length scale $1/m = \hbar/M$, i.e. the Compton wavelength. The rules of quantum theory supplemented by relativistic invariance imply a limitation on the localizability of a single particle, to a distance of order its Compton wavelength.

One can consider the matrix element $\langle 0|\phi(x)\phi(y)|0\rangle$ further: we note that quantum field fluctuations in the vacuum, i.e.

$$\begin{aligned}
&\langle 0|\phi(x)\phi(y)|0\rangle - \langle 0|\phi(x)|0\rangle\langle 0|\phi(y)|0\rangle \\
&\equiv \langle 0|\phi(x)\phi(y)|0\rangle
\end{aligned}$$

actually diverge quadratically as $x \to y$. To control this divergence, it is necessary to define averaged fields over space-time

$$\overline{\phi} = \frac{1}{VT} \int_V d^3 x \int_{-T/2}^{T/2} dt\, \phi(\vec{x}, t) \tag{56}$$

so that $\langle 0|\overline{\phi}^2|0\rangle$ is finite for finite V, T: only measurements of fields averaged both in space and time can have any physical meaning. Another interpretation of the matrix element $\langle 0|\phi(x)\phi(y)|0\rangle$, for a specified time ordering such that $x^0 > y^0$, is the probability amplitude that an initial state of a particle $\phi(y)|0\rangle$, not quite localized at $\vec{y}$, will end up as $\phi(x)|0\rangle$ at a later time, not quite localized at $\vec{x}$: to allow for the possibility that $x^0 < y^0$, one defines the Feynman propagator so that the above interpretation always holds.

$$i\Delta_F(x - y) = \langle 0|T\phi(x)\phi(y)|0\rangle \tag{57}$$

where the time ordering puts "later on the left", i.e.

$$T\phi(x)\phi(y) = \theta(x^0 - y^0)\phi(x)\phi(y) + \theta(y^0 - x^0)\phi(y)\phi(x) \tag{58}$$

where $\theta(x) = 1$ if $x > 0$, 0 if $x < 0$: it is often convenient to define $\theta(0) = 1/2$.

The usual expression for $i\Delta_F(x - y)$ follows easily from the previous work and the integral representation of the step function: one easily gets

$$i\Delta_F(x - y) = \hbar \int \frac{d^4k}{(2\pi)^4}\ e^{-ik\cdot(x-y)}\ \frac{i}{k^2 - m^2 + i\epsilon} \tag{59}$$

We conclude this lecture by addressing an important point mentioned earlier: how the quantization rule for the scalar field (23), involving a commutator, implies Bose-Einstein statistics for the free particle states comprising our basis. We will see how the connection is established, not in full generality, but by considering an illuminating special case. Look at the matrix element of the commutator (23) between the vacuum and a two-particle state, i.e.

$$\langle 0|[\phi(x), \phi(y)]|\vec{k}_1, \vec{k}_2\rangle = 0 \tag{60}$$

This vanishes because the commutator is an ordinary function, and the two states are orthogonal. This can be written as

$$\begin{aligned} \sum_{\vec{k}} \langle 0|\phi(x)|\vec{k}\rangle\langle\vec{k}|\phi(y)|\vec{p}_1\vec{p}_2\rangle \\ - \langle 0|\phi(y)|\vec{k}\rangle\langle\vec{k}|\phi(x)|\vec{p}_1\vec{p}_2\rangle = 0 \end{aligned} \tag{61}$$

since only the single particle state contributes to the completeness relation here. Now it follows after some simple calculations that this can hold, only if

$$\begin{aligned} \langle\vec{k}|\phi(y)|\vec{p}_1, \vec{p}_2\rangle =& \langle\vec{k}|\vec{p}_1\rangle\langle 0|\phi(y)|\vec{p}_2\rangle \\ &+ \langle\vec{k}|\vec{p}_2\rangle\langle 0|\phi(y)|\vec{p}_1\rangle \end{aligned} \tag{62}$$

The crucial point is the "+" relative sign, which follows from the fact of the symmetry under the interchange of $\vec{p}_1$ and $\vec{p}_2$: for Fermi-Dirac statistics, this would have been replaced by a "$-$" sign, reflecting the antisymmetry under the interchange of $\vec{p}_1$ and $\vec{p}_2$, and under this assumption, Eq. (61) would never be fulfilled, contradicting our assumptions. As a bonus, we also check that the assertion following Eq. (46) does hold true: the proportionality constant is indeed $\langle 0|\phi|\vec{p}\rangle$. This can be established quite easily for the general case by considering appropriate matrix elements of Eq. (23).

We are now ready to turn our attention to the problem of interacting quantum field theory, having set up the necessary tool kit for our investigations.

IV. QUANTUM THEORY OF THE SELF-INTERACTING SCALAR FIELD

Now consider the theory of a quantum scalar field satisfying the equation of motion

$$\Box\phi(x) = -\mu^2\phi(x) - \lambda\phi^3(x) \tag{1}$$

In the classical case, this equation can be derived from the least action principle, given the Lagrangian density

$$\mathcal{L}(\phi, \partial_\mu\phi) = \frac{1}{2}\,\partial_\mu\phi\partial^\mu\phi - V(\phi) \quad, \tag{2}$$

$$V(\phi) = \frac{1}{2}\,\mu^2\phi^2 + \frac{1}{4}\,\lambda\phi^4 \tag{3}$$

We note immediately that the commutation relation (III.23) postulated for the free field case ($\lambda = 0$) no longer holds for arbitrary space-time points x and y: indeed, one can no longer assume that the function $F(x, y)$ in (III.15) is simply an ordinary function, because of the presence of the cubic term in Eq. (1). Nevertheless, the quantum equation of motion can still be derived from the Hamiltonian

$$H = \int d^3y\; \frac{1}{2}\,(\dot{\phi}^2(y) + (\vec{\nabla}\phi(y))^2 + \mu^2\phi^2(y) + \frac{1}{2}\,\lambda\phi^4(y)) \tag{4}$$

when we note that it is sufficient to know the field commutation relations at equal times: one can then simply take Eq. (III.26) as a basic postulate, namely

$$\left[\phi\left(\vec{y}, x^0\right), \dot{\phi}(\vec{x}, x^0)\right] = i\hbar\delta^3\left(\vec{x} - \vec{y}\right) \tag{5}$$

together with

$$[\phi\left(\vec{y}, x^0\right), \phi\left(\vec{x}, x^0\right)] = \left[\dot{\phi}\left(\vec{y}, x^0\right), \dot{\phi}\left(\vec{x}, x^0\right)\right] = 0 \tag{6}$$

It is then a simple exercise to show that Eq. (1) follows from the Heisenberg equation of motion, as was done in the previous lecture.

The theory defined by Eq. (1), unlike that described in the previous lecture, is not exactly soluble: in particular, as we shall see, there is no exact formula relating the parameters appearing in the equation of motion (1) and the physical mass of the particle corresponding to the field $\phi(x)$: this observation should be remembered, as it will have important consequences as regards the interpretation of the parameters μ and λ.

We assume the existence of a ground state $|0\rangle$, as before, and now one has

$$
\begin{aligned}
\langle 0|\Box\phi(x)|0\rangle &= 0 \\
&= -\mu^2\langle 0|\phi(x)|0\rangle - \lambda\langle 0|\phi^3(x)|0\rangle
\end{aligned} \tag{7}
$$

Thus, one can no longer conclude that the vacuum expectation value of $\phi(x)$ vanishes: from translational invariance one has

$$
\begin{aligned}
\langle 0|\phi(x)|0\rangle &= \langle 0|e^{iP\cdot x/\hbar}\phi(0)e^{-iP\cdot x/\hbar}|0\rangle \\
&= \langle 0|\phi(0)|0\rangle
\end{aligned} \tag{8}
$$

so the assumption that the ground state is an eigenstate of P^μ implies that the vacuum expectation value (vev) is a constant; we shall write

$$
\langle 0|\phi(x)|0\rangle \equiv v \tag{9}
$$

We shall continue the analysis by defining a quantum field $\hat{\phi}$ whose vev vanishes, i.e. write

$$
\phi(x) = \hat{\phi}(x) + v \tag{10}
$$

so

$$
\langle 0|\hat{\phi}(x)|0\rangle \equiv 0 \tag{11}
$$

Eq. (7) then takes the form

$$
\begin{aligned}
\mu^2 v + \lambda v^3 &+ 3\lambda v\langle 0|\hat{\phi}^2(x)|0\rangle \\
&+ \lambda\langle 0|\hat{\phi}^3(x)|0\rangle = 0
\end{aligned} \tag{12}
$$

For a general potential $V(\phi)$, with the equation of motion

$$
\Box\phi(x) = -\frac{dV(\phi)}{d\phi} \equiv -V'(\phi(x)) \tag{13}
$$

one gets, expanding $V'(\phi(x))$,

$$
\begin{aligned}
V'(\phi) =&\, V'(v) + \hat{\phi}(x)V''(v) \\
&+ \frac{1}{2}\,\hat{\phi}^2(x)V'''(v) + \frac{1}{6}\,\hat{\phi}^3(x)V^{(iv)}(v)
\end{aligned} \tag{14}
$$

where we have assumed $V(\phi)$ to be at most quartic in $\phi(x)$, the following equation

$$
\begin{aligned}
V'(v) &+ \frac{1}{2}\,V'''(v)\langle 0|\hat{\phi}^2(x)|0\rangle \\
&+ \frac{1}{6}\,V^{(iv)}(v)\langle 0|\hat{\phi}^3(x)|0\rangle = 0
\end{aligned} \tag{15}
$$

If quantum fluctuation effects are completely neglected, one gets simply

$$
V'(v) = 0 \tag{16}
$$

or, in the present case

$$\mu v^2 + \lambda v^3 = 0 \tag{17}$$

which is the classical equation for a constant scalar field. We see that (17) admits several solutions: $v = 0$ is apparently always possible, although if $\mu^2 < 0$ (and given that $\lambda > 0$ so that the potential is bounded from below), then one has the possibility

$$v^2 = -\mu^2/\lambda \tag{18}$$

In fact, by calculating $V''(v)$, one sees that when $\mu^2 < 0$, the solution $v = 0$ is unstable, corresponding to a local maximum of $V(\phi)$.

To go beyond this simple classical approximation we need an·approximation scheme: the simplest possibility is to calculate the vacuum matrix elements in (12) or (15) by taking $\hat{\phi}(x)$ to behave like a free field, i.e. that its non-vanishing matrix elements correspond to states differing by only one particle, as seen in the previous chapter, all of which can be related to the basic matrix element

$$\langle 0|\hat{\phi}(x)|\vec{p}\rangle = \sqrt{\frac{\hbar}{2\omega_k V}}\, e^{-ik\cdot x} \tag{19}$$

cf. Eq. (III.51), where $\omega_k = \sqrt{\vec{k}^2 + m^2}$. The question is then what is the appropriate mass to take in Eq. (19). To get the leading $O(\hbar)$ quantum correction, it will suffice to use m^2 as determined to $O(1)$, i.e. neglecting quantum fluctuations.

Once again, assuming that single particle states $|\vec{p}\rangle$ exist with mass M, i.e.

$$P^2|\vec{p}\rangle = M^2|\vec{p}\rangle$$

it follows that,

$$\begin{aligned}
\langle 0|\Box\phi(x)|\vec{p}\rangle &= \langle 0|\Box\hat{\phi}(x)|\vec{p}\rangle \\
&= -M^2/\hbar^2 \langle 0|\hat{\phi}(x)|\vec{p}\rangle
\end{aligned} \tag{20}$$

and now from the equation of motion (1), writing $m^2 = M^2/\hbar^2$,

$$\begin{aligned}
m^2\langle 0|\hat{\phi}(x)|\vec{p}\rangle =&\, \mu^2\langle 0|\hat{\phi}(x)|\vec{p}\rangle \\
&+ 3\lambda v^2\langle 0|\hat{\phi}(x)|\vec{p}\rangle \\
&+ 3\lambda v\langle 0|\hat{\phi}^2(x)|\vec{p}\rangle \\
&+ \lambda\langle 0|\hat{\phi}^3(x)|\vec{p}\rangle
\end{aligned} \tag{21}$$

For a general quartic potential, one gets

$$\begin{aligned}
m^2\langle 0|\hat{\phi}(x)|\vec{p}\rangle =&\, V''(v)\langle 0|\hat{\phi}(x)|\vec{p}\rangle \\
&+ \frac{1}{2}\, V'''(v)\langle 0|\hat{\phi}^2(x)|\vec{p}\rangle \\
&+ \frac{1}{6}\, V^{(iv)}(v)\langle 0|\hat{\phi}^3(x)|\vec{p}\rangle
\end{aligned} \tag{22}$$

One easily sees that to $O(1)$, i.e. ignoring quantum fluctuations, the mass m^2 is given by

$$m^2 = V''(v) = \mu^2 + 3\lambda v^2 \tag{23}$$

The classical requirement of stability $V''(v) > 0$ is mirrored in the requirement that the mass of the particle in the quantum theory be real.

With Eq. (23), we can go back to the vacuum equations (12) or (15): in the free-field approximation one sees that $\langle 0|\hat\phi^3(x)|0\rangle$ vanishes. One is left with the $O(\hbar)$ equation

$$\mu^2 v + \lambda v^3 + 3\lambda v \langle 0|\hat\phi^2(x)|0\rangle = 0 \tag{24}$$

The matrix element can be defined as the limit $x \to y$ of the free field Feynman propagator, Eq. III.58, i.e.

$$\langle 0|\hat\phi^2(x)|0\rangle \equiv \lim_{x \to y} i\Delta_F(x - y; m^2(v))$$
$$= \hbar \int \frac{d^4k}{(2\pi)^4} \frac{i}{k^2 - m^2(v) + i\epsilon} \tag{25}$$

where $m^2(v)$ is given by (23).

This is clearly divergent, and this problem will have to be addressed soon: a particularly useful concept here is that of the effective potential. The idea here is to consider the vacuum expectation value of the equation of motion, as the minimization equation of an "effective potential", which is a function of a classical field ϕ_c, i.e. identify

$$0 = \langle 0|\frac{dV(\phi)}{d\phi}|0\rangle\bigg|_{\langle 0|\phi|0\rangle=v} \equiv \frac{dV_{\text{eff}}(\phi_c)}{d\phi_c}\bigg|_{\phi_c=v} \tag{26}$$

Eq. (24) can be interpreted in this way, given Eq. (25), provided we identify

$$\frac{dV_{\text{eff}}(\phi_c)}{d\phi_c} = \mu^2\phi_c + \lambda\phi_c^3 + 3\lambda\phi_c \int \frac{d^4k}{(2\pi)^4} \frac{i\hbar}{k^2 - m^2(\phi_c) + i\epsilon} \tag{27}$$

with a ϕ_c-dependent mass

$$m^2(\phi_c) = \mu^2 + 3\lambda\phi_c^2 \tag{28}$$

Noting that $dm^2(\phi_c) = 6\lambda\phi_c d\phi_c$, Eq. (27) can be formally integrated to yield

$$V_{\text{eff}}(\phi_c) = \frac{1}{2}\mu^2\phi_c^2 + \frac{1}{4}\lambda\phi_c^4$$
$$- \frac{i\hbar}{2}\int \frac{d^4k}{(2\pi)^4} \ln\left(m^2(\phi_c) - k^2 - i\epsilon\right) \tag{29}$$

a very famous result due to S. Coleman and E. Weinberg. Another way of writing $V_{\text{eff}}(\phi_c)$ follows if we simply use the equivalent definition

$$\langle 0|\hat\phi^2(x)|0\rangle = \lim_{x \to y} \langle 0|\hat\phi(x)\hat\phi(y)|0\rangle$$
$$= \hbar \int \frac{d^3k}{(2\pi)^3 2\omega_k} \tag{30}$$

[cf. Eq. (III.54)], where $\omega_k = \sqrt{\vec{k}^2 + m^2(\phi_c)}$, so that $dm^2(\phi_c) = 2\omega_k d\omega_k$ which leads to

$$V_{\text{eff}}(\phi_c) = \frac{1}{2}\,\mu^2\phi_c^2 + \frac{1}{4}\,\lambda\phi_c^4 + \frac{\hbar}{2}\int\frac{d^3k}{(2\pi)^3}\,\sqrt{\vec{k}^2 + m^2(\phi_c)} \tag{31}$$

and is very suggestive, in that the $O(\hbar)$ term looks like the vacuum energy density of a free scalar field with a ϕ_c-dependent mass. Indeed, it is a simple exercise to show this by calculating the quantity $\langle 0|T^{00}(x)|0\rangle$ for the free field case explained in the previous lecture. Eq. (31) incidentally suggests the correct generalization of these results to a more general theory in which vector and fermion fields are coupled to a scalar field, in such a way that they obtain masses proportional to v, as in the standard model: let these particles be labeled i, so that $m_i(\phi_c) = g_i\phi_c = m_i(v)\phi_c/v$. Then, the effective potential to $O(\hbar)$ is given by a sum of vacuum energy densities,

$$
\begin{aligned}
V_{\text{eff}}(\phi_c) = {}& \frac{1}{2}\,\mu^2\phi_c^2 + \frac{1}{4}\,\lambda\phi_c^4 + \frac{\hbar}{2}\int\frac{d^3k}{(2\pi)^3}\,\sqrt{\vec{k}^2 + m^2(\phi_c)} \\
& + \frac{\hbar}{2}\sum_i C_i\int\frac{d^3k}{(2\pi)^3}\sqrt{\vec{k}^2 + m_i^2(\phi_c)}
\end{aligned}
\tag{32}
$$

where C_i is a counting factor, given by

$$C_i = (2S_i + 1)(-1)^{2S_i}2^{\lambda_i} \tag{33}$$

where S_i is the spin of particle i, and where $\lambda_i = 0, 1$ according to whether i is identical to its antiparticle or not. Fermions contribute to the sum with negative weight, corresponding to the negative energy density of the Dirac sea.

We now return to Eq. (27): we realize that μ^2 and λ as parameters appearing in the equation of motion (1) do not directly correspond to physical quantities, but rather enter into expressions for physical quantities calculated in some approximation. Let us relabel them μ_0^2 and λ_0, respectively: we can now try to interpret them in the context of our simple approximation scheme in which only the lowest order quantum correction is kept.

Since the divergence difficulties appear in $O(\hbar)$, let us further define

$$\mu_0^2 = \mu^2 - \delta\mu^2 \tag{34}$$

$$\lambda_0 = \lambda - \delta\lambda \tag{35}$$

in which $\delta\mu^2$ and $\delta\lambda$ are of $O(\hbar)$ and serve to absorb the divergences. We will see by explicit calculation that this will suffice here, although in general one should also rescale the fields themselves. The parameters μ^2 and λ are finite, and must be defined in terms of conditions on physical quantities, known as renormalization conditions. Write Eq. (27) as follows, neglecting terms of $O(\hbar^2)$ and higher,

$$V'_{\text{eff}}(\phi_c) = \mu^2\phi_c + \lambda\phi_c^3 - \delta\mu^2\phi_c - \delta\lambda\phi_c^3 + 3\hbar\lambda\phi_c\Delta(m^2(\phi_c)) \tag{36}$$

which also serves to define $\Delta(m^2(\phi_c))$. This can also be expressed as a sum of terms of different order in $\hbar$, namely

$$V'_{\text{eff}}(\phi_c) = V^{(0)'}(\phi_c) + \Delta V^{(1)'}(\phi_c) \tag{37}$$

with

$$V^{(0)}(\phi_c) = \frac{1}{2}\,\mu^2\phi_c^2 + \frac{1}{4}\,\lambda\phi_c^4 \tag{38}$$

At the physical minimum $\phi_c = v$, one has of course

$$V'_{\text{eff}}(v) = 0 \tag{39}$$

We shall choose to fix the parameters μ^2 and λ by requiring that

$$\Delta V^{(1)'}(\phi_c = v) = 0 \tag{40}$$

$$\Delta V^{(1)''}(\phi_c = v) = 0 \tag{41}$$

Other prescriptions are possible, and we will soon return to the consequences of this freedom of choice in renormalization conditions.

As a consequence of the two conditions (40) and (41), the minimization condition and mass formula to $O(\hbar)$ will be of the same form as at tree-level (i.e. in the absence of quantum fluctuations) namely

$$V'_{\text{eff}}(v) = \mu^2 v + \lambda v^3 = 0 \tag{42}$$

$$V''_{\text{eff}}(v) = m^2 = \mu^2 + 3\lambda v^2 = 2\lambda v^2 \tag{43}$$

In fact, strictly speaking, the quantity m^2 defined by Eq. (43) is not quite the physical mass $M^2/\hbar^2$ of the scalar particle, but differs from it by a finite calculable amount: we will not discuss this additional subtlety further here. It is sufficient that the renormalized parameters μ^2 and λ do completely specify the physical properties of the quantum theory to $O(\hbar)$. Eq. (40) now yields

$$-\delta\mu^2 \cdot v - \delta\lambda \cdot v^3 + 3\hbar\lambda v \Delta(m^2(v)) = 0 \tag{44}$$

while Eq. (41) becomes

$$\begin{aligned} -\delta\mu^2 - 3\delta\lambda \cdot v^2 + 3\hbar\lambda\Delta(m^2(v)) \\ + 18\hbar\lambda^2 v^2 \Delta'(m^2(v)) = 0 \end{aligned} \tag{45}$$

where $\Delta'(m^2) = d\Delta(m^2)/dm^2$. We recall the definition of $\Delta(m^2)$, (cf. Eq. 36),

$$\Delta(m^2) = \int \frac{d^4k}{(2\pi)^4}\, \frac{i}{k^2 - m^2 + i\epsilon} \tag{46}$$

One can solve (44) and (45) to obtain

$$\delta\mu^2 = 3\hbar\lambda\Delta(m^2(v)) - 9\hbar\lambda^2 v^2 \Delta'(m^2(v)) \tag{47}$$

$$\delta\lambda = 9\hbar\lambda^2 \Delta'(m^2(v)) \tag{48}$$

so that, using Eq. (28),

$$
\begin{aligned}
V'_{\text{eff}}(\phi_c) = {} & \mu^2\phi_c + \lambda\phi_c^3 \\
& + 3\hbar\lambda\phi_c\{\Delta(m^2(\phi_c)) - \Delta(m^2(v)) \\
& - (m^2(\phi_c) - m^2(v))\Delta'(m^2(v))\}
\end{aligned} \tag{49}
$$

This is in fact completely finite! To see this, note that one can evaluate the integral in Eq. (46) by making a so-called Wick rotation, changing variables from k^0 to ik_4, with the result that $k^2 = k^{0^2} - \vec{k}^2$ becomes $k^2 = -k_E^2 \equiv -k_4^2 - \vec{k}^2$ and the integral is now over a four-dimensional Euclidean volume: thus,

$$\Delta(m^2) = \int \frac{d^4 k_E}{(2\pi)^4} \frac{1}{k_E^2 + m^2} \tag{50}$$

Using the fact that the total solid angle in four dimensions is $2\pi^2$ (exercise!) one can go to hyperspherical polar coordinates and write

$$
\begin{aligned}
\Delta(m^2) &= \int_0^\Lambda \frac{2\pi^2 k_E^3 dk_E}{16\pi^4} \frac{1}{k_E^2 + m^2} \\
&= \frac{1}{16\pi^2} \left[\Lambda^2 - m^2 \ln \frac{\Lambda^2}{m^2} + 0\left(\frac{1}{\Lambda^2}\right) \right]
\end{aligned} \tag{51}
$$

where a momentum space (actually, a wave-number space!) cut-off Λ has been introduced. One clearly sees the quadratic and logarithmic divergences appearing when this cut-off is taken to infinity. Using this explicit expression, one easily checks that all cut-off dependence disappears from the term in curly brackets in Eq. (49), and the result is

$$
\begin{aligned}
V'_{\text{eff}}(\phi_c) = {} & \mu^2\phi_c + \lambda\phi_c^3 \\
& + \frac{3\hbar\lambda\phi_c}{16\pi^2} \{m^2(\phi_c) \ln m^2(\phi_c)/m^2(v) - m^2(\phi_c) + m^2(v)\}
\end{aligned} \tag{52}
$$

This is easily integrated to yield the renormalized effective potential, with $m^2(\phi_c) = \mu^2 + 3\lambda\phi_c^2$,

$$
\begin{aligned}
V_{\text{eff}}(\phi_c) = {} & \frac{1}{2} \mu^2\phi_c^2 + \frac{1}{4} \lambda\phi_c^4 \\
& + \frac{\hbar}{64\pi^2} \{m^4(\phi_c) \ln m^2(\phi_c)/m^2(v) \\
& - \frac{3}{2} m^4(\phi_c) + 2m^2(\phi_c)m^2(v)\}
\end{aligned} \tag{53}
$$

Again, one can easily write down the analogous formula for the renormalized effective potential in the standard model: to Eq. 53 add the contribution of other physical fields, i.e. with $m_i(\phi_c) = g_i \phi_c$,

$$V_{\text{eff}}(\phi_c) = \{\text{RHS of (53)}\}$$
$$+ \sum_i C_i \frac{\hbar}{64\pi^2} \frac{m_i^4}{v^4} \left\{ \phi_c^4 \ln \phi_c^2 / v^2 - \frac{3}{2} \phi_c^4 + 2\phi_c^2 v^2 \right\} \tag{54}$$

The crucial quantity is, in the standard model,

$$B \equiv \sum_i C_i \frac{\hbar}{64\pi^2 v^4} m_i^4$$
$$= \frac{\hbar}{64\pi^2 v^4} \left(6m_W^4 + 3m_Z^2 - 4\sum_f m_f^2 \right) \tag{55}$$

where the sum is over quark and lepton types, not forgetting a factor of 3 for color in the quark sum. In practice, only the top quark is heavy enough to contribute significantly, so

$$B \simeq \frac{\hbar}{64\pi^2 v^4} \left(6m_W^4 + 3m_Z^2 - 12m_t^4 \right) \tag{56}$$

The expressions (54) and (56) have been used as starting points for the derivation of bounds on scalar and fermion masses. Interested readers are referred to the literature for further details: a good starting point is the review by M. Sher referred to in the preface.

To conclude this lecture, we discuss how the results so far obtained provide indications that the $\lambda\phi^4$ theory may be trivial in the infinite cut-off limit, and how this might lead to an upper bound on the scalar particle mass. From (48) and (51), we deduce that

$$\lambda_0 = \lambda + \frac{9\hbar\lambda^2}{16\pi^2} \ln \Lambda^2 / em^2 \tag{57}$$

where $\ln e = 1$ and $m^2 \equiv m^2(v)$. One refers to λ_0 as the bare, and to λ, as the renormalized, coupling constant. Renormalizability of the theory implies that given a value for λ, a change in cut-off can be compensated entirely by a change in λ_0, such that

$$\frac{\partial \lambda_0}{\partial \ln \Lambda} = \frac{9\hbar\lambda^2}{8\pi^2} = \frac{9\hbar\lambda_0^2}{8\pi^2} \tag{58a}$$

accurate to $O(\hbar)$. One also notices that Eq. (57) would still hold, had we used a different renormalization prescription to define $\delta\mu^2$ and $\delta\lambda$: indeed, all that was required was the cancellation of Λ^2 and $\ln \Lambda^2$ terms in $V_{\text{eff}}(\phi_c)$. More generally, we expect a relation of the form

$$\lambda_0 = \lambda + \frac{9\hbar\lambda^2}{16\pi^2} \ln \Lambda^2 / M^2 \tag{58b}$$

where M^2 is some chosen momentum (wave-number!) scale, not to be confused with the physical mass of the scalar particle discussed earlier in this lecture. The renormalized coupling constant will now depend on this scale, while the bare coupling clearly does not, so

$$\frac{\partial \lambda_0}{\partial \ln M} = 0 = \frac{\partial \lambda}{\partial \ln M} - \frac{9\hbar\lambda^2}{8\pi^2} + O(\hbar^2) \tag{59}$$

i.e. to $O(\hbar)$,

$$\frac{\partial \lambda}{\partial \ln M} \equiv \beta(\lambda) = \frac{9\hbar\lambda^2}{8\pi^2} \tag{60}$$

This is the beta function for $\lambda\phi^4$ theory, which describes how the renormalized coupling constant changes with the renormalization scale M. With the boundary condition $\lambda(M = \Lambda) = \lambda_0$, Eq. (60) can be easily integrated to give

$$\frac{1}{\lambda_0} = \frac{1}{\lambda(M)} - \frac{9\hbar}{8\pi^2} \ln \Lambda/M \tag{61}$$

or

$$\lambda_0 = \frac{\lambda(M)}{1 - \frac{9\hbar\lambda(M)}{8\pi^2} \ln \Lambda/M} \tag{62}$$

If one holds $\lambda(M)$ fixed while trying to take the limit $\Lambda \to \infty$, one sees that the bare coupling λ_0 diverges at a critical value of the cut-off, Λ_C, with

$$\frac{\Lambda_C}{M} = \exp\left(8\pi^2/9\hbar\lambda(M)\right) \tag{63}$$

This is known as a Landau ghost pole at Λ_C, for reasons that will not be explained further. The only way to push this pole to infinity and so define a reasonable renormalized theory is apparently to have $\lambda(M) = 0$, i.e. to have the renormalized coupling vanish in the infinite cut-off limit: this, if true, would make the renormalized theory trivial. Indications that this is indeed what happens have been found in lattice simulations.

What would the implications of the triviality of the self-interacting scalar field theory be for the Standard Model? In practice, one could still have the Higgs mechanism for mass generation, but the scalar theory should only be viewed as an effective low-energy description, with an explicit upper cut-off Λ to its range of validity. This in turn implies that the scalar particle mass should be less than this cut-off: if one takes this mass to be given by $m_H^2 = 2\lambda(m)v^2$, say, one would be led to the rough upper bound

$$\begin{aligned} m_H^2 &< \frac{16\pi^2 v^2}{9\hbar \ln \Lambda/m_H} \\ &< \frac{16\pi^2 v^2}{9\hbar} \end{aligned} \tag{64}$$

that follows by requiring avoidance of the Landau pole and the conservative bound $\ln \Lambda/m_H > 1$. With $\hbar^{1/2}v \simeq 246$ GeV the Higgs vev in the Standard Model, we get the often quoted upper bound on the Higgs boson mass,

$$\hbar m_H < O(1) \text{ TeV} \quad . \tag{65}$$

Lattice simulations typically give a slightly lower value around 700 GeV or so. It should nevertheless be remembered that triviality of $\lambda\phi^4$ theory, although nearly proven, has not in fact been rigorously established as yet.

V. RELATIVISTIC COSMOLOGY AND SCALAR FIELDS

In this final lecture, we will see how classical scalar fields have come to play a dominant role in recent investigations of cosmology, insofar as they can lead to inflation in the early universe. To explain these concepts, it will be necessary to first set the stage with a mini-course of relativistic cosmology as formulated in the framework of Einstein's general relativity. The dynamical equations of the expanding universe will be written down in a physically motivated way without recourse to the full mathematical machinery of general relativity. After a discussion of some of the successes of Big Bang cosmology will come a short treatment of one of its shortcomings known as the flatness problem. This will set the stage for a short discussion of inflation, how it resolves the flatness problem, and of the role of scalar fields in the early universe in allowing for inflation.

BASICS OF RELATIVISTIC COSMOLOGY

Relativistic cosmology is based on the general theory of relativity proposed by Albert Einstein in 1915, who gave us a revolutionary view of gravitation and space-time. Instead of having matter move through a passive space-time, Einstein asserted that the presence of matter distorted space-time itself so that the interval between two events becomes

$$ds^2 = g_{\mu\nu}(x)dx^\mu dx^\nu \tag{1}$$

that is, a space-time dependent metric $g_{\mu\nu}(x)$ replaces the usual constant Minkowski metric $\eta_{\mu\nu}$. Particles then move from point A to point B along the analog of straight lines in this curved space-time grid (known as geodesics), according to the variational equation

$$\delta \int_A^B ds = 0 \tag{2}$$

The central problem is to determine the form of $g_{\mu\nu}(x)$ given a distribution of matter/energy characterized by an energy-momentum tensor $T_{\mu\nu}$. Einstein guessed the following beautiful equation relating the geometry of space-time and the distribution of matter,

$$R_{\mu\nu} - \frac{1}{2}\, g_{\mu\nu} R = 8\pi G T_{\mu\nu} \tag{3}$$

where $R_{\mu\nu}$ and R are constructed in terms of derivatives of $g_{\mu\nu}(x)$ and are a geometric measure of space-time curvature (they are known as the Ricci tensor and scalar curvature, respectively, and are in turn related by the process of tensor contraction, using $g_{\mu\nu}(x)$, from a rank-four tensor known as the Riemann curvature). In Eq. (3), G is Newton's gravitational constant: the stress tensor $T_{\mu\nu}$ no longer satisfies the simple conservation law $\partial^\nu T_{\mu\nu} = 0$, which is now replaced by a different one involving a generalized derivative known as a covariant derivative. Note that for $T_{\mu\nu} = 0$ corresponding to empty space, the usual Minkowski metric $\eta_{\mu\nu}$ is a solution of (3).

In this framework, the dynamical equations of cosmology follow from a number of hypotheses. The first is the cosmological principle, which asserts that space is isotropic and homogeneous, namely that there are no preferred directions or locations. Given the existence of galaxies as aggregates of matter, it is clear that this assumption should hold over suitably large distances so that one can speak of a constant average matter density.

The second assumption is that the energy-momentum tensor of matter/energy is that of a perfect fluid,

$$T^{\mu\nu} = (p + \rho)u^\mu u^\nu - p g^{\mu\nu} \tag{4}$$

The general form of the line-element (1) follows from the cosmological principle,

$$ds^2 = dt^2 - S^2(t)d\sigma^2 \tag{5}$$

Here, one uses a simple system of coordinates in which clocks at rest with respect to elements of the cosmological fluid keep time at the same rate independently of their position and of elapsed time: such coordinates are known as comoving coordinates, and one speaks of a cosmic time t. In Eq. (5), the "space" part of the interval, $d\sigma^2$, must then be time independent and of the form

$$d\sigma^2 = h_{ij}(\vec{x})dx^i dx^j \tag{6}$$

with $h_{ij} = \delta_{ij} f^{(i)}(r)$ diagonal, because of the assumed homogeneity and isotropy of space.

To determine the precise form of (6), one imagines that the homogeneous and isotropic space is embedded in a fictitious four dimensional space, in much the same way that the line-element or distance on the surface of a sphere is embedded in three-dimensional space. Introducing a fictitious fourth space coordinate w, we consider the volume defined by the equation

$$x^2 + y^2 + z^2 \pm w^2 = \pm a^2 \tag{7}$$

where a is the radius of curvature. This covers the three possible cases known to exist for homogeneous spaces: if $a \to \infty$, one gets flat three dimensional space (think of the analog in 3 dimensions: for example, the surface of a sphere of infinite radius is flat!). For finite a^2, the geometry is spherical if the positive sign is chosen, corresponding to positive curvature: in the opposite case of negative curvature, the geometry is hyperbolic. Writing $\rho^2 = x^2 + y^2 + z^2$, the spatial interval in the three dimensional space of (x, y, z), subject to the constraint (7) is

$$
\begin{aligned}
d\sigma^2 &= dx^2 + dy^2 + dz^2 \pm dw^2 \\
&= d\rho^2 + \rho^2 d\Omega^2 \pm dw^2
\end{aligned}
\tag{8}
$$

where we have gone to spherical coordinates $(x, y, z) \to (\rho, \theta, \phi)$ and $d\Omega^2 = d\theta^2 + \sin^2\theta d\phi^2$. To get dw^2, we use (7) to get

$$
\rho d\rho \pm w dw = 0
\tag{9}
$$

so that, squaring, and using (7) again,

$$
dw^2 = \frac{\rho^2 d\rho^2}{a^2 \mp \rho^2}
\tag{10}
$$

Thus, in all,

$$
d\sigma^2 = \frac{d\rho^2}{1 \mp \rho^2/a^2} + \rho^2 d\Omega^2
\tag{11}
$$

which is indeed of the form (6), and if $a^2 \to \infty$ this corresponds to flat space $d\sigma^2 = dx^2 + dy^2 + dz^2$. One can define a dimensionless coordinate r via $\rho = ra$, and correspondingly the dimensionful factor $R(t) = S(t)a$ in (5) to finally get the so-called Lemaitre-Friedmann-Robertson-Walker line element or metric,

$$
ds^2 = dt^2 - R^2(t) \left[\frac{dr^2}{1 - kr^2} + r^2 d\Omega^2 \right]
\tag{12}
$$

Here, $k = 0$ corresponds to flat space (but not in general flat <u>space-time</u> because of the presence of the factor $R(t)$), while $k = +1$ (-1) correspond to the spherical (hyperbolic) geometry for space.

The interpretation of this line-element is as follows: the particles of the cosmological fluid have <u>fixed</u>, dimensionless, coordinate positions (r, θ, ϕ) on a curved 3-D space grid, with the <u>spacing</u> between points changing with time according to the scale factor $R(t)$. To take a 2-D example, consider the surface of an expanding balloon: the $R(t)$ is the expanding radius of curvature of the balloon into a fictitious 3-D space. In particular, $\dot{R}(t) = dR/dt$ does <u>not</u> measure an expansion into an "outside" region of space: space (and space-time) just <u>is</u>, there is no "outside".

We can discuss at this point the origin of cosmological red-shifts. Consider a galaxy at coordinate r_e and cosmic time t_e emitting light that arrives at us at the later time t_0, our coordinate being set to $r_0 = 0$: light still travels along null intervals $ds^2 = 0$ so one has

$$\int_{t_e}^{t_0} \frac{dt}{R(t)} = \int_0^{r_e} \frac{dr}{\sqrt{1 - kr^2}} \equiv \xi \tag{13}$$

the next crest of the light-wave reaches us at $t_0 + \Delta t_0$, having left at $t_e + \Delta t_e$ and still

$$\int_{t_e + \Delta t_e}^{t_0 + \Delta t_0} \frac{dt}{R(t)} = \xi \tag{14}$$

Since $R(t)$ hardly changes during the intervals Δt_0 and Δt_e, it follows that

$$\frac{\Delta t_e}{R(t_e)} = \frac{\Delta t_0}{R(t_0)} \tag{15}$$

Since the period $T(= \Delta t_0$ or $\Delta t_e)$ is equal to the wavelength λ, or the inverse of frequency ν, one has

$$\frac{\lambda_0}{\lambda_e} = \frac{R(t_0)}{R(t_e)} \tag{16}$$

or

$$\nu(t)R(t) = \text{ const.} \tag{17}$$

One usually defines the red-shift

$$z = \frac{\lambda_0 - \lambda_e}{\lambda_e} = \frac{R(t_0)}{R(t_e)} - 1 \tag{18}$$

This will indeed be a red-shift if $R(t_0) > R(t_e)$. One can expand

$$R(t_e) = R(t_0) - (t_0 - t_e)\dot{R}(t_0) + \frac{1}{2}(t_0 - t_e)^2 \ddot{R}(t_0) + \ldots \tag{19}$$

Keeping only terms to lowest order in $(t_0 - t_e)$, the red-shift can be written as

$$z \simeq (t_0 - t_e)\frac{\dot{R}(t_0)}{R(t_0)} + \ldots \tag{20}$$

In the same approximation, the dimensionless coordinate distance ξ is given by

$$\xi \approx \frac{t_0 - t_e}{R(t_0)} + \ldots \tag{21}$$

Now we recall that to transform a coordinate distance into a physical distance, one multiplies by the scale factor at the appropriate time, so in fact

$$r(t_0) = \xi R(t_0) \approx (t_0 - t_e) + \ldots \tag{22}$$

and so one has the approximate result

$$z \simeq \frac{\dot{R}(t_0)}{R(t_0)} r(t_0) + \ldots \tag{23}$$

In this limit, $z \ll 1$ and one can identify its value with the velocity of the source, so

$$v(t_0) \simeq H(t_0) r(t_0) \tag{24}$$

for small red-shifts. Equations (23) and (24) are the famous Hubble law of cosmological expansion, whereby observed red-shifts are attributed to a universal expansion of galaxies, with the famous Hubble "constant":

$$H(t_0) \equiv H_0 = 100 \, h_0 \, km \sec^{-1} Mpc^{-1} \tag{25}$$

where $\frac{1}{2} \lesssim h_0 < 1$ reflects our uncertain knowledge of this crucial parameter. We see that H_0 is constant in the sense that it is independent of spatial position, but it does change with time: in general,

$$H(t) = \dot{R}(t)/R(t) \tag{26}$$

The dynamical equations of cosmology will determine the behavior of $R(t)$ given a distribution of matter/energy characterized by $T_{\mu\nu}$. One set of equations will follow from (3), given the LFRW metric used to calculate the left-hand side, and the perfect fluid tensor (4) on the right-hand side. Another set of equations follows from the more general covariant conservation law for $T_{\mu\nu}$. In fact, the assumed homogeneity and isotropy of space allows for a very simple derivation of the resulting equations for $R(t)$ as we now indicate.

THE DYNAMICAL EQUATIONS OF COSMOLOGY

One exploits the homogeneity and isotropy of space to derive equations that will hold locally, where the effects of curvature can be correctly described by Newtonian gravitation. A first equation follows from the local conservation law for the perfect fluid tensor $T^{\mu\nu}$, with $g_{\mu\nu}(x)$ simply replaced by $\eta_{\mu\nu}$: one has, as seen in the second lecture,

$$\partial_\nu T^{\mu\nu} = 0 = \partial_\nu \left[(p + \rho) u^\mu u^\nu - p \eta^{\mu\nu} \right] \tag{27}$$

where p and ρ refer to the frame in which the fluid is momentarily at rest. We get,

$$u^\mu \partial_\nu [(p + \rho) u^\nu] + (p + \rho) u^\nu \partial_\nu u^\mu - \partial^\mu p = 0 \tag{28}$$

Multiplying by u_μ, and using the fact that since $u_\mu u^\mu = 1$, $u_\mu \partial_\nu u^\mu = 0$, we find

$$\partial_\nu \left[(p + \rho) u^\nu \right] - u_\mu \partial^\mu p = 0 \tag{29}$$

or in fact

$$(p + \rho) \partial_\nu u^\nu + u^\nu \partial_\nu \rho = 0 \tag{30}$$

In the frame in which the fluid is momentarily at rest, one has $u^0 = 1$, $\vec{u} = 0$; this also means that $\partial_\nu u^0 = 0$, so $\partial_\nu u^\nu = \vec{\nabla} \cdot \vec{u}$. Eq. (30) finally becomes

$$\frac{\partial}{\partial t} \rho + (p + \rho)\vec{\nabla} \cdot \vec{v} = 0 \tag{31}$$

because $\vec{u} = u^0 \vec{v}$ where $\vec{v}$ is the usual velocity, and $\vec{\nabla} u^0 = 0$. From the assumed homogeneity and isotropy of the fluid, it follows that ρ and p are only functions of time, $\rho = \rho(t)$, $p = p(t)$. Thus, $\vec{\nabla} \cdot \vec{v}$ must also only be a function of time:

$$\vec{\nabla} \cdot \vec{v} = \frac{\partial v_x}{\partial x} + \frac{\partial v_y}{\partial y} + \frac{\partial v_z}{\partial z} \equiv 3\, H(t) \tag{32}$$

since all terms in the middle of (32) must be equal. Integrating (32) leads to the local form of the Hubble law,

$$\vec{v} = H(t)\, \vec{r} \tag{33}$$

so by Eq. (24) $H(t)$ is identified with $\dot{R}/R$. Eq. (31) now takes the form

$$\dot{\rho} + 3\, H(t)(p + \rho) = 0 \tag{34}$$

This is essentially the equation $dU = -pdV$ for the change in internal energy due to work against pressure, as is easily verified by writing $U = \rho V$ and remembering that volumes scale like $R^3(t)$.

To get a second equation, we consider a small test particle of mass μ at a distance $R(t)\xi = r(t)$ from the origin: if the volume $\frac{4\pi}{3}\, R^3(t)\xi^3$ is sufficiently small, we can use Newtonian mechanics to write for the total energy of the particle

$$\frac{1}{2}\, \mu v^2 - \frac{1}{R\xi}\, G\mu \cdot \frac{4\pi}{3}\, \rho R^3 \xi^3 \equiv E \tag{35}$$

and where we have used that only the matter/energy interior to the volume contributes to the gravitational potential energy. Using (33), we get

$$H^2(t)R^2(t) - \frac{8\pi G\rho}{3}\, R^2(t) = 2E/\mu\xi^2 \tag{36}$$
$$\equiv \text{const.}$$

There are three possibilities for the constant on the right-hand side of Eq. (36): it can be positive, negative or zero, and in fact is to be identified with $-k$, where $k = 0, \pm 1$ is the parameter determining the geometry of space in the LFRW line element.

Thus one recovers the second dynamical equation of relativistic cosmology,

$$H^2(t) - \frac{8\pi G}{3}\, \rho(t) = -\frac{k}{R^2(t)} \tag{37}$$

A third equation can be derived by taking the time derivative of (37) and using (34) to eliminate $\dot{\rho}$, namely

$$\ddot{R}(t) = -\frac{4\pi}{3}\,(\rho + 3p)R(t) \tag{38}$$

If it is assumed that the equation of state $p = p(\rho)$ is such that $\rho + 3p > 0$ (this is clearly the case for radiation and ordinary matter, but see later in this lecture!), this means that $\ddot{R}(t) < 0$, i.e. that the expansion of the universe is decelerated by the attractive gravitational forces. Looking to the past, this also means that $\dot{R}(t)$ was larger at earlier times than it is now: thus, at some finite time in the past, set at $t = 0$, $R(t = 0)$ must have been zero and the density infinite! Thus this model is the called Big Bang cosmology.

Note also that if $k = -1$ or 0, Eq. (37) shows that $\dot{R}(t)$ can never vanish and so the universe must expand forever. If $k = +1$ on the other hand, $\dot{R}$ will vanish when $\rho R^2 = 3/8\pi G$: after that, the universe will contract since $\ddot{R} < 0$, and will eventually collapse back to $R = 0$ (Big Crunch!).

Can we distinguish between these three cases on the basis of present-day observations? Write equation (37) in the form

$$k = \frac{8\pi G}{3}\,\rho R^2 \left(1 - \frac{1}{\Omega(t)}\right) \tag{39}$$

where we have introduced the density parameter $\Omega(t)$,

$$\Omega(t) = \frac{8\pi G\rho(t)}{3H^2(t)} \tag{40}$$

Note that k is a constant throughout the evolution of the universe: no classical process can change the geometry of space.

It is clear that a measurement of the present-day value Ω_0 of the density parameter serves to distinguish the three cases: if $\Omega_0 > 1$, then $k = +1$ and the universe will eventually collapse, while if $\Omega_0 \leq 1$ one has either $k = 0$ or -1 and the universe will expand forever. The present-day density parameter can be written as

$$\Omega_0 = \rho_0/\rho_c \tag{41}$$

where the critical density ρ_c can be calculated from the Hubble parameter (Eq. 25) to be

$$\rho_c = \frac{3H_0^2}{8\pi G} = 1.88 \times 10^{-29} h_0^2 g/cm^3 \tag{42}$$

Now the value of Ω_0 is inferred from velocity measurements on scales greater than $100\ kpc$, which are all consistent with $0.1 \lesssim \Omega_0 \lesssim 0.4$. Conservative bounds however allow larger values, so the possible range is more like $0.05 \leq \Omega_0 \leq 4$. The portion of Ω_0 in luminous matter is much smaller, $5 \times 10^{-3} \leq \Omega_{lum} \leq 0.02$: the excess of Ω_0 over Ω_{lum} leads to the inference that most of the matter in the universe is nonluminous, "dark" matter, but that is another story.

It is remarkable that we are unable to determine whether space is open, closed or flat: but as we shall see later, it is even more remarkable that Ω_0 is close to unity! This is at the heart of the so-called flatness problem, a possible resolution of which appeals to the physics of scalar fields leading to what is known as "inflation", as we shall see.

THERMAL HISTORY OF THE UNIVERSE

We should now discuss some consequences of the dynamical equations of cosmology: assume that at some period in the past, matter and radiation were at equilibrium at some temperature T. Then we can take ($\hbar = c = 1$),

$$
\begin{aligned}
\rho_r &= \frac{\pi^2}{30} \, g_{\text{eff}}(T)T^4 \\
\rho_m &= n_B m_N + \frac{3}{2} \, n_B T \\
p_r &= \frac{1}{3} \, \rho_r \\
p_m &= n_B T
\end{aligned}
\tag{43}
$$

the usual expressions for the density and pressure, where n_B is the baryon number density, m_N is the nucleon rest mass and Boltzmann's constant k has been set to unity: this means that temperature is measured in energy units, the equivalence being

$$
1 \text{ eV} = 1.16 \times 10^4 \; K
\tag{44}
$$

The factor $g_{\text{eff}}(T)$ counts the number of degrees of freedom of particles (bosons and fermions) with masses considerably smaller than T: the precise expression is $g_{\text{eff}} = g_B + \frac{7}{8} \, g_F$ where $g_\gamma = 2$, etc. We will not worry too much about this factor for our crude numerical estimates. Using Eq. (43) in Eq. (34) yields

$$
\left(4 \, \rho_r + \frac{3}{2} \, n_B T \right) \frac{\dot{T}}{T} = - (4 \, \rho_r + 3 \, n_B T) \, \frac{\dot{R}}{R}
\tag{45}
$$

where we have used the fact that the total baryon number $N_B \propto n_B R^3$ is conserved, so that $\dot{n}_B = -3 n_B \dot{R}/R$. Now consider the quantity $3 \, n_B T/4 \, \rho_r$: the number density for radiation is given by the formula

$$
n_r = \frac{\zeta(3)}{\pi^2} \left(g_B + \frac{3}{4} \, g_F \right) T^3
\tag{46}
$$

where g_B and g_F are again numbers of degrees of freedom of massless bosons and fermions: the effective total number in (46) differs from that in ρ_r because the integrals over Bose-Einstein and Fermi-Dirac distributions are different in the two cases; $\zeta(3) = 1.202\ldots$ is the Riemann zeta-function of argument three. We see that

$$\rho_r/n_r \approx 3\ T \approx \rho_\gamma/n_\gamma \tag{47}$$

so that one can write, roughly

$$\frac{3n_B T}{4\rho_r} \approx \frac{n_B}{n_r} \approx \frac{n_B}{n_\gamma} \tag{48}$$

where $n_B/n_\gamma \equiv \eta$ is the ratio of baryon – to photon number densities. Eq. (45) now takes the form

$$\left(1 + \frac{\eta}{2}\right) \frac{\dot{T}}{T} = -(1+\eta)\ \frac{\dot{R}}{R} \tag{49}$$

We see that if $\eta \ll 1$, one has $R\dot{T} + \dot{R}T = 0$ so

$$R(t)T(t) = \text{const.} \tag{50}$$

throughout the evolution of a universe with matter and radiation. Further, in that case, since $n_B \propto R^{-3}$, η itself is constant ($\eta \propto 1/R^3 T^3$). Also, and this is very important, by combining (50) with (17) we see that a black-body distribution of photons with frequency $\nu(t)$ remains a black-body distribution since then $\nu(t)/T(t)$ = constant also! We conclude that if $\eta \ll 1$, there should be a relic black-body distribution of photons permeating the universe!

It was George Gamow, in the late forties, who provided the first estimate of the present-day temperature of this universal black-body background of photons. Before giving his reasoning, we shall need some preliminary information. Under the assumption $RT = \text{const.}$, we note that at early times corresponding to higher temperatures, the radiation energy density will dominate and further by Eq. (37), the curvature term proportional to T^2 will be negligible as compared to the T^4 dependence of the term involving ρ_r. Thus a knowledge of k is unnecessary for an exploration of the early universe, and one can safely set $k = 0$ in (37) for all practical purposes.

In that case, with the equation of state in the simple form

$$p = (\gamma - 1)\rho \tag{51}$$

where $\gamma = 4/3$ for radiation, $\gamma \cong 1$ for matter, one can easily solve the dynamcial equations for $R(t)$. From (34) we get

$$\dot{\rho} + 3\gamma\rho\dot{R}/R = 0 \tag{52}$$

so one has

$$\rho R^{3\gamma} = \text{const.} \equiv A \tag{53}$$

as the appropriate equation for different choices of γ: when radiation dominates, this is $\rho R^4 = $ const. while for matter dominance $\rho R^3 = $ const.: in the first case, one has $RT= $ const. again, while in the second case $\rho R^3 = $ const. reduces to the statement of baryon number conservation.

Putting (53) into (37) now yields

$$\dot{R} R^{\frac{3\gamma}{2}-1} = \left(\frac{8\pi GA}{3}\right)^{1/2} \tag{54}$$

or

$$\frac{2}{3\gamma} \frac{d}{dt} R^{3\gamma/2} = \left(\frac{8\pi GA}{3}\right)^{1/2} \tag{55}$$

which is easily integrated subject to $R(t = 0) = 0$:

$$R(t) = \left(6\pi\gamma^2 GA\right)^{1/3\gamma} t^{2/3\gamma} \tag{56}$$

whence

$$H(t) = \frac{\dot{R}(t)}{R(t)} = \frac{2}{3\gamma t} \tag{57}$$

$$\rho(t) = \frac{1}{6\pi\gamma^2 Gt^2} \tag{58}$$

The observables H and ρ are seen not to depend on the choice of A or of R. When radiation dominates, one can use (58) and (43) to write for the age of the universe

$$t = \left(\frac{45}{16\pi^3 G}\right)^{1/2} \frac{g_{\text{eff}}^{-1/2}}{T^2} \tag{59}$$

It is a useful exercise to put in units in this equation, to obtain

$$t(\text{sec}) \cong \frac{2g_{\text{eff}}^{-1/2}(T)}{T^2(\text{MeV})} \tag{60}$$

with t in second and T in MeV: thus, at $T \simeq 1$ MeV, the universe was about 1 second old.

We can now proceed with Gamow's argument. One starts with the observation that helium comprises about 25% of the observed mass of the universe, a number too large to have resulted from the burning of stars (which do in fact turn hydrogen to helium, but not nearly enough). Gamow concluded that most of the helium must have been produced in the early universe. Many nuclear reactions participate in this process, but the crucial one is the production of deuterons $n+p \rightarrow D+\gamma$. Once deuterons are produced via this electromagnetic process, they can combine in further reactions not involving photons and quickly convert into helium. Now the inverse reaction of photodissociation $\gamma + D \rightarrow n + p$ will keep the number of deuterons at a rather low equilibrium figure, but if the universe has cooled enough, there won't be enough photons with energy larger than the binding energy of the deuteron, about 2 MeV, to effect this dissociation: a crude estimate of the temperature at which this will occur is a fraction of this binding energy, say $T_* \simeq 100$ keV, corresponding to an age of the universe of about 100 sec, or a few minutes (hence the title of Steven Weinberg's celebrated book on cosmology "The First Three Minutes": highly recommended). Now the rate of formation of deuterons depends on $(\sigma v_{\text{rel}})_{D\gamma}$ the cross-section for $n + p \rightarrow D + \gamma$ times the relative velocity and the density n_B of baryons: this rate should be such that the reaction occurs within the age $t_* \simeq 100$ sec of the universe: given an estimate per (σv), this gives us an idea of the primordial baryon density $n_B(t_*)$, via the estimate

$$(\sigma v_{\text{rel}})_{D\gamma} n_B(t_*) \sim t_*^{-1} \tag{61}$$

We can guess $\sigma_{D\gamma} \sim O(10^{-3}) \times 10^{-26}$ cm^2 because this is an electromagnetic process while for $T_* \sim 100$ keV, $v^2/c^2 \approx 10^{-4}$ for nucleons, so $(\sigma v_{\text{rel}})_{D\gamma} \approx 10^{-20}$ cm^3 sec^{-1}. From (61) one then gets the estimate, $t_* \sim 10^2$ sec,

$$n_B(t_*) \simeq 10^{18} \text{ cm}^{-3} \tag{62}$$

and using Eq. (46) with $T_* \simeq 100$ keV one finds with (62) that

$$\eta = \frac{n_B(t_*)}{n_\gamma(t_*)} \simeq 10^{-10} \tag{63}$$

which is indeed much less than one. Then, using the fact that $RT = $ const. one can write in terms of present-day values (recall $n_B \propto R^{-3}$)

$$\frac{T_0}{T_*} \approx \left[\frac{n_{B,0}}{n_B(t_*)}\right]^{1/3} \tag{64}$$

The present number density of baryons is estimated at about 10^{-7} cm^{-3} so

$$T_0 \approx 5 \times 10^{-9} T_*$$
$$\approx 6°K \tag{65}$$

This remarkable prediction of the Big Bang cosmology was borne out by the discovery of the microwave background with $T_0 \simeq 2.73°K$ by Penzias and Wilson in 1965. The value of 10^{-10} for η deduced in (63) is also remarkably close to the accepted range of values for this parameter.

We can use η to estimate the age of the universe at which it became matter dominated, by writing

$$T_{eq}^4 \approx n_B(t_{eq}) m_N \tag{66}$$

i.e.

$$T_{eq} \approx \eta \, m_N \approx 10^{-1} \text{ eV} \tag{67}$$

so (with (60))

$$t_{eq} \approx O(10^6) \text{ years} \tag{68}$$

Given that the age of the universe is evidently in excess of the age of the earth at 4.5×10^9 yrs., we see that for most of its evolution, the universe was matter dominated. One can then use Eq. (57) with $\gamma = 1$ to estimate the age of the universe at

$$t_0 \cong \frac{2}{3H_0} = \frac{6.7 \times 10^9 \text{ yrs}}{h_0} \tag{69}$$

corresponding to the assumption of flat ($k = 0$) space. Nuclear dating and the inferred ages of old globular clusters lead to the estimate of $15 - 20 \times 10^9$ yrs. for t_0. If h_0 is in fact close to one, the assumption $k = 0$ i.e. $\Omega_0 = 1$ exactly would seem to be untenable. However, there are many ways to remedy this problem: we will not discuss this further.

The thermal history of the early universe can be followed given the Standard Model of particle physics, as well as nuclear physics inputs, to allow for an accurate calculation of relative nuclear abundances, which led to an early bound on the number of effectively massless neutrinos (via $g_{\text{eff}}(T)$), and many other successes. But some aspects of the Big Bang cosmology remain mysterious, especially as regards the initial conditions required to lead to some of the properties of the present day universe.

Particle physics can offer some help: for example, the fundamental parameter η can be calculated given a model of particle physics with baryon number violation and C- and CP-violation. η is then seen as the asymmetry between baryons and antibaryons produced in out-of-equilibrium processes in the very early universe, and is even calculable entirely in terms of particle physics parameters. In what follows, we shall explain another difficulty with initial conditions known as the flatness problem, and show how a hypothetical homogeneous classical scalar field playing a dominant role at some stage in the history of the early universe can resolve this problem through a mechanism known as inflation.

THE FLATNESS PROBLEM

It is a puzzle that Ω_0, while uncertain, is still of order unity. To see why, recall the fundamental equation (37),

$$H^2(t) - \frac{8\pi G \rho(t)}{3} = -\frac{k}{R^2(t)}$$

During a period of expansion dominated by a fluid with equation of state $p = (\gamma - 1)\rho$ in the early universe, we saw that $\rho R^{3\gamma} = \text{const}$. For $3\gamma > 2$, we notice that the ρ-dependent term decreases faster than the curvature-dependent term as the universe expands and cools, and it could be expected that by the present-day the universe's expansion should be curvature-dominated: this is the flatness puzzle, namely why is the present-day universe consistent with the flat space geometry, $k = 0$. Note that if $k = 0$ as an initial condition, there is no problem! However, the data do allow the other two possibilities as well, and this is indeed remarkable, as we shall now explain.

Let us estimate the value of $\Omega(t_i)$ at some early time t_i during the era of radiation dominance, which ended at t_{eq} (Eq. 68): from Eq. (39) we have

$$
\begin{aligned}
\frac{1}{\Omega(t_i)} - 1 &= \left(\frac{1}{\Omega(t_{eq})} - 1 \right) \frac{\rho_{eq} R_{eq}^2}{\rho_i R_i^{2}} \\
&= \left(\frac{1}{\Omega(t_{eq})} - 1 \right) \left(\frac{R_i}{R_{eq}} \right)^2
\end{aligned}
\tag{70}
$$

where we have used $\rho R^4 = $ const. in the radiation dominated-era (this relation holds strictly speaking for $k = 0$ only: this minor inconsistency can be remedied and does not change the conclusion we shall reach). One can now do the same for the matter-dominated from t_{eq} to the present-day t_0:

$$
\begin{aligned}
\frac{1}{\Omega(t_{eq})} - 1 &= (\frac{1}{\Omega_0} - 1) \frac{\rho_0 R_0^{2}}{\rho_{eq} R_{eq}^2} \\
&= (\frac{1}{\Omega_0} - 1) \frac{R_{eq}}{R_0}
\end{aligned}
\tag{71}
$$

where now we have $\rho R^3 = $ const. Putting (70) and (71) together we get

$$
\begin{aligned}
\frac{1}{\Omega(t_i)} - 1 &= \left(\frac{1}{\Omega_0} - 1 \right) \left(\frac{R_i}{R_{eq}} \right)^2 \left(\frac{R_{eq}}{R_0} \right) \\
&\equiv \left(\frac{1}{\Omega_0} - 1 \right) \left(\frac{T_{eq}}{T_i} \right)^2 \left(\frac{T_0}{T_{eq}} \right)
\end{aligned}
\tag{72}
$$

using $RT = $ const. Rearranging, we see that

$$
\left(\frac{T_{eq}}{T_0} \right) \left(\frac{T_i}{T_{eq}} \right)^2 \left| \frac{1}{\Omega(t_i)} - 1 \right| = \left| \frac{1}{\Omega_0} - 1 \right| \simeq O(1)
\tag{73}
$$

The factor multiplying the density parameter on the left-hand side will depend on t_i: we know that $T_{eq}/T_0 \simeq 10^3$ from previous considerations. If we consider taking $T_i \simeq O(1)$ MeV, then the factor is 10^{17}: this means that the kinetic energy of expansion and the gravitational attraction at $T_i \sim 1$ MeV had to be matched and cancel to an accuracy of 10^{-17} or better for Ω_0 to be of order unity now.

If one chose an earlier time or temperature, this fine-tuning gets even more extreme: choosing $T_i \simeq 10^{15}$ GeV increases the overall-factor to 10^{53}!! It is only because of this extreme fine-tuning that the universe is as it seems at present: any failure to exquisitely fine-tune would have led to a universe which would have lasted only a mere fraction of a second before recollapsing, or would have expanded and cooled much too fast for any of the physical processes that led to our universe as we know it to have occurred.

A possible solution to this difficulty is easily guessed: we saw at the beginning of this paragraph that one should expect a problem when $3\gamma - 2 > 0$, i.e. when $\rho + 3p > 0$ for $\rho > 0$. Conversely, a period of expansion with $3\gamma - 2 < 0$, i.e. $\rho + 3p < 0$, $\rho > 0$ should alleviate the problem. We shall first see how this works, and then see how a classical scalar field can have such a peculiar equation of state.

Suppose that the very early universe is radiation dominated from $t = 0$ to $t = t_i$, at which time it enters a period of expansion dominated by a fluid with $3\gamma - 2 < 0$, which ends at $t = t_f$; the universe is then once again radiation-dominated from t_f to t_{eq}, and matter dominated from t_{eq} to the present-day t_0. A similar reasoning to the one that led us to Eq. (72) will now give the following equation of the $\Omega(t)$ at some time $t < t_i$:

$$
\begin{aligned}
\Omega(t)^{-1} - 1 &= (\Omega_0^{-1} - 1)\left(\frac{\rho_0 R_0^{\,2}}{\rho_{eq} R_{eq}^{\,2}}\right)\left(\frac{\rho_{eq} R_{eq}^{\,2}}{\rho_f R_f^{\,2}}\right)\left(\frac{\rho_f R_f^{\,2}}{\rho_i R_i^{\,2}}\right)\left(\frac{\rho_i^{\,2} R_i^{\,2}}{\rho(t) R^2(t)}\right) \\
&= (\Omega_0^{-1} - 1)\left(\frac{T_0}{T_{eq}}\right)\left(\frac{T_{eq}}{T_f}\right)^2 \left(\frac{R_f}{R_i}\right)^{2-3\gamma}\left(\frac{T_i}{T(t)}\right)^2
\end{aligned}
\tag{74}
$$

where we have used $\rho R^{3\gamma} = $ const. and $RT = $ const. in eras of radiation and matter dominance, leaving γ open during the new era of expansion between t_i and t_f. Note this expression reduces to (72) when $t = t_i = t_f$.

It is now clear that with $R_f > R_i$, if one chooses $2 - 3\gamma > 0$ as mentioned before, one can arrange for the new factor in (74) to be large enough to cancel out the very small result obtained by taking the product of the remaining factors, thus resolving the fine-tuning problem. Turning this around, one sees that in fact given any reasonable initial condition, the existence of this new era of expansion in the early universe would lead us to expect that independently of the value of k, one would predict that the present-day value of density parameter Ω_0 should be unity, to great accuracy.

Why is the choice $2 - 3\gamma > 0$ so effective? We have already noted that with $\rho > 0$ this corresponds to $\rho + 3p < 0$, which in view of the dynamical equation (37) corresponds to <u>accelerated</u> expansion $\ddot{R} > 0$. It is conceivable that under these conditions the argument given earlier for a singularity at $t = 0$ could be obviated, but we will not discuss this further.

The question is now what kind of new physics could lead to an equation of state with $\gamma < 2/3$ so that a period of accelerated expansion, that is, a period of <u>inflation</u>, could occur in the Early Universe. We will now see that we have learned about classical scalar fields could provide the answer.

CLASSICAL SCALAR FIELDS AND INFLATION

We recall from the second lecture that the energy-momentum tensor of a classical scalar field with a Lagrangian density $\mathcal{L}$

$$\mathcal{L} = \frac{1}{2}\, \partial_\mu \phi \partial^\mu \phi - V(\phi) \tag{75}$$

is of the form (II.23):

$$T^{\mu\nu} = \partial^\mu \phi \partial^\nu \phi - \eta^{\mu\nu}\mathcal{L} \tag{76}$$

To see that this can be written in the form (II.36) of a perfect fluid energy-momentum tensor, introduce a unit time-like four-velocity u^μ via

$$u^\mu = \partial^\mu \phi / u_\nu \partial^\nu \phi \tag{77}$$

which also implies that

$$\partial^\mu \phi \partial_\mu \phi = (u \cdot \partial \phi)^2 \tag{77'}$$

With the identification (77) and (77'), and by comparison with (II.36) we see that a classical scalar field behaves like a perfect fluid with

$$\rho = \frac{1}{2}\, \partial_\mu \phi \partial^\mu \phi + V(\phi) \tag{78}$$

$$p = \frac{1}{2}\, \partial_\mu \phi \partial^\mu \phi - V(\phi) \tag{79}$$

both quantities being Lorentz invariant as they must.

In cosmological applications we shall be only interested in uniform, time dependent fields so in fact, in the comoving frame,

$$\rho = \frac{1}{2}\, \dot{\phi}^2 + V(\phi) \tag{80}$$

$$p = \frac{1}{2}\, \dot{\phi}^2 - V(\phi) \tag{81}$$

We see that the equation of state for the classical scalar field allows for two extreme values, $\gamma = 2$ for $p = \rho$ when $\dot{\phi}^2 \gg V(\phi)$, and $\gamma = 0$ for $p = -\rho$, when $\dot{\phi}^2 \ll V(\phi)$. This latter case obviously satisfies the inflationary constraint $\gamma < 2/3$, and we will now examine it a little more closely.

When $\gamma = 0$, the perfect fluid energy-momentum tensor reduces to ($\rho = $ const.)

$$T^{\mu\nu} = \rho\, \eta^{\mu\nu} \tag{82}$$

which as we have seen in (III.30) is characteristic of the vacuum expectation value of the energy momentum tensor of a quantum field: thus, the equation of state $p = -\rho$ is often referred to as that of the vacuum. We can go back to the dynamical equations of cosmology and now examine their behaviour when $\gamma = 0$, i.e. vacuum dominance, in the early universe when curvature effects can be neglected. One easily sees that the discussion leading to the rule $RT = $ const. in Eq. (50) is unmodified, as is that leading to $\rho R^{3\gamma} = $ const., which now is just $\rho = $ const. Eq. (54) now simply reads

$$\frac{\dot{R}}{R} = H = \left(\frac{8\pi G\rho}{3} \right)^{1/2} = \text{ const.} \tag{83}$$

so the universe undergoes exponential expansion (inflation!) during a period of vacuum dominance:

$$R \propto e^{Ht} \tag{84}$$

with constant H given in (83). Because $RT = $ const. still holds, the universe is very cold at the end of an inflationary period, and a physical mechanism must be built-in for reheating the universe: this is an essential part of any model of inflation, and the universe is typically reheated to a temperature close to that at the beginning of inflation. This is definitely not an adiabatic process!

If the inflation starts at t_i at the end of a radiation dominated period, one will have by the continuity of $H(t)$ the relation

$$H = 1/2 \; t_i \tag{85}$$

Referring back to (74) with $\gamma = 0$ we see that

$$\left(\frac{R_f}{R_i} \right)^2 = e^{2(t_f - t_i)/t_i} \tag{86}$$

If $T_i \simeq 10^{14}$ GeV, say, corresponding to $t_i \simeq 10^{-35}$ sec, an expansion R_f/R_i by a factor of order 10^{35} is feasible if $t_f \approx 4 \times 10^{-34}$ sec, a duration of the merest fraction of a second! Of course, a mechanism for ending the period of vacuum dominance and inflation, while initiating the process of reheating must exist and is an important part of models of inflation. Inflation by these kinds of amounts also solves other conceptual difficulties of the Big Bang cosmology such as the horizon problem, which will not be further discussed here.

To see how the equation of state $p = -\rho$ could effectively hold in the course of the evolution of a classical scalar field in the early universe, we can use (80) and (81) in Eq. (34) to obtain the equation for the scalar field

$$\ddot{\phi} + 3 \; H(t)\dot{\phi} + V'(\phi) = 0 \tag{87}$$

If, say as a result of initial conditions on the scalar field, there is a period of vacuum dominance so that $H = $ const., we can see how this regime can be self-consistently maintained over a certain period. Take as the simplest example the potential $V(\phi) = \frac{1}{2} \; m^2 \phi^2$ so that $V'(\phi) = m^2 \phi$. Then (87) is easily solved,

$$\phi(t) = \phi_1 e^{\lambda_1 t} + \phi_2 e^{\lambda_2 t} \tag{88}$$

where t is measured from the beginning of the vacuum dominance regime, and with

$$\lambda_{1,2} = -\frac{3H}{2} \pm \sqrt{\frac{9H^2}{4} - m^2} \qquad (89)$$

If $m^2/H^2 \ll 1$, $\lambda_1 \simeq -m^2/3H$ and $\lambda_2 \simeq -3H$. We see that $\phi(t)$ quickly tends to a nearly constant value, so that $T^{\mu\nu} \simeq \eta^{\mu\nu}V(\phi)$ and one has self-consistency. These ideas underlie all models of inflation, for example the chaotic inflationary model of Linde.

Thus we see that a period of inflation can take place if a classical scalar field (known sometimes as the "inflaton") comes to dominate the evolution of the very early universe. It has been shown that the quantum fluctuations of such a field in an exponentially expanding background could give rise to density perturbations which at a much later time could lead to the formation of galaxies. All inflationary models predict on this basis that $\Omega_0 = 1$ to an accuracy of typically 10^{-4}, although it is not possible to go into details here.

It should be emphasized that no scalar field in any of the many postulated extensions of the Standard Model (i.e. motivated by particle physics considerations) has the properties expected of an inflaton. A new, very weakly interacting scalar must be invented purely for the purpose of generating inflation.

THE COSMOLOGICAL CONSTANT PROBLEM

We close with a few remarks on the cosmological constant problem. The original equation suggested by Einstein for relativistic cosmology was in fact of the form (cf. (3))

$$R_{\mu\nu} - \frac{1}{2}\, g_{\mu\nu}R = 8\pi G T_{\mu\nu} + \Lambda g_{\mu\nu}$$

with Λ known as the cosmological constant. We see that it can be understood, given (82), as being due to a contribution to $T_{\mu\nu}$ of a universal constant energy density

$$\epsilon_{\text{VAC}} = \Lambda/8\pi G \qquad (90)$$

The present bounds on Ω_0 lead to a bound

$$\epsilon_{\text{VAC}} \lesssim O\left(10^{-46}\right)\ \text{GeV}^4 \qquad (91)$$

which is an incredibly small number on the scale of particle physics: for example, one usually characterizes the vacuum of quantum chromodynamics as having an energy density of order $-(200\ \text{MeV})^4$ or so, which is fully 42 orders of magnitude too large. How to reconcile the cosmological constraint (91) with the expectations of particle physics as regards the properties of the physical vacuum is an unsolved problem, a discussion of which is worthy of another set of lectures! The interested reader is referred to an excellent (though quite technical in places) review, emphasizing the point of view of the particle physicist: S. Weinberg, Reviews of Modern Physics, volume 61, p.1, 1989.

ACKNOWLEDGEMENTS

I wish to thank Professor T. Ferbel for his kind invitation to lecture in St. Croix and for so ably organizing a very pleasurable and enriching summer school, as well as for his patience in enduring the long delays surrounding the production of this written version of the lectures. I am especially indebted to Ms. Judy Mack for her tireless efforts in typing up the manuscript. The preparation of these lecture notes was supported in part by the Department of Energy, under contract number DE-AC02-83ER40105.

THE QUARK MIXING MATRIX AND CP VIOLATION

C. Jarlskog

Department of Physics
University of Stockholm
Stockholm, Sweden

An introduction is given to the quark mixing matrix and CP violation. The topics discussed include Classification of Symmetries, the Discrete P, C and CP Symmetries, the Electroweak Model, the Lagrangian of the Electroweak Model, the Quark Mass Terms and the Origin of CP Violation, Spontaneous Symmetry Breaking, the Quark Mixing Matrix, the Number of Families, CP Violation with Three Families, the KM Type Parametrizations of the Quark Mixing Matrix, Unitarity Triangles and the CP Violation Area, the Commutator Formalism for CP Violation, Manifestations of CP Violation, and Comments on CP Violation in the Beauty Sector.

1 Introduction

The most important single concept in theoretical particle physics in the past decades has been that of SYMMETRY. It has indeed served as the "password" of the theorist into the realm of "new physics". You, being young experimentalists, may wonder if there has also been a corresponding single "password" in experimental physics. Words such as accelerators and detectors come to the mind but I don't know the answer.

Historically, given a fundamental equation one could find its symmetries and use them, for example, in order to simplify calculations. More importantly, symmetries have always led to a better understanding of what is going on and thus to enlightenment. With time the theorists have become braver in employing symmetries. Symmetries have been

Techniques and Concepts of High Energy Physics VII
Edited by T. Ferbel, Plenum Press, New York, 1994

postulated to ensure the validity of one's (preconceived) notations on what the laws of nature are allowed to do or not to do. For instance, it was believed that the laws of nature must be symmetric under the exchange of left and right and therefore only theories which obeyed that symmetry were considered to be "reasonable". Finally, as the latest step in this development the symmetries have been promoted to be the "mother" of the laws of nature. Indeed the laws of nature are nowadays believed to be "born" out of symmetry requirements. This does not mean that we understand the origin of the symmetries themselves. The question remains why some symmetries and not others are "chosen". For instance, why is the symmetry of the Standard Model $SU(3) \times SU(2) \times U(1)$ and why not some other symmetry? Because of the very important role played in the past by symmetries, in the next section I will list some of these concepts. I will not go through them in these notes as the material is easily found in many books and articles, for example those listed in [1].

2 Classification of Symmetries

Since symmetries have been very essential in the development of the modern particle physics they have gotten specific labels on them specifying what they are and what they are not. One often meets such antonyms as

- space-time versus internal symmetry

- continuous symmetry versus discrete symmetry

- exact versus approximate symmetry

- global symmetry versus local symmetry

- abelian symmetry versus non-abelian symmetry

- gauge versus non-gauge symmetry

- manifest versus hidden or spontaneously broken symmetry

- accidental versus fundamental symmetry

It is essential to understand what the above concepts mean as they belong to the everyday vocabulary of the theoretical particle physicists. Throughout these notes, we shall see several examples of the above symmetries. It is generally (but not always) believed that the only good symmetry is a gauge symmetry, preferably based on a simple non-abelian symmetry group. The reason is that then there is only one single "unified" force and the number of free parameters in the theory is minimal. However, we should not forget that beliefs are not necessarily facts. Perhaps our prejudice that the gauge symmetries are the only really good symmetries is not true. It may turn out that a discrete-looking symmetry is actually the footprint of a subtle, perhaps gauge-looking, symmetry at a more fundamental level. That is why we should treat all, specially badly understood, symmetries with great respect and hope that someday we will understand them. Another point to be emphasized is that some of the above antonyms are unified in a number of symmetries employed by nature. For example the CPT symmetry "unifies" parity (P) and time-reversal (T) which are space-time symmetries with charge-conjugation symmetry (C) which is an "internal symmetry, i.e., it has nothing to do with space-time.

⋆ **Exercise**

Give examples of the above symmetries. Of course, I don't expect you to be sure that your examples of "accidental versus fundamental symmetry" are truly so. A symmetry believed to be accidental may turn out to be a fundamental one, within an extended framework.

Many theorists have a "dream" (sometimes referred to as Einstein's dream) that they will someday discover that all forces in nature are unified in just one single unified force. Although we can't be sure that such a dream has anything to do with reality we are encouraged by the past achievements. The path in the past has been that of unification, the last step being that we now have a unified-looking theory [2] of weak and electromagnetic interactions and there is a good chance that a Grand Unified Theory of electroweak and strong interactions will eventually be found. However, that may take quite some time. The electroweak theory is not all that unified - it has two fundamental coupling constants instead of just one which is what one would expect from a truly unified theory. What is much worse is that the electroweak model predicts the existence of "Higgs forces". We don't usually think of them as new forces but that is exactly what they are, if one takes the model seriously. These hypothetical "forces" resemble the "good old" low energy hadronic interactions, such as the pion-nucleon forces and because of that they are often referred to as the Yukawa sector of the model. This state of affairs is considered to be very unsatisfactory by most physicists. One tends to believe that there is a much more elegant underlying theory of which the electroweak model is but the "tip of the iceberg". At present we don't know what there is below the "tip". There is, on the theoretical scene no sufficiently attractive theory that we can get very excited about and say to ourselves that "it must be it!". Indeed the question of what lies beyond the standard model is a very open one.

3   The Discrete Symmetries P, C and CP

These symmetries have been with us for a long time. Historically, they were found to be symmetries of the Schrödinger and the Dirac equations. It was firmly believed by physicists that the laws of nature must be symmetric under parity and charge conjugation symmetries. Thus the experimental verification, in 1957 [3] of the ideas of Lee and Yang [4] that P and C need not be good symmetries came as a shocking surprise to the physics community. After all how could the laws of nature not be left-right symmetric? We still lack the answer to that question but have no other choice than to follow nature in formulating our theories. That is why the Standard Electroweak Model [2] postulates violation of parity and charge-conjugation symmetries by introducing left-handed doublets but right-handed singlets for quark and lepton families. We still hope that someday the "P and C puzzle" will be understood. The "founding fathers" such as Pauli who did not like the violation of P and C symmetries found some consolation in the fact that the combined CP symmetry seemed to be respected in nature. Pauli was satisfied with what he believed to be a fact that God has the particles in His left hand and the antiparticles in His right hand. The antiparticle could then be identified with the CP-image of the particle. The CP "mirror" looked beautiful. In it left and right were interchanged and the signs of the internal charges were reversed and that sufficed to turn a particle into its antiparticle. By the time this beautiful idea got generally accepted came the second chocking surprise. CP was not a good symmetry! The discovery of CP violation [5] meant that the CP mirror was cracked. It could not be used to define the antiparticle. One had now to rely on a more complex mirror, the CPT mirror, for that purpose. An additional "mysterious" feature of CP violation was that it was "tiny". Parity and C symmetry were manifestly maximally violated but apparently not CP! What could be the origin of such a tiny effect? Why did it only appear in the decays of the long-lived neutral kaon? As soon as the physics community got over the CP-"shock" it tried to understand and explain it. In fact, in the early years of CP there was a great deal of theoretical as well as experimental interest in it and the origin of CP violation was much debated. Was it due to strong interactions?, electromagnetic?, weak interactions?, superweak interactions? For example, one could argue that since α/π and the measured CP violation parameter e were not so different the origin of CP violation was most probably electromagnetic. Some of the models proposed attributed CP violation to the existence of:

- a long-range (cosmic) force

- an octet of strongly interacting vector bosons

- new particles called the a-particles or chimerons

- mirror fermions

- leptoquarks

- gravitational dipole, etc.

That period was a very exciting one both for theorists, who could freely speculate, and experimentalists who could do a new class of experiments. A large amount of work was done on improving the early results obtained in the neutral kaon system as well as on searches for CP violation in many other processes. Gradually, however, it was found that there was no evidence for CP violation anywhere except in some decay modes of the K_L. Unfortunately, the situation has not changed since then and there is still no evidence for CP violation anywhere else. For a review of the theoretical as well the experimental situation of CP during those early years see, for example, the book by Kabir [6].

By 1974 the experiments on CP violation had become quite accurate and within the experimental uncertainties one had $\eta_{+-} = \eta_{00}$. Here η_{ij} denotes the ratio of the amplitudes $A(K_L \to \pi^i \pi^j)/A(K_S \to \pi^i \pi^j)$. All results were in agreement with Wolfenstein's superweak model [7]. A substantial improvement of the experimental results appeared to be so difficult that the subject of CP violation looked like a closed chapter in the history of physics. See, for example, the 1974 review talk by Kleinknecht [8]. Ten years later, however, the situation had changed dramatically and CP violation had once again become a central issue in particle physics. This was partially due to the fact that a new generation of measurements of η_{+-} and η_{00}, at CERN and at Fermilab., were planned with an unprecedented accuracy. A second reason for the renewed interest in CP violation had to do with the discovery of new fundamental hadrons, charm and beauty, which called for an extension of the earlier works and promised new phenomena. In fact nowadays the major item on the shopping list of the future beauty factories is the search for CP violation. A third important factor had to do with the entrance of cosmology and the early universe into the lives of particle physicists. It is believed that the observed baryon asymmetry of the universe has to do with CP violation. This subject has been pursued for more than a decade by many physicists but unfortunately there is as yet no real understanding of how the necessary CP violation is generated in the early universe. To all this we should also add the discovery of the so called strong CP problem also called the "θ puzzle" in QCD and its possible solutions to which a great deal of theoretical work has been devoted.

By far the most serious "Model" of CP violation in our time, i.e., almost three decades after the discovery of the phenomenon, is the minimal version of the Standard Electroweak

Model [2]. Although the Standard Model is not expected to be the "final theory" it does seem to be a very good perhaps only "effective theory" and is highly successful in describing all relevant data. The Standard Electroweak Model [2] with only two families of quarks and leptons is automatically CP conserving (see below). In order to incorporate CP violation one way was through extension to 3 families, as proposed by Kobayashi and Maskawa [9]. However it is generally believed that the Minimal Standard Model does not generate enough CP violation to explain the baryon asymmetry of the universe. In addition, the solution of the "θ puzzle" also seems to lie outside the realm of the Minimal Standard Model. Thus other sources of CP violation seem to be badly needed. Some such possible sources have been often discussed in the literature since a long time ago and because of various reasons not necessarily related to cosmology or the theta puzzle. For instance, the possibility of having spontaneous CP violation (i.e., the hypothesis that CP violation is due to relative phases of vacuum expectation values of different Higgs fields) was put forward at a very early stage. For an early review see Ref. [10]. Models for CP violation within the Standard Model but with an extended Higgs sector were also proposed [11]. Some authors proposed that the origin of CP violation be sought in extended gauge theories such as the left-right symmetric models. For a review see, for example, Ref. [12].

3.1 Parity in Field Theory

In the language of local relativistic quantum field theory, which up to now has been the most successful universal language in particle physics phenomenology, the P and C symmetries are not intuitively obvious symmetries. They have to be "hammered" into us. This is specially the case for fermions. For example, the parity operation of a particle with spin 1/2 (Dirac) particle is somewhat involved. Suppose that $\psi(\overline{x}, t)$ is a solution to the Dirac equation, viz.,

$$(i\gamma^\mu \partial_\mu - m)\psi(\overline{x}, t) = 0$$

where our notations here are as in Ref. [13]. We see immediately that the above equation does not imply that $\psi(-\overline{x}, t)$ is also a solution to the Dirac equation. In fact it is not because of the partial derivative; the space and time components in it transform with opposite signs. The solution is to define the parity image to be (proportional to) $\gamma_0 \psi(-\overline{x}, t)$ which automatically satisfies the original Dirac equation, viz.,

$$(i\gamma^\mu \partial_\mu - m)\gamma_0 \psi(-\overline{x}, t) = 0$$

Thus $\gamma_0 \psi(-\overline{x}, t)$ multiplied by an arbitrary phase is the parity transformed Dirac spinor. In the momentum space what happens is that the "small components" of the spinor change sign. In other words, the momentum of the particle reverses its direction, exactly as expected from a parity tranformation. For a scalar field, say $\phi(\overline{x}, t)$, however, the parity transformation is much simpler, viz., $\phi(\overline{x}, t) \rightarrow \phi(-\overline{x}, t)$, up to a phase. You may wonder why the spinor case is more complicated than say the scalar one. The answer is because $\psi(\overline{x}, t)$ is not directly measurable. Measurables are the bilinears (current-like objects) for example of the form $\overline{\psi}(\overline{x}, t) \mathbf{O} \psi(\overline{x}, t)$, where $\mathbf{O}$ is one of the usual five "operators", scalar, pseudoscalar, vector, axialvector and tensor. These have indeed simple and intuitively transparent transformation properties under parity. It is a useful exercise to compute the transformation properties of the bilinears, i.e., to do the following exercise.

⋆ **Exercise**

Assuming that the $\psi_j(\overline{x}, t) \rightarrow exp(i\phi_j)\gamma_0\psi_j(-\overline{x}, t)$, where $j = 1, 2$ and ϕ_j are arbitrary phases, show that the bilinears transform as follows:

	$(\vec{x}, t)$	**P** $\rightarrow$	$(-\vec{x}, t)$
scalar	$\overline{\psi}_1\psi_2$	$\rightarrow$	$\overline{\psi}_1\psi_2$
pseudoscalar	$\overline{\psi}_1\gamma_5\psi_2$	$\rightarrow$	$-\overline{\psi}_1\gamma_5\psi_2$
vector	$\overline{\psi}_1\gamma_\mu\psi_2$	$\rightarrow$	$\overline{\psi}_1\gamma^\mu\psi_2$
axialvector	$\overline{\psi}_1\gamma_\mu\gamma_5\psi_2$	$\rightarrow$	$-\overline{\psi}_1\gamma^\mu\gamma_5\psi_2$
tensor	$\overline{\psi}_1\sigma_{\mu\nu}\psi_2$	$\rightarrow$	$\overline{\psi}_1\sigma^{\mu\nu}\psi_2$

where ψ_1 and ψ_2 are two different fields (for example, the fields of an up quark and a down quark respectively) and for simplicity we take the "parity phases" of the two fields to be equal, $\phi_1 = \phi_2$.

The transformation of the bilinears, in the above exercise, also tell us how scalars, pseudoscalars, vectors, axial vectors, and tensors transform under parity. Note that since we are using the Feynman metric (as for example in the book by Bjorken and Drell [13]) $Q^k = -Q_k$ where k=1,2,3 and $Q_0 = Q^0$, for any four vector Q^μ. Thus the spatial components and the time component of the above bilinears transform with opposite signs, under parity, as is intuitively expected.

A point to be kept in mind is that the parity of a free field, by itself, is devoid of physical significance (a noninteracting particle is not even observable). For example, consider a free charged scalar field $\phi(\vec{x}, t)$ described by the free Lagrangian density

$$L_0(\vec{x}, t) = \partial^\mu \phi^+(\vec{x}, t) \, \partial_\mu \phi(\vec{x}, t) - m^2 \phi^+(\vec{x}, t)\phi(\vec{x}, t)$$

where m is the mass. The equation of motion reads $(\partial^\mu\partial_\mu + m^2)\phi(\vec{x},t) = 0$. We may define the parity operation via $\phi(\vec{x},t) \rightarrow \phi^P(\vec{x},t) = exp(i\eta)\phi(-\vec{x},t)$, where η is an arbitrary real number (phase). By definition, the parity transformed field, $\phi^P(\vec{x},t)$, is constructed such that it satisfies the same equation of motion as $\phi(\vec{x},t)$. Furthermore, $L_o(\vec{x},t) \rightarrow L_o(-\vec{x},t)$ which means that the corresponding action, i.e., the integral over space-time of the Lagrangian, $\int d^4x L_{int}$, is invariant (simply verified by changing the dummy variable $\vec{x}$ to $-\vec{x}$, in the action integral). However, the parity of $\phi(\vec{x},t)$ is not fixed because η is arbitrary. The above argument is also valid if there are several free fields. It is the **interaction** which fixes the "relative" parities of the different fields, provided that parity is a good symmetry. If parity is not a good symmetry then there is no way of choosing the phases (the η's) such that $L(\vec{x},t) \rightarrow L(-\vec{x},t)$. Let us demonstrate this by giving a very simple example, as follows.

Consider a spin-1/2 (Dirac) particle interacting with a real (neutral) spin-0 object. Let us call them "nucleon" and "meson" respectively and take the Lagrangian density to be

$$
\begin{aligned}
L(\vec{x},t) &= i\bar{\psi}(\vec{x},t)\gamma^\mu\partial_\mu\psi(\vec{x},t) - m\bar{\psi}(\vec{x},t)\psi(\vec{x},t) + \frac{1}{2}\partial^\mu\phi(\vec{x},t)\,\partial_\mu\phi(\vec{x},t) - V \\
&\quad \bar{\psi}(\vec{x},t)\,(a + ib\gamma_5)\psi(\vec{x},t)\,\phi(\vec{x},t)
\end{aligned}
\tag{1}
$$

Here V is a "potential", which is assumed to depend on $(\phi(\vec{x},t))^2$, being such that our meson has a well defined mass. The last term in the above Lagrangian describes the interactions of the nucleon with the meson . These interactions are scalar (with coupling constant a) and pseudoscalar (with coupling constant b). These constants are real due to the requirement that L be hermitian (without the i in front of the γ_5 the constant b would have been purely imaginary). The hermiticity of the Lagrangian is necessary in order to have a unitary transition operator (S-matrix), viz., $S = T(exp(i\int d^4x L_{int}))$, where T indicates time-ordering. Under parity, we have

$$
\begin{array}{lll}
\text{scalar density} & \bar{\psi}(\vec{x},t)\psi(\vec{x},t) & \rightarrow \quad \bar{\psi}(-\vec{x},t)\psi(-\vec{x},t) \\
\text{psuedoscalar density} & \bar{\psi}(\vec{x},t)\gamma_5\psi(\vec{x},t) & \rightarrow \quad -\bar{\psi}(-\vec{x},t)\gamma_5\psi(-\vec{x},t)
\end{array}
$$

These relations do not change if we define $\psi^P(\vec{x},t) = exp(i\eta)\,\psi(-\vec{x},t)$, η being an arbitrary real number. Thus, we see that the two pieces (scalar and pseudoscalar) in the last term require respectively $\phi(\vec{x},t) \rightarrow \phi(-\vec{x},t)$ and $\phi(\vec{x},t) \rightarrow -\phi(-\vec{x},t)$ in order to yield parity conservation, i.e., $L(\vec{x},t) \rightarrow L(-\vec{x},t)$. These two requirements being incompatible parity is violated, in the model considered. How can such a parity violation in the Lagrangian give a physically measurable effect? For that to happen we need to find a process $i \rightarrow f$, for which the two terms with opposite parity contribute such that we get a transition probability of the form

$$\Gamma(i \rightarrow f) = |A_1 + A_2|^2 \tag{2}$$

Since the two amplitudes A_1 and A_2 have opposite parities then going to the parity conjugate process we would get

$$\Gamma(i_P \rightarrow f_P) = |A_1 - A_2|^2 \tag{3}$$

In such a case the transition probablities for the process and its parity image will, in general, be different indicating that parity is violated. This simple example also demonstrates that in order to get a large effect it is desirable to choose transitions such that $|A_1| \simeq |A_2|$, The parity asymmetry is defined by

$$a_P = \frac{\Gamma - \bar{\Gamma}}{\Gamma + \bar{\Gamma}} \tag{4}$$

where Γ and $\bar{\Gamma}$ denote the rates for the process and its parity image respectively. Note that $-1 \leq a_P \leq +1$. A few comments are now in order.

First of all, note that from the transformation properties of the fields one may derive the transformation properties of the annihilation and the creation operators and use them to define how the states transform (see for example Ref.[1]).

In spite of the fact that parity is not always conserved in Nature, it is of paramount importance in physics, because it is conserved in strong and electromagnetic interactions. Furthermore, it is useful for "classification" purposes, even in cases where it is not conserved.

It is interesting to note that there is an additional "complication" when defining the parity of a free Majorana particle (i.e., a spin-1/2 particle which is its own antiparticle). We shall briefly discuss this point in the subsection on CP-transformations.

3.2   C-Conjugation

At the level of free fields, the operation of C-conjugation is rather simple to understand. A free field has a Fourier decomposition of the symbolic form $\phi(x) = \sum \{(\ldots)\, a + (\ldots) b^\dagger\}$ where $a(a^\dagger)$ and $b(b^\dagger)$ are respectively the annihilation (creation) operators for the "particle" and its "antiparticle". Here, we have suppressed the plane waves as well as indices, such as momenta, spin, colour, etc. Under C-conjugation the role of the operators a and b is interchanged. C-conjugation flips the signs of internal charges, such as the electric charge, baryon number, strangeness, etc. However, we should keep in mind that C is not a good symmetry, in Nature. Therefore, the C-conjugation does not change a particle into its antiparticle. At the level of free fields it does interchange the a and b but these need not correspond to the operators for the physical particles. [In order to define the antiparticle one must use the CPT operation, where T denotes the time-reversal operation.] Nevertheless, just like parity and for the same reasons, C-conjugation is a very useful concept. We shall now define the action of C-conjugation for the free fields and examine its consequences for the interaction pieces in the Lagrangian.

Under C, the free fields transform as follows

$$
\begin{array}{cccc}
 & & \mathbf{C} & \\
\text{scalar field} & \phi(x) & \rightarrow & \phi^\dagger(x) \\
\text{Dirac spinor} & \psi(\vec{x}, t) & \rightarrow & C\bar{\psi}^T(\vec{x}, t) \\
 & \bar{\psi}(\vec{x}, t) & \rightarrow & -\psi^T(\vec{x}, t)\, C^{-1} \\
\text{vector field} & V_\mu(\vec{x}, t) & \rightarrow & -V_\mu^\dagger(\vec{x}, t) \\
\text{Axial field} & A_\mu(\vec{x}, t) & \rightarrow & A_\mu^\dagger(\vec{x}, t)
\end{array}
$$

Here C is a 4 by 4 unitary matrix which satisfies the condition $C^{-1}\gamma_\mu C = -\gamma_\mu^T$, where the letter T means "transpose" (should not be confused with T-reversal). This relation for the C is obtained by requiring that the C-conjugated Dirac field satisfies the same free field equation of motion as the field itself. The matrix C is usually taken to be $C = i\gamma^2\gamma^0$. Again, as in the case of parity, we may include arbitrary phases in the above definitions and, for example take $\phi \rightarrow exp(i\eta)\phi^\dagger$, where η is an arbitrary phase. We shall discuss such (sometimes necessary) modifications whenever needed. Let us , through a simple example, demonstrate how one does such C transformations. Take two Dirac fields ψ_1 and ψ_2 interacting with a vector field V_μ via

$$
L_{int} = g \left\{ \bar{\psi}_2 \gamma^\mu \psi_1\, V_\mu + \bar{\psi}_1 \gamma^\mu \psi_2\, V_\mu^\dagger \right\}
$$

where the second term is the hermitian conjugate of the first one . It must be there to ensure the hermiticity of the Lagrangian. Furthermore, g is a real coupling constant. Under C we get

$$
\begin{aligned}
\bar{\psi}_2 \gamma^\mu \psi_1 &\rightarrow -\psi_2^T\, C^{-1} \gamma^\mu C\, \bar{\psi}_1^{\,T} = \psi_2^T\, (\gamma^\mu)^T\, \bar{\psi}_1^{\,T} \\
&= -[\bar{\psi}_1 \gamma^\mu \psi_2]^T = -\bar{\psi}_1 \gamma^\mu \psi_2
\end{aligned}
$$

where the minus signs, on the last line, come from changing the order of the fields ψ_1 and ψ_2. These, being fermion fields, anticommute with each other. Note that the quantity in the last backets is a one by one "matrix". Therefore, the superscript T can be omitted. Comparing the above two equations, we see that the Lagrangian will be invariant provided we take $V_\mu \to -V_\mu^\dagger$, under the C-operation. Thus C conjugation has simply interchanged the two terms in our Lagrangian. This is a general feature i.e., the C-conjugation takes a term in the Lagrangian into its hermitian conjugate (up to a sign and modulus the coupling constant in front which is not affected). Since V_μ annihilates a vector particle, say V^- and creates its antiparticle V^+ and $V^\dagger$ does just the opposite operations, we see that in this example the roles of the particle and its antiparticle have been interchanged.

For a real vector field such as the photon, the invariance under C gives, $A_\mu \to -A_\mu$. This is no surprise. The photon field couples to the electromagnetic current $\bar{\psi}\gamma^\mu\psi$. Since the current changes sign the photon field must also do so in order to ensure the invariance of the electromagnetic interactions. The photon has, therefore, $C = -1$ which is a very useful information to know as it has far reaching consequences (selection rules). Finally, in this section, we quote the transformation properties of the spinor bilinears, under C

$$
\begin{array}{lccc}
& & \textbf{C} & \\
& (\vec{x}, t) & \to & (\vec{x}, t) \\
\text{scalar} & \bar{\psi}_1\psi_2 & \to & \bar{\psi}_2\psi_1 \\
\text{pseudoscalar} & \bar{\psi}_1\gamma_5\psi_2 & \to & \bar{\psi}_2\gamma_5\psi_1 \\
\text{vector} & \bar{\psi}_1\gamma_\mu\psi_2 & \to & -\bar{\psi}_2\gamma_\mu\psi_1 \\
\text{axialvector} & \bar{\psi}_1\gamma_\mu\gamma_5\psi_2 & \to & \bar{\psi}_2\gamma_\mu\gamma_5\psi_1 \\
\text{tensor} & \bar{\psi}_1\sigma_{\mu\nu}\psi_2 & \to & -\bar{\psi}_2\sigma_{\mu\nu}\psi_1
\end{array}
$$

Here again, for simplicity, we have taken the C-phases of the two fields to be the same. The proof of the above equations is again left as an exercise. From these transformations, we see that the "pion-nucleon" interaction Lagrangian, Eq. (1) conserves C.

3.3 CP Transformations

Finally, the operations CP or PC are obtaind by performing the indicated operations in sequence. The spinor bilinears transform under CP according to

$$
\begin{array}{lccc}
& & \textbf{CP} & \\
& (\vec{x}, t) & \to & (-\vec{x}, t) \\
\text{scalar} & \bar{\psi}_1\psi_2 & \to & \bar{\psi}_2\psi_1 \\
\text{pseudoscalar} & \bar{\psi}_1\gamma_5\psi_2 & \to & -\bar{\psi}_2\gamma_5\psi_1 \\
\text{vector} & \bar{\psi}_1\gamma_\mu\psi_2 & \to & -\bar{\psi}_2\gamma^\mu\psi_1 \\
\text{axialvector} & \bar{\psi}_1\gamma_\mu\gamma_5\psi_2 & \to & -\bar{\psi}_2\gamma^\mu\gamma_5\psi_1 \\
\text{tensor} & \bar{\psi}_1\sigma_{\mu\nu}\psi_2 & \to & -\bar{\psi}_2\sigma^{\mu\nu}\psi_1
\end{array}
$$

As an example of the application of the above CP transformations, consider again our "pion-nucleon" interaction, Eq. (1). It is CP violating, because it violates P but conserves C. In fact Lagrangians similar to it have been used to describe CP violation in various processes, for example, in the decay $K_L \rightarrow \mu^+\mu^-$. As in the case discussed in connection with parity, here we will find that the probability of the above transition and its CP conjugate reaction (in which $\mu^+ \leftrightarrow \mu^-$ and the directions of the momenta are reversed but the spins are not affected) are not the same. In order to see such an effect one needs to measure the polarization of at least one of the muons which is a very tough experiment. As in the case of parity, we may define the CP asymmetry as

$$a_{CP} = \frac{\Gamma - \bar{\Gamma}}{\Gamma + \bar{\Gamma}} \tag{5}$$

where now Γ and $\bar{\Gamma}$ denote respectively the rate for the process and its *CP conjugate* reaction. As before this asymmetry also lies between -1 and $+1$ and the conditions for getting a large asymmetry is similar to what we had for parity (see the discussion after Eq. (3)).

Let us now briefly discuss the "slight complication" in the case of Majorana particles. A Majorana particle is a spin-1/2 particle which is its own antiparticle. This means that ψ and $\bar{\psi}$ are not independent, i.e., they contain the same annihilation and creation operators. This is because the particle is its own antiparticle whereby a and b, which were the annihilation operators for the particle and antiparticle respectively, are equivalent. Let us try to take the relation between ψ and $\bar{\psi}$ to be of the form

$$\psi = \psi^C \tag{6}$$

Such a relation is called "the Majorana condition". Now, we try to use our previous definition that the parity transformation of our fermion field is given by

$$\psi^P(\vec{x}, t)) = \gamma^0 \psi(-\vec{x}, t) \tag{7}$$

Taking the C-transform of this equation we find

$$\{\psi^P(\vec{x}, t))\}^C = \{\gamma^0 \psi(-\vec{x}, t)\}^C = -\gamma^0 \{\psi(-\vec{x}, t)\}^C = -\gamma^0 \psi(-\vec{x}, t) = -\psi^P(\vec{x}, t) \tag{8}$$

The reason for getting the minus sign is that the matrices C and γ_0 (corresponding to charge conjugation and parity operations, respectively) anticommute. Comparing Eqs. (6) and (8) we see that the Majorana conditions for the field and its parity image have opposite signs. This we don't like at all. We would like to have a "universal" Majorana condition for our field. You may convince yourself that putting in arbitrary phases in the above definitions of the Majorana condition and P transformation (i.e., $\psi = exp(i\theta^C)\psi^C$ and $\psi^P(\vec{x},t)) = exp(i\theta^P)\gamma^0\psi(-\vec{x},t))$ and requiring the Majorana condition to be the same for the field and its parity image yields that $exp(2i\theta^P) = -1$. In other words, the parity phase of a Majorana particle is $\pm\pi/2$. Thus $\psi^P(\vec{x},t)) = \pm i\gamma^0\psi(-\vec{x},t)$. Since parity and charge conjugation, in general, are not good symmetries one may repeat the above discussion for their product CP with the conclusion that the CP phase of a Majorana particle is $\pm i$.

4   The Electroweak Model

The modern framework for discussing CP violation is the Standard Model. The reason is that the Model is so successful that one must take it very seriously and understand it thoroughly, in order to be able to detect deviations from its predictions.

The Standard Model has two distinct parts, the Electroweak Model [2] and QCD. Accordingly, one distinguishes between two types of CP violation. The CP violation originating from QCD, usually referred to as the strong CP "problem" or the θ puzzle which does not concern us here (for a review see, for example Ref. [14]). Here, we shall consider some aspects of the electroweak CP violation. In the context of field theories the two are, of course, related to each other at some level due to quantum (higher order) corrections. The Electroweak Model had three distinct sectors

1. The gauge sector, here denoted by G.

2. The sector containing the fundamental fermions (families), denoted by f

3. The Higgs sector, denoted by H.

The Electroweak Model is a gauge theory which means that its forces are "dictated" by symmetry principles. These forces are mediated by the intermediate vector (spin one) bosons, called gauge bosons. In the Electroweak Model one assumes that the relevant symmetry is a $SU(2) \times U(1)$. This means that there are four intermediate bosons in the theory. These make up two charged W-bosons (W^+ and W^-) and two neutral ones (the photon γ and the Z). It is the experiment which has guided the theory, indicating the symmetry to be chosen. There is no a priori understanding of why the symmetry group should actually be $SU(2) \times U(1)$. If it had been a different group (for example $SU(3)$)

then the spectrum of the particles of force would have been different (for example eight intermediate bosons for $SU(3)$). The gauge sector of the Electroweak Model contains one free parameter, the coupling constant g of the SU(2) group. The SU(2) group is more beautiful than the U(1) because its intermediate bosons have self-interactions while the U(1) part by itself would have been "empty" - its gauge boson does not interact with itself or with the W's.

Next there is the **fermion sector**, denoted by f. There, the fermions (quarks and leptons) are introduced in left-handed doublets and right-handed singlets under the SU(2) symmetry group. What this means is that the right-handed constituents of matter, by construction, do not interact with the W's and thus there are no right-handed charged currents. Thus, parity and charge conjugation asymmetries, observed in nature, are introduced into the Electroweak Model by hand; they are not "explained", as we shall see soon (after Eq. (10)). This is one of the reasons for interest in more symmetric generalizations of the Electroweak Model, i.e., the left-right models. In such models there are more gauge bosons. These, if they exist, must be quite heavy since they have not been seen yet. At present we have only (model dependent) lower limits on their masses (roughly 800 GeV), from precision tests of the Standard Model. Future high-energy accelerators could look for such objects.

In the Electroweak Model, the fermions interact with the gauge sector via a "link", $L(f, G)$, which we shall introduce soon. There, a new coupling constant, g', enters as well as a number of hidden parameters, the weak hypercharges Y. These are chosen to satisfy the relations $Q = I_3 + Y$, where Q is the electric charge of the fermion f. The I and Y refer respectively to the weak isospin and the weak hypercharge of that fermion. It is true that there are relations among the Y's if one requires that the anomalies cancel [15]. But such relations should come out by themselves without having to put in extra assumptions, such as the cancellation of anomalies and the number of colours, as is required in the Electroweak Model. Therefore, the fermionic sector of the model is less beautiful than its gauge sector. In extensions of the standard model the link $L(f, G)$ would, in general, be different. The fundamental fermions would then have new interactions which have not been detected yet.

If we had only the above two sectors (f and G) of the Electroweak Model all fermions and gauge bosons would have been massless. This is because the symmetry structure of the Electroweak Model, with its left-handed doublets and right-handed singlets, does not allow mass terms for these particles. Thus, the symmetry is somehow "broken" in the real world where all the above particles (gauge bosons, quarks and leptons) are known to be massive with the exception of the photon and the neutrinos.

The Higgs sector, denoted by H, is introduced to solve the mass problem. This is done by introducing a number of spin zero particle degrees of freedom (scalar particles) into the theory, all along respecting the symmetry $SU(2) \times U(1)$. The theory then contains what is called a Higgs potential, $V(H)$, as well as the "links" $L(G,H)$ and $L(H,f)$. Once the isospins and the hypercharges of the Higgs multiplet(s) are specified there are no new parameters in the link $L(G,H)$. The number of parameters in the Higgs potential $V(H)$ depends on how the Higgs multiplets are chosen. In the Minimal Electroweak Model the Higgs is a doublet and there are only two free parameters in the Higgs potential. The link $L(H,f)$ is the "ugliest" feature of the Electroweak Model. First of all it involves solely non-gauge interactions. In fact every term in $L(H,f)$ is invariant by itself, under the action of the symmetry group, and, therefore, its coefficient is not constrained by other terms in the Lagrangian. These terms give interactions which remind one of the old-fashioned pion-nucleon interactions and therefore they are usually called the Yukawa terms. In the Minimal Electroweak Model, with three families and massless neutrinos, there are fifteen free parameters in $L(H,f)$. Thus from the seventeen free parameters of the minimal electroweak model fifteen of them are due to the Higgs! With massive neutrinos the number of parameters in the model can be much larger. Unfortunately the Electroweak Model throws no light on the question of the neutrino mass. There is no a priori reason why the neutrino should be massless but the model is also perfectly consistent with neutrinos being massless. The same could have been said about the quarks and leptons as well. The model would have not minded at all if say the electron had been massless, but we know that it is not, through experiment.

Before ending this section a comment is in order. In spite of the fact that the Electroweak Model has been very successful in describing all the observed couplings of fermions to the gauge bosons (currents) one should not forget that there are still major missing or untested pieces in the model, viz.,

- The top quark has not been found yet.

- The Higgs particle remains to be discovered.

- The triple and quartic gauge boson couplings have not been seen yet. These are $WW\gamma$, WWZ, $WWWW$, $WWZZ$, $WW\gamma\gamma$, $WWZ\gamma$. The model also predicts that there are no ZZZ or $ZZZZ$ couplings. The observation of the triple and quartic couplings would provide the proof, so far missing, that the theory is nonabelian.

- The CP violation, as we shall see soon, is supposed to be due to the quark mixing matrix of the Electroweak Model. This claim of the model is badly tested. We still do not know whether the so called "direct" CP violation predicted by the model is there or not.

- Our knowledge of the neutrino sector of the Electroweak Model is very poor.

Except for the discovery of the top quark, which may come quite soon, all the above tests of the Electroweak Model will probably take some time.

5 The Lagrangian of the Electroweak Model

The Lagrangian density of the Electroweak Model is, symbolically of the form

$$L = L(f, G) + L(H, f) + L(G, H) + L(G) - V(H) \tag{9}$$

$$
\begin{aligned}
f &= \text{fermions (quarks, leptons)} \\
G &= \text{gauge bosons } (\vec{W} \text{ and } B) \\
H &= \text{the Higgs doublet}
\end{aligned}
$$

This Lagrangian is constructed so that it is invariant under the local (space-time dependent) symmetry group $SU(2) \times U(1)$. Consider first the hadronic sector of the theory. The leptonic sector may be treated in a similar way, if the neutrinos are Dirac particles. The case where the neutrinos are their own antiparticles (Majorana particles) is far more complicated to treat, specially if CP is violated. Here, we shall only be concerned with the hadronic sector of the Electroweak Model.

Under SU(2), the quark fields are assumed to transform as doublets ,if they are left-handed, and as singlets, if they are right-handed. In other words, one introduces the multiplets

$$\begin{pmatrix} q_{jL} \\ q'_{jL} \end{pmatrix} \qquad q_{jR}, \qquad q'_{jR}, \qquad j = 1, 2, ... N$$

where $q \equiv \psi_q(\vec{x}, t) \equiv q(\vec{x}, t)$ and

$$q_R = \frac{1 + \gamma_5}{2} q \qquad q_L = \frac{1 - \gamma_5}{2} q$$

Here the index j is the *family* index, and N denotes the number of families. In the Standard Model the number N is arbitrary. One just sums over the families, treating them as copies. At our present state of knowledge, it is inappropriate to refer to the families as *generations* because, in the Standard Model, there is no *mother - daughter* relationship between the copies. The situation may turn out to be different if ,e.g., the quarks would turn out to be composite objects.

Consider now the Lagrangian in Eq. (9). The hadronic part of the first term, on the RHS, is given by

$$
\begin{aligned}
L(f,G) \;=\; & \sum_{j=1}^{N} \Big\{ \overline{(q,q')_{jL}} \; i\gamma^{\mu} [\partial_{\mu} I - ig_2 \frac{\vec{\sigma}}{2}.\vec{W}_{\mu} - ig_1(\frac{1}{6})B_{\mu} I \,] \begin{pmatrix} q_{jL} \\ q'_{jL} \end{pmatrix} \\
& + \; \overline{q_{jR}} \; i\gamma^{\mu} [\partial_{\mu} - ig_1(\frac{2}{3}) \; B_{\mu}]q_{jR} + \overline{q'_{jR}} \; i\gamma^{\mu} [\partial_{\mu} - ig_1(\frac{-1}{3}) \; B_{\mu}]q'_{jR} \Big\}
\end{aligned}
\tag{10}
$$

Here I denotes the unit matrix and the numbers in the parenthesis are the eigenvalues of the weak hypercharge Y (the generator of the U(1) group). They are chosen such that $Q \equiv I_3 + Y$ has the eigenvalues 2/3 and -1/3 for the up-type and the down-type quarks respectively; $Q \; q_L = (I_3 + Y) \; q_L = (\frac{1}{2} + \frac{1}{6}) \; q_L = 2/3 \; q_L$, etc.

The Lagrangian density, Eq. (10) *formally* violates both parity and C-conjugation symmetry. Let us check this for one of the terms, say

$$
E(\vec{x},t) \equiv \overline{q_{jR}} \; \gamma^{\mu} [\partial_{\mu} - ig_1(\frac{2}{3}) \; B_{\mu}]q_{jR} = \bar{q}_j \gamma^{\mu} [\partial_{\mu} - ig_1(\frac{2}{3}) \; B_{\mu}] \frac{1+\gamma_5}{2} \; q_j
\tag{11}
$$

$$
E(\vec{x},t) \xrightarrow{P} \bar{q}_j(-\vec{x},t)\gamma_0\gamma^{\mu} [\partial_{\mu} - ig_1(\frac{2}{3}) \; B^{\mu}(-\vec{x},t)] \frac{1+\gamma_5}{2} \; \gamma_0 \; q_j(-\vec{x},t),
\tag{12}
$$

where we have used that under parity $q_j \to \gamma_0 q_j$ and $B_{\mu} \to B^{\mu}$ as well as $(\vec{x},t) \to (-\vec{x},t)$ according to our earlier results (see the section on parity) We may now use the identities $\frac{1+\gamma_5}{2}\gamma_0 = \gamma_0\frac{1-\gamma_5}{2}$, $\gamma_0\gamma^{\mu}\gamma_0 = \gamma_{\mu}$ and $\gamma_0\gamma^{\mu}\gamma_0\partial_{\mu} = \gamma_{\mu}\partial^{\mu} = \gamma^{\mu}\partial_{\mu}$, taken at $(-\vec{x},t)$ to show that

$$
E(\vec{x},t) \xrightarrow{P} \bar{q}_j\gamma_{\mu}[\partial^{\mu} - ig_1(\frac{2}{3}) \; B^{\mu}] \frac{1-\gamma_5}{2} q_j = \bar{q}_L \gamma_{\mu}[\partial^{\mu} - ig_1(\frac{2}{3}) \; B^{\mu}]q_L
$$

Here the fields are taken at the space-time point $(-\vec{x},t)$. Similarly, we can show that in the other terms also what happens is that, under P, left (L) and right (R) are interchanged, and the space coordinates are flipped. Thus what we find is that the left-handed quarks which originally interacted with the W no longer do so. Now instead the right-handed ones are interacting with the W. Thus parity is, formally, violated. We say formally because the fields in L(f,G) are not yet the physical fields, and what matters is whether physics is invariant under P or not. In the Electroweak Model, the noninvariance under P remains even after the Lagrangian is rewritten in terms of the physical fields (see below). The proof of the violation under C follows the same pattern. It is due to the simultaneous presence of vector and axial vector interactions, because, under C, the vector density is "odd" but the axial vector density is "even" . However, the above Lagrangian, L(f,G) is invariant under CP because we may choose the phases such that $L(\vec{x},t) \to L(-\vec{x},t)$, as follows. First we consider the term proportional to the quantity

$$E_N \equiv \overline{(q,q')} \, \gamma^\mu \sigma_3 W_\mu^3 (1 - \gamma_5) \begin{pmatrix} q \\ q' \end{pmatrix} = \bar{q}\gamma^\mu W_\mu^3(1 - \gamma_5)q - \bar{q}'\gamma^\mu W_\mu^3(1 - \gamma_5)q' \qquad (13)$$

$$\xrightarrow{CP} -\bar{q}\gamma_\mu W_\mu^{3'}(1 - \gamma_5)q - [q \to q'] \qquad (14)$$

where $(W_\mu^3)'$ is the CP transformed of W_μ^3. Since we had $W_\mu^3 \xrightarrow{P} W^{3\mu}$, we define the C-conjugation to give $W_\mu^3 \to -W_\mu^3$, whereby the above term is symmetric under CP, i.e., $E_N(\vec{x},t) \to E_N(-\vec{x},t)$. Now we consider the charged current terms, i.e.,

$$E_C \equiv \bar{q}\gamma^\mu[W_\mu^1 - iW_\mu^2](1 - \gamma_5)q' + \bar{q}'\gamma^\mu[W_\mu^1 + iW_\mu^2](1 - \gamma_5)q$$

$$\xrightarrow{CP} -\bar{q}'\gamma_\mu[W_\mu^1 - iW_\mu^2]'(1 - \gamma_5)q - \bar{q}\gamma_\mu[W_\mu^1 + iW_\mu^2]'(1 - \gamma_5)q'$$

where again $(\vec{x},t)$ is replaced with $(-\vec{x},t)$, in the arguments of the fields which we have not written out, and the primes on the W's denote CP tranformations, to be determined now. Under parity, we had $W_\mu^j(\vec{x},t) \to W^{\mu j}(-\vec{x},t)$. Therefore, CP is conserved, because we can define

$$W_\mu^1 \xrightarrow{C} -W_\mu^1 \qquad W_\mu^2 \xrightarrow{C} +W_\mu^2$$

whereby $E_C(\vec{x},t) \to E_C(-\vec{x},t)$, under CP. To summarize, we have

$$W_\mu^j \xrightarrow{P} W^{j\mu}$$

$$(W_\mu^1, W_\mu^2, W_\mu^3) \xrightarrow{C} -(W_\mu^1, -W_\mu^2, W_\mu^3)$$

$$(W_\mu^1, W_\mu^2, W_\mu^3) \xrightarrow{CP} -(W^{1\mu}, -W^{2\mu}, W^{3\mu}) \qquad (15)$$

Note that the extra minus sign in front of the W_μ^2 also ensures that each of the three components of the field tensor has a well defined transformation property, under C. The field tensor is defined by

$$W_{\mu\nu}^i = \partial_\mu W_\nu^i - \partial_\nu W_\mu^i + g_2 \, \epsilon^{ijk} \, W_\mu^j W_\nu^k$$

where $i, j, k = 1, 2, 3$. Thus in spite of the fact that the field tensor contains both linear and quadratic terms in W it transforms rather simply, viz.,

$$(W^1_{\mu\nu}, W^2_{\mu\nu}, W^3_{\mu\nu}) \xrightarrow{P} (W^{1\mu\nu}, W^{2\mu\nu}, W^{3\mu\nu})$$

$$(W^1_{\mu\nu}, W^2_{\mu\nu}, W^3_{\mu\nu}) \xrightarrow{C} -(W^1_{\mu\nu}, -W^2_{\mu\nu}, W^3_{\mu\nu})$$

Since $L(W) \equiv -\frac{1}{4} W^i_{\mu\nu} W^{i\mu\nu}$ it is symmetric under P as well as C. This fact is not trivial because $L(W)$ is not the free Lagrangian for the W but also contains interaction terms.

6 The Quark Mass Terms, the Origin of CP Violation

We now return to the Lagrangian of the Standard Model, Eq. (9). In the previous section, we showed that the first term (on the RHS) which gives the interactions of the fermions with the gauge bosons, is CP conserving. We showed that L(W) conserves CP as well. Furthermore, we may show that $L(G) \equiv L(W) + L(B)$ as well as all the remaining terms, in Eq. (9), with the exception of the term $L(H, f)$, conserve both C and P. The demonstration that $L(G, H)$ is symmetric under P and C is quite similar to what was done at the end of the last section. We know how the gauge bosons transform and can show that it is possible to define the appropriate transformation formulae, also for components of the Higgs doublet, such that $L(G, H)$ has the correct transformation properties under P and C.

All CP violation, in the Electroweak Model, originates from the term $L(H, f)$ which describes the interactions of the fermions with the Higgs doublet. The hadronic part of this term is given by

$$L(H, f) = \sum_{j,k=1}^{N} \left\{ Y_{jk} \overline{(q, q')_{jL}} \begin{pmatrix} \phi^{(0)*} \\ -\phi^{(-)} \end{pmatrix} q_{kR} + Y'_{jk} \overline{(q, q')_{jL}} \begin{pmatrix} \phi^{(+)} \\ \phi^{(0)} \end{pmatrix} q'_{kR} + h.c. \right\} \quad (16)$$

Here Y_{jk} and Y'_{jk} are coupling constants, sometimes referred to as the Yukawa couplings, in analogy with pion-nucleon coupling constants. These coupling constants are, within the Standard Model, arbitrary complex numbers. The reason being that each term (corresponding to arbitrary j and k) in Eq. (16) is invariant (see below). The scalar fields appearing in Eq. (16) are

$$H \equiv \begin{pmatrix} \phi^{(+)} \\ \phi^{(0)} \end{pmatrix} \quad and \quad H^C \equiv \begin{pmatrix} \phi^{(0)*} \\ -\phi^{(-)} \end{pmatrix}$$

and denote the Higgs doublet and its C-conjugate respectively. Note that the Higgs doublet represents four real scalar fields, which may be introduced as $\phi^{(+)} = \frac{1}{\sqrt{2}}(\phi_1 + i\phi_2)$

and $\phi^{(0)} = \frac{1}{\sqrt{2}}(\phi_0 + i\phi_3)$ The point is that both H and $H^C \equiv i\sigma_2 H^*$ transform as doublets under SU(2), whereby the Lagrangian (16) is manifestly invariant under SU(2). To see this note that under the action of SU(2) the doublet H, by definition, transforms as

$$H \rightarrow exp(i\vec{\alpha}.\vec{\sigma})\ H \equiv U(\vec{\alpha})\ H$$

where $\vec{\alpha}$ are the three (real) parameters of the SU(2) group. Therefore,

$$\overline{(q,\ q')_{jL}}\ H \rightarrow \overline{(q,\ q')_{jL}}\ U^\dagger(\vec{\alpha})U(\vec{\alpha})\ H = \overline{(q,\ q')_{jL}}\ H$$

However, the H^* does not transform as a doublet, viz., $H^* \rightarrow U^*(\vec{\alpha})\ H^* \neq U(\vec{\alpha})\ H^*$. That is why we can not use it but instead use H^C which has the correct transformation property, viz., $H^C \rightarrow exp(i\vec{\alpha}.\vec{\sigma})\ H^C = U(\vec{\alpha})\ H^C$. The invariance under U(1) is ensured by requiring that the electric charge be conserved.

The above Lagrangian, $L(H, f)$ involves scalar and pseudoscalar interactions and is, therefore, as far as the symmetries go, rather similar to the interaction term in the Lagrangian of Eq. (1). We shall discuss its symmetry properties after it undergoes spontaneous symmetry breakdown, as follows.

7   Spontaneous Symmetry Breaking

Under the spontaneous symmetry breaking, the field ϕ_0 is shifted, viz., $\phi_0 \rightarrow \phi_0 + v$, where v is a real number called the vacuum expectation value of the field ϕ_0, and the three fields ϕ_j , j=1,2,3, are "eaten" by the $W^{(\pm)}$ and the Z bosons which become massive. We shall not go into the details of the spontaneous symmetry breaking mechanism but refer the interested reader to the literature [16]. The essential point is that what remains, from Eq. (16), is only the terms involving the field ϕ_0, i.e.,

$$L(f, H) \overset{SSB}{\longrightarrow} -\sum_{j,k=1}^{N} \{m_{jk}\ \overline{q_{jL}}\ q_{kR} + m'_{jk}\overline{q'_{jL}}q'_{kR}\ + h.c.\}(1 + \frac{1}{v}\phi_0) \tag{17}$$

where the quantities

$$m_{jk} = -\frac{v}{\sqrt{2}}\ Y_{jk}, \qquad m'_{jk} = -\frac{v}{\sqrt{2}}\ Y'_{jk}$$

are called the quark mass matrices. Since j,k =1,2,..,N, the mass matrices are N by N matrices, N being the number of families. Again, formally, the Lagrangian (17) violates the discrete symmetries P, C, and CP (or T). Consider, for example, the terms in it having to do with the up-kind quarks, i.e.,

$$E \equiv m_{jk}\overline{q_{jL}}\ q_{kR} +\ h.c. = \frac{1}{2}[\ m_{jk}\bar{q}_j(1 + \gamma_5)q_k + m^*_{jk}\bar{q}_k(1 - \gamma_5)q_j]$$

$$= \frac{1}{2} \bar{q} \left[(m + m^\dagger) + (m - m^\dagger)\gamma_5 \right] q$$

Thus parity is conserved if the mass matrix is hermitian, because then the pseudoscalar terms, in Eq. (17), would be absent. Under C-conjugation, using the transformation properties of the spinors under C, we have that

$$E \xrightarrow{C} \bar{q} \left[(m + m^\dagger)^T + (m - m^\dagger)^T \gamma_5 \right] q.$$

where T now refers to transposition. Thus the mass matrix m must be symmetric in order for C to be a good symmetry. Finally, under the CP operation we have

$$E \xrightarrow{CP} \bar{q} \left[(m + m^\dagger)^T - (m - m^\dagger)^T \gamma_5 \right] q.$$

where, in addition, $\vec{x} \rightarrow -\vec{x}$. Thus CP invariance requires the mass matrices to be real, viz.,

$$m = m^* \quad m' = m'^*$$

We should keep in mind that we are, so far, dealing with nonphysical fields, as we have not yet written the theory in terms of the physical fields. However, we shall do so now.

In order to find the physical fields we must diagonalize the quark mass matrices, m and m'. From theory of matrices, it is known that any square matrix (hermitian or not) can be diagonalized with the help of two unitary matrices, which in our case implies that we can find four matrices such that

$$U_L \, m \, U_R^\dagger = D \equiv Diag.(m_u, m_c, m_t, ...) \tag{18}$$

$$U_L' \, m' \, U_R'^\dagger = D' \equiv Diag.(m_d, m_s, m_b, ...) \tag{19}$$

Here the U_L denotes the unitary matrix (i.e., $U_L U_L^\dagger = 1$) which multiplies the mass matrix m from the left, etc. The daggers on the matrices, on the RHS, are introduced for the sake of notational convenience. D is a diagonal N by N matrix and the quantities m_u, m_d, etc., denote the eigenvalues, of the mass matrices, i.e., the quark masses. To find the matrix U_L we multiply Eq. (18) with its hermitian conjugate and obtain

$$U_L m U_R^\dagger \, U_R m^\dagger U_L^\dagger = U_L m m^\dagger U_L^\dagger = D^2 = Diag.(m_u^2, m_c^2, m_t^2, ...)$$

Thus we see that the matrix U_L diagonalizes the *hermitian* matrix $mm^\dagger$. Similarly, the matrix U_R diagonalizes the hermitian matrix $m^\dagger m$, a procedure familiar from elementary quantum mechanics where one often has to diagonalize the perturbation Hamiltonian matrix, which is usually hermitian. Similarly, we can derive analogous relations for the down-kind sector by simply putting primes on the U's and taking the appropriate mass eigenvalues. Substituting Eq. (18) into Eq. (17) gives, for the up-kind quarks

$$\overline{q_{jL}} \, m_{jk} q_{kR} = \overline{q_L} \, m q_R = \overline{q_L} \, U_L^\dagger U_L m U_R^\dagger U_R q_R \tag{20}$$

$$= \overline{U_L q_L} \, D \, U_R q_R = \overline{U_L q_L} \begin{pmatrix} m_u & 0 & 0 & \cdots \\ 0 & m_c & 0 & \cdots \\ 0 & 0 & m_t & \cdots \\ \cdot & \cdot & \cdot & \cdots \\ \cdot & \cdot & \cdot & \cdots \end{pmatrix} U_R q_R. \tag{21}$$

Treating the down type quarks in the same way, we find that the physical fields are

$$q_L^{phys} = U_L q_L = U_L \begin{pmatrix} u_L \\ c_L \\ t_L \\ \vdots \end{pmatrix} \qquad q_L^{\prime phys} = U_L' q_L' = U_L' \begin{pmatrix} d_L \\ s_L \\ b_L \\ \vdots \end{pmatrix} \tag{22}$$

Similar relations are also valid for the R-handed quarks, where we simply replace the index L by R, in Eq. (22). Substituting Eq. (22), and its analog for the right-handed quarks, into Eq. (16) gives us the Lagrangian in terms of the physical fields, viz.,

$$L(H, f) = -(1 + \frac{\phi_0}{v}) \Big[m_u \bar{u} \frac{(1 + \gamma_5)}{2} u + m_c \bar{c} \frac{(1 + \gamma_5)}{2} c + \cdots$$
$$+ m_d \bar{d} \frac{(1 + \gamma_5)}{2} d + m_s \bar{s} \frac{(1 + \gamma_5)}{2} s + \cdots \Big] + h.c.$$

The pseudoscalar terms here cancel due to the contribution coming from the Hermitian conjugate and we get

$$L(H, f) = -(1 + \frac{\phi_0}{v}) [m_u \bar{u} u + m_c \bar{c} c + \dots + m_d \bar{d} d + m_s \bar{s} s + \cdots]. \tag{23}$$

The essential feature of $L(H, f)$ is that it conserves, seperately, P , C and thus also CP as well as T. This demonstrates that the apparent symmetry properties of a Lagrangian need not have anything to do with *physics*, if the Lagrangian in question is expressed in terms of unphysical fields. Let us summarize: given the mass matrices m and m' we may find the four unitary matrices (U, U', with subscripts L and R) needed to diagonalize them. At the same time we find the relations between the physical and the unphysical fields (Eq. (22) and its analog for the right-handed quarks). In rewriting the $L(H, f)$ in terms of the physical fields we find that there is no trace of the U and U' matrices left. The interactions of the Higgs with the fermions is automatically P and C conserving, as well as flavour conserving. In extensions of the Model these properties will, in general, not be valid.

Now that we have the term $L(H, f)$ in terms of the physical quark fields, we turn our attention to the only other term in the Lagrangian of the Standard Model, Eq. (9), which needs to be expressed in terms of the physical quark fields, i.e., the term $L(f, G)$. Consider first any "neutral current term", i.e., any term which involves only either the up-kind quarks or the down-type quarks but not both, for example the term which we had in Eq. (11)

$$E(\vec{x}, t) \equiv \bar{q}_{j_R} \gamma^\mu [\partial_\mu - ig_1(\frac{2}{3})\ B_\mu] q_{j_R} = \bar{q}_R \gamma^\mu [\partial_\mu - ig_1(\frac{2}{3})\ B_\mu] q_R$$

$$= \overline{q_R^{phys}} U_R \gamma^\mu [\partial_\mu - ig_1(\frac{2}{3})\ B_\mu] U_R^\dagger q_R^{phys}$$

$$= \overline{q_R^{phys}} \gamma^\mu [\partial_\mu - ig_1(\frac{2}{3})\ B_\mu] q_R^{phys} \tag{24}$$

where we have used the identity $U_R U_R^\dagger = 1$, whereby the matrix U_R again "disappears". Thus the neutral current term considered is flavour conserving, viz.,

$$\overline{q_R^{phys}}(...)q_R^{phys} = \bar{u}(...)u + \bar{c}(...)c + \bar{t}(...)t + ...$$

where (...) stands for the factor involving the derivative and the B-field, in Eq. (11). Evidently, the above calculation can be extended to all the neutral current terms in the Lagrangian $L(f, G)$. In such terms several "miracles" happen. The neutral currents all conserve CP (the proof is similar to the one presented in Eq. (14)). In addition, there are no flavour changing neutral currents because of two facts, (i) these terms are helicity conserving (i.e., they involve either the index L or R but not both) and (ii) the four matrices U and U' (with indices L or R) are unitary. The absence of flavour changing neutral currents means that there are no factors of the type $\bar{u}\gamma^\mu(a + b\gamma_5)c$, in the Lagrangian of the Standard Model. This absence of the flavour changing neutral currents is a very important feature of the Model. It is usually referred to as the Glashow-Iliopoulos-Maiani mechanism or simply the GIM mechanism [17]. Note that the interactions of the physical vector bosons A (the photon) and the Z are then automatically flavour conserving because the photon and the Z are linear combinations of W^3 and B. As there are stringent limits, from experiment, on flavour changing neutral currents, the absence of such currents is one of the great successes of the Model. It should be noted that in the "extensions" of the Standard Model one generally encounters difficulties in this respect.

Having treated the $L(H, f)$ and the neutral current terms (as well as the kinetic terms) in $L(f, G)$, what remains is the charged current terms, which we shall discuss now. We see that such terms are helicity conserving but *do mix* the up-type and the down-type quarks, simply because $W^\pm$ carry one unit of charge. Since the charged currents only involve left-handed quarks the matrices with the subscript R do not enter at all and the U with the index L gets multiplied by $U'^\dagger$ with the same index. But $UU'^\dagger$ need not be equal to unity. All we know is that it is a unitary matrix. Dropping the numerical factors and the coupling constant, the charge current terms in Eq. (10) are given by

$$X_C \equiv [W_\mu^1 - iW_\mu^2] \, \overline{q_L} \, \gamma^\mu q_L' \, + h.c. = [W_\mu^1 - iW_\mu^2] \, \overline{q_L^{phys}} \, \gamma^\mu U_L U_L'^\dagger q_L'^{phys} + h.c.$$

$$= [W_\mu^1 - iW_\mu^2] \, \overline{q_L^{phys}} \, \gamma^\mu V q_L'^{phys} \, + h.c. \equiv [W_\mu^1 - iW_\mu^2] \, J_C^\mu \, + h.c. \tag{25}$$

where $V \equiv U_L U_L'^\dagger$ and J_C^μ denotes the charged current

$$J_C^\mu \equiv \overline{(u,c,t,...)_L} \, \gamma^\mu \begin{pmatrix} V_{ud} & V_{us} & V_{ub} & \cdots \\ V_{cd} & V_{cs} & V_{cb} & \cdots \\ V_{td} & V_{ts} & V_{tb} & \cdots \\ . & . & . & \cdots \\ . & . & . & \cdots \\ . & . & . & \cdots \end{pmatrix} \begin{pmatrix} d_L \\ s_L \\ b_L \\ \vdots \end{pmatrix} \tag{26}$$

The matrix V is the so called quark mixing matrix.

8   The Quark Mixing Matrix

We shall now, in more detail, consider the charged currents, the expression X_C, where all the CP violation in the Standard Model resides. X_C is maximally P and C violating, due to the equal strengths of the vector and axial vector interactions and the fact that these currents transform with opposite signs so well under P as under C. Under CP , using our table in the section on CP transformations and the Eq. (15) and omitting the superscript $phys$, we have

$$X_C = (W_\mu^1 - iW_\mu^2)\overline{q_j}\gamma^\mu V_{jk}(1 - \gamma_5)q_k' + (W_\mu^1 + iW_\mu^2)\overline{q_k'}\gamma^\mu V_{jk}^*(1 - \gamma_5)q_j$$

$$\xrightarrow{CP} (W_\mu^1 + iW_\mu^2)\overline{q_k'}\gamma^\mu V_{jk}(1 - \gamma_5)q_j + (W_\mu^1 - iW_\mu^2)\overline{q_j}\gamma^\mu V_{jk}^*(1 - \gamma_5)q_k'$$

to be supplemented with $(\vec{x}, t) \to (-\vec{x}, t)$. Thus we find that CP conservation requires the matrix V to be real. The reality requirement means that the matrix must be real, modulus unmeasurable phases, as shall be described now. We know from quantum mechanics that the phase of a wave function is not a measurable quantity. In other words a wave function ψ and $exp(i\eta)\psi$, where η is a real number, are physically equivalent. The situation is the same in field theory. What matters is not the absolute phases but the relative phases of different fields. Therefore, we must examine which phases in V are measurable and which ones are not. In the charged current J_C^μ, Eq. (26), all the fields are left-handed ones. The phases of these fields are arbitrary, in other words not measurable quantities. Therefore, we may redefine them by letting

$$u_L \rightarrow e^{i\phi(u)}u_L, \quad c_L \rightarrow e^{i\phi(c)}c_L, \quad etc.$$

$$d_L \rightarrow e^{i\phi(d)}d_L, \quad s_L \rightarrow e^{i\phi(s)}s_L, \quad etc.$$

Here the quantities $\phi(f)$, $f = u, c, ..; d, s, ...$ are arbitrary real numbers. There are thus 2N such quantities, if there are N families. Under the above phase transformation we have, for example for the case of 3 families

$$V \rightarrow \begin{pmatrix} e^{-i\phi(u)} & 0 & 0 \\ 0 & e^{-i\phi(c)} & 0 \\ 0 & 0 & e^{-i\phi(t)} \end{pmatrix} \begin{pmatrix} V_{ud} & V_{us} & V_{ub} \\ V_{cd} & V_{cs} & V_{cb} \\ V_{td} & V_{ts} & V_{tb} \end{pmatrix} \begin{pmatrix} e^{i\phi(d)} & 0 & 0 \\ 0 & e^{i\phi(s)} & 0 \\ 0 & 0 & e^{i\phi(b)} \end{pmatrix}$$

Therefore, for 3 families, as well as for *any number of families* what happens is that

$$V_{\alpha j} \rightarrow exp[i((\phi(j) - \phi(\alpha))]V_{\alpha j} \tag{27}$$

where α and j denote an up-kind and a down-kind quark, respectively. Before going on, we should examine how this rephasing affects the remaining terms in the Lagrangian of the Model. Consider first the neutral current in $L(f, G)$ expressed in terms of the physical fields, for example the quantity in Eq. (11). Such terms are manifestly invariant under the above rephasing because they are flavour as well as helicity conserving. However the $L^{phys}(H, f)$ is affected because it connects L and R fields (e.g., $\overline{u_R}u_L$) and we are only rephasing the L fields. This can, however, be remedied by rephasing any right-handed quark field exactly with the same phase as the corresponding left-handed one whereby $L(f, G)$ has the same form when expressed in terms of the physical fields as it had when written in terms of the unphysical fields.

Let us now count the number of parameters of the quark mixing matrix. A general unitary N by N matrix has N^2 parameters (for parametrizations of unitary matrices see, e.g., Ref. [18]). $N(N-1)/2$ of these parameters may be taken as Euler angles which one introduces in dealing with rotations in N dimensional (Euclidean) space. The remaining parameters are called phases. We see from Eq. (27) that $2N - 1$ of these phases are not measurables. This comes about because we had $2N$ unmeasurable phases, $\phi(j)$ and $\phi(\alpha)$, but in Eq. (27) only the phase differences appear and there are $2N - 1$ such quantities. Therefore, V has $N^2 - (2N-1) = (N-1)^2$ parameters among which $N(N-1)/2$ are the rotation angles. Therefore, the number of phases in V is $(N-1)(N-2)/2$. For the case of 2 families we find 1 rotation angle and no phases. The quark mixing matrix is then of the form

$$\begin{pmatrix} cos\theta & sin\theta \\ -sin\theta & cos\theta \end{pmatrix} \qquad (28)$$

This matrix, which in the literature is usually called the Cabibbo matrix [19] emerged gradually, out of much effort. What made it "compulsory" was that it could, due to being an orthogonal matrix, explain the absence of the flavour changing neutral currents (the GIM mechanism, [17]). The essential point to note about the matrix V, for 2 families, is that it is automatically real. In other words, CP is automatically conserved, in the Standard Model with 2 families. This fact was a serious problem for the Standard Model at that time. A possible solution was to extend the Higgs sector of the Model [11], or more generally associate CP violation to the process of spontaneous symmetry breaking. In another class of possible solutions it was proposed that one should go beyond the Standard Model and consider, for example left-right models with two families. Finally, the problem could also be solved by introducing a third family, as was first suggested by Kobayashi and Maskawa [9]. Since we know now that there are (at least) 3 families the KM mechanism (i.e., the proposal that the observed CP violation is due to the existence of 3 families) can not be circumvented. Furthermore, we know a great deal more about the number of families than we knew in 1970's thanks to results from LEP (See, for example Ref. [20]).

Before closing this section it is amusing to note that, for the leptonic sector, the above two-by-two matrix was written down [21] as early as in 1962, in connection with neutrino mixing.

9   The Number of Families

In the Minimal Electroweak Model the number of families is given by the number of the neutrinos. These are all massless, by definition. The number of families is then computed from the "invisible" width of the Z, viz.,

$$\Gamma_{inv} \equiv \Gamma_Z - (\Gamma_h + \Gamma_e + \Gamma_\mu + \Gamma_\tau) \qquad (29)$$

$$\mathbf{N_{fam}} = N_\nu = \Gamma_{inv}/\Gamma_\nu \qquad (30)$$

where Γ_Z denotes the total width, the subscript h refers to hadrons and Γ_ν is the theoretical width (166 MeV) for one massless neutrino. Γ_e is the width of the Z decay into an electron-positron pair, etc.

The recent results from LEP give [20], for the number of families a result very close to three, the most recent result being 3.04 ± 0.04. Note that we don't understand why the number of standard families is 3. Some years ago there were hopes that the number of families could be computed from first principles (geometry of compactified manifolds). But the hopes did not materialize. Due to LEP results, below, we shall restrict our discussions to the case of 3 families. Before doing so, however, it is important to keep in mind that the determination of the number of families is model-dependent. The LEP result depends on assuming the validity of the Minimal Electroweak Model. Otherwise, for example, the measured number of families need not even be an integer. It could be, for example less than three even if there are three families. All one needs to do to get such a result is to extend the minimal standard model just a "tiny bit", by allowing right-handed neutrinos. In fact there is no known reason why we should not have such objects. All other fermions have right-handed partners, why then not the neutrino? It is true that right-handed neutrinos introduce, in general, a large number of new parameters into the theory but we get something in return. Massive neutrinos can oscillate. They can, in some cases, take part in double-beta decay. They could perhaps be a substantial component of the so called dark matter in the universe.

One would naively guess that adding right-handed neutrinos would increase the invisible width of the Z because there are more channels into which the Z can decay. But just the opposite thing happens [22]). The right-handed neutrinos have vanishing isospins and hypercharges. Therefore, they do not couple to the gauge bosons. However, they can mix with the left-handed neutrinos and weaken their couplings to the Z. Such mixings can only decrease the invisible width - never increase it. There is a theorem which summarizes this result, as follows.

Take the Minimal Standard Model with N_{fam} families. Add to it an arbitrary number of right-handed neutrinos. Then, the effective number of families can only be smaller than (or equal to) the true number of families [22]

$$N_{fam}^{eff} \equiv \Gamma_{inv}/\Gamma_\nu \leq N_{fam} \tag{31}$$

The upshot of the theorem is that LEP alone cannot measure the number of families, unless one assumes that the Minimal Electroweak Model is correct. It is important to keep this fact in mind. There could be much more structure at higher energies and it is vital not to be biased by the successes of the Electroweak Model.

10 CP Violation with 3 Families

Consider now the case of 3 families. The number of parameters in V is $(3-1)^2 = 4$, i.e., 3 rotation angles and one phase. The first example of such a matrix was given by Kobayashi and Maskawa [9]. We shall return to it in the next section. The angles and the phase occuring in the quark mixing matrix, for 3 families, are generally referred to as either the KM angles and the KM phase, or the (quark) mixing angles and the CP phase. Unfortunately there are many different-looking parametrizations of the KM matrix. Therefore, in the next section we shall give the totality of these parametrizations so that the reader can make her/his own choice. Before doing so let us note that in the limit where two quarks *with the same charge* are degenerate in mass there is no CP violation, in the 3 family case. Let us take, as an example, the s and the b quarks to be degenerate, whereby they would be indistinguishable. Then there is an extra symmetry in the Model, i.e., unitary rotations in the space spanned by the d and s quarks leave the Lagrangian invariant. Therefore we could take our "new" s quark to be proportional to the linear combination $V_{us}\, s + V_{ub}\, b$, whereby, as seen from Eq. (26), the up quark only couples to two down-type quarks. It does not couple to b^{new}. We see immediately that the element V_{13} of our new KM matrix is zero and therefore the first row, in the mixing matrix may be written as $(cos\theta, sin\theta, 0)$. By use of unitarity we can construct the rest of the matrix and find

$$\begin{pmatrix} cos\theta & sin\theta & 0 \\ -sin\theta cos\phi & cos\theta cos\phi & sin\phi \\ sin\theta sin\phi & -cos\theta sin\phi & cos\phi \end{pmatrix} \tag{32}$$

This matrix leads to a CP conserving theory because it is real. The above argument can be repeated for any pair of up-type or down-type quarks. Therefore, we have the following six necessary (but not sufficient) conditions for CP violation

$$m_u \neq m_c, \quad m_c \neq m_t, \quad m_t \neq m_u, m_d \neq m_s, \quad m_s \neq m_b, \quad m_b \neq m_d, \tag{33}$$

11 The KM Type Parametrizations of the Quark Mixing Matrix

The first observation to be made about the KM parametrization is that it can be written as product of three "rotation" matrices where one of them involves a phase, viz.,

$$V_{KM} = R_{23}(\theta_3, \delta) R_{12}(\theta_1, 0) R_{23}(\theta_2, 0) \tag{34}$$

where

$$R_{12}(\theta, \phi) \equiv \begin{pmatrix} cos\theta & sin\theta e^{i\phi} & 0 \\ -sin\theta e^{-i\phi} & cos\theta & 0 \\ 0 & 0 & 1 \end{pmatrix} \quad R_{23}(\theta, \phi) \equiv \begin{pmatrix} 1 & 0 & 0 \\ 0 & cos\theta & sin\theta e^{i\phi} \\ 0 & -sin\theta e^{-i\phi} & cos\theta \end{pmatrix} \quad (35)$$

where $R_{ij}(\theta, \phi)$ denotes a unitary rotation in the *ij-plane* by the angle θ and the phase ϕ. We have here spelled out two of the three independent such matrices. The remaining one R_{31} is obtained by cyclic permutation. It should be mentioned that angles are limited to the range $0 \leq \theta_j \leq \pi/2$ and $0 \leq \delta \leq 2\pi$. From Eq. (34), we see immediately that the general structure of this type of parametrization is of the form

$$V = R_{\bullet\bullet}(\ ,\ ) \, R_{\bullet\bullet}(\ ,\ ) \, R_{\bullet\bullet}(\ ,\ ) \quad (36)$$

We shall refer to these matrices as the left, right and the middle matrices. The dots, as indices of the R's, indicate all possible choices of the plane of the rotation, and empty spaces are introduced to be filled with appropriate angles (θ, ϕ). Which possible choices do we have? Consider first the middle matrix. There are 3 possible choices for its indices, i.e., 12, 23 and 31. Suppose that we choose the combination 12. Then we are not allowed to use the indices 12 any more, for the left or right matrices, because then we do not obtain the most general matrix. However, we may take the left and right to have both the indices 23 or 31, or let one of them have 23 and the other 31. The left and right indices may be taken to be equal (the usual Euler construction in classical mechanics) because rotation matrices do not commute. Thus, with 12 in the middle there are 4 possibilities. Repeating the same procedure for the indices 23 and 31 gives us 3x4=12 such possibilities. Furthermore we can put the phase in three different slots. Therefore, we have altogether 3x12=36 different looking parametrizations which are all equivalent to the KM parametrization.

Let us consider the conditions for having CP violation coming from the quark mixing matrix. These conditions are to be added to our previous conditions, on non-degeneracy of the quarks (Eq. (33)). Take the most general quark mixing matrix, as indicated by Eq. (36), e.g., the KM matrix $V_{KM} = R_{23}(\theta_3, \delta)R_{12}(\theta_1, 0)R_{23}(\theta_2, 0)$. None of the rotation angles θ_j is allowed to be 0 or $\pi/2$ if CP is to be violated. For example if θ_3 vanishes, the matrix R_{23} becomes the unit matrix and V becomes real. Similarly, is θ_1 vanishes V becomes a rotation matrix in the 23 space. In fact the mixing matrix has then 4 zeros as the first family does not mix with the others, and there is no CP violation, etc. Finally, even if all the rotation angles are different from 0 and $\pi/2$, in order to have a nonreal V, we must require $\delta \neq 0, \pi$. In summary, there are eight necessary conditions on the angles and the phase, in order to have CP violation, viz.,

$$\theta_j \neq 0, \frac{\pi}{2} \quad \delta \neq 0, \pi, \quad j = 1, 2, 3 \tag{37}$$

Thus there are altogether 14 necessary conditions for having CP violation, in the Standard Model, with 3 families. We shall see later that these 14 conditions are nicely unified in a single relation, found by the present author ([23], [24]) as I shall discuss soon.

One important feature of the 3 family case is that in many computations, in the past, of the CP violating effects, the authors who did not make small angle approximations, were finding that the effects are proportional to the quantity

$$J \equiv sin^2\theta_1 sin\theta_2 sin\theta_3 cos\theta_1 cos\theta_2 cos\theta_3 sin\delta \tag{38}$$

This quantity , up to a sign, is obtained from the quark mixing matrix as follows ([23], [24], [25]). Take the matrix V; cross out one row and column. A 2 by 2 matrix is obtained. Put stars on the diagonal (or the off-diagonal) elements of the 2 by 2 matrix. For example, by crossing out the third row and the third column, and putting the stars along the anti-diagonal, we find

$$\begin{pmatrix} V_{11} & V_{12}^* \\ V_{21}^* & V_{22} \end{pmatrix} \tag{39}$$

Now multiply the 4 elements of this matrix and take the imaginary part of the product. The quantity J is obtained, up to a sign, viz., $J = Im[V_{11}V_{12}^*V_{21}^*V_{22}]$. The exact formula reads

$$Im[V_{\alpha j} V_{\beta k} V_{\alpha k}^* V_{\beta j}^*] = J \sum_{\gamma,l} \epsilon_{\alpha\beta\gamma} \, \epsilon_{jkl} \tag{40}$$

where the sums go from 1 to 3 and the Greek (Latin) indices stand for the up-(down-) kind quarks. For example, $\alpha = 1$ denotes the up quark and j=2 refers to the strange quark, etc. Thus the quantity J is a universal quantity, in the sense that it does not depend on the parametrization. Note that J is phase convention independent. For example, if we rephase the up quark, then due to the fact that one of the elements, V_{uj}, appearing in J is complex conjugated, while the other one is not, J does not change, etc. In fact the "invariants" of the quark mixing matrix are the moduli of its elements $|V_{\alpha j}|$ and the quantity J. These quantities are invariant in the sense that they do not depend on the parametrization chosen. Thus one may use them in order to give a parametrization which is angle and phase independent, as we shall do in the next section.

Finally, a very useful phenomenological parametrization of the quark mixing matrix was suggested by Wolfenstein [26]. It is much used in the literature because it is convenient and easy to remember. It is given by

$$\begin{pmatrix} 1 - \frac{1}{2}\lambda^2 & \lambda & A\lambda^3(\rho - i\eta) \\ -\lambda & 1 - \frac{1}{2}\lambda^2 & A\lambda^2 \\ A\lambda^3(1 - \rho - i\eta) & -A\lambda^2 & 1 - \frac{1}{2}\lambda^2 \end{pmatrix} \qquad (41)$$

Here λ, ρ, η and A are phenomenological parameters. The above parametrization of the quark mixing matrix is obtained [26] by expanding in powers of the empirical parameter $\lambda \simeq .22$ and neglecting terms of order λ^4 and higher. The empirical values of these parameters was discussed, at this School by Danilov [27].

12   Unitarity Triangles and the CP Violation Area

As noted, in the previous section, the invariant measurables of the quark mixing matrix are the moduli of its elements i.e., the quantities $|V_{\alpha j}|$, and the quantity J. The angle and phase independent parametrization is based on the observation [28] that even the quantity J, up to a sign, is a function of the moduli of the quark mixing matrix. In fact it is given by

$$4J^2 = -\lambda(a^2, b^2, c^2)$$

where the a, b, c are products of moduli of the elements of the quark mixing matrix, as we shall see soon. Let us first remind ourselves that the function λ is the triangular function sometimes also refered to as the Källén λ function. It should be familiar to the reader from two body kinematics. This function is given by [29]

$$\lambda(x, y, z) = x^2 + y^2 + z^2 - 2xy - 2yz - 2zx$$

How do we choose the a, b, c? We take the quark mixing matrix and cross out either one row or one column. Let us assume that we cross out the second column. We then obtain the matrix

$$\begin{pmatrix} V_{11} & V_{13} \\ V_{21} & V_{23} \\ V_{31} & V_{33} \end{pmatrix}$$

Now a equals the modulus of the product of the two elements on the first row, b and c are the corresponding quantities for the second and the third row, i.e.,

$$a = |V_{11}V_{13}| \qquad b = |V_{21}V_{23}| \qquad c = |V_{31}V_{33}|$$

Similarly, one can cross out one row and construct the corresponding a, b, c from the moduli of the products of the elements on the same column. The conclusion to be drawn is that even J, up to a sign, is a function of the moduli. Since there are 4 independent moduli the matrix, up to a two-fold ambiguity, may be parametrized by using 4 independent moduli (i.e., the choice must not include moduli of 3 elements from the same row or the same column). We shall now give a simple derivation of the above result.

Let us consider [28] the quark mixing matrix as a "lattice", the elements $V_{\alpha j}$ being the vertices. Let us denote the vertices with dots and introduce oriented horizontal "links" on the lattice, as shown below,

$$
\begin{array}{ccccc}
\bullet & \xrightarrow{a_1} & \bullet & \xrightarrow{b_1} & \bullet \\[4pt]
\bullet & \xrightarrow{a_2} & \bullet & \xrightarrow{b_2} & \bullet \\[4pt]
\bullet & \xrightarrow{a_3} & \bullet & \xrightarrow{b_3} & \bullet
\end{array}
$$

Here, the a_1 link or arrow connects the two elements V_{11} and V_{12}. We define the a_1 link to be the product of the first factor times the complex conjugate of the second one (to which the arrow points). Thus

$$
a_1 = V_{11} V_{12}^* \qquad a_2 = V_{21} V_{22}^* \qquad a_3 = V_{31} V_{32}^*
$$

Thus the a_j, j=1,2,3, are in general complex numbers. Unitarity tells us that the sum of these three complex numbers must vanish,

$$
a_1 + a_2 + a_3 = 0 \tag{42}
$$

Therefore, the three links a_j form a triangle, in the complex plane. We shall refer to this triangle as the a-unitarity triangle. Suppose that we change the phase of one of the quarks, entering in the a links, for example the phase of the strange quark. We see, from Eq. (42) that the whole triangle rotates with that phase angle but its form as well as its orientaion are invariants. If CP is conserved, then due to the fact that the quark mixing matrix is real, the triangle collapses to a line. Conversely, if any measurement would establish that any of the angles in the triangle is nonzero then we know that CP must be violated. From Eq. (40) we see that the quantity J is given by

$$
J = Im(a_1 \, a_2^*) = |a_1 a_2| \, sin\phi_{12}^a
$$

where, ϕ_{12}^a is the oriented angle between the vectors $\vec{a}_2$ and $\vec{a}_1$. Thus J is simply twice the area of the unitarity triangle that we have been considering. We may repeat the above argument for the b links. We will obtain the corresponding b-triangle. Furthermore, we can introduce the long horizontal, say c links, connecting the third and the first columns. Again we will obtain a c-triangle. Finally, there are 3 vertical link vectors, connecting the first and the second rows $(\vec{d})$, the second and the third rows $(\vec{e})$ and the third and the first rows $(\vec{f})$, viz.,

$$
\begin{array}{ccc}
\bullet & \bullet & \bullet \\
\downarrow d_1 & \downarrow d_2 & \downarrow d_3 \\
\bullet & \bullet & \bullet \\
\downarrow e_1 & \downarrow e_2 & \downarrow e_3 \\
\bullet & \bullet & \bullet
\end{array}
$$

Thus we have altogether 6 unitarity triangles, viz.,

$$\sum_{j=1}^{3} a_j = \sum_{j=1}^{3} b_j = \sum_{j=1}^{3} c_j = \sum_{j=1}^{3} d_j = \sum_{j=1}^{3} e_j = \sum_{j=1}^{3} f_j = 0$$

All these triangles collapse to lines if CP is conserved. The most interesting feature of them is that all these 6 triangles, in spite of having different shapes, have the **same area**. It is easy to check this fact, which is a consequence of unitarity, by using Eq.(40),

$$
\begin{aligned}
J = 2(area) = & \ Im(a_1 a_2^*) = Im(b_1 b_2^*) = Im(c_1 c_2^*) \\
= & \ Im(d_1 d_2^*) = Im(e_1 e_2^*) = Im(f_1 f_2^*)
\end{aligned}
$$

In conclusion, we have six unitarity triangles. The sides of these triangles are determined by the moduli of the links, or in other words by the moduli of the elements of the quark mixing matrix. Triangles are very special. Knowing the sides of a triangle, tells us everything about the triangle (the angles) except its overall orientation. The two fold ambiguity is due to this orientation (the sign of J). Thus we can parametrize the quark mixing matrix, including its CP properties, using four moduli and the sign of J. This fact may seem a bit strange. Afterall, $|V_{\alpha j}|$ are related to the probability that $W \to \alpha + \bar{j}$, i.e., they are not CP odd quantities. Within the Model, however, these probabilities are constrained, and give us information about CP violation in much the same way as one says that one tests parity violation by comparing the cross sections for neutrino and antineutrino neutral current reactions, where a difference between the cross sections proves that parity is violated *provided* one assumes vector and axial vector interactions.

13   The Commutator Formalism for CP Violation

In the previous sections we have seen that, in the Standard Model with 3 families, 14 conditions must be satisfied in order to have CP violation. Once again, these conditions were

$$m_u \neq m_c, \ m_c \neq m_t, \ m_t \neq m_u, \ m_d \neq m_s, \ m_s \neq m_b, \ m_b \neq m_d,$$

$$\theta_j \neq 0, \frac{\pi}{2}, \quad \delta \neq 0, \pi, \quad j = 1, 2, 3$$

where the last line is for the KM parmetrization, but in any other parametrization there will be a similar set of conditions. It is interesting to note that these conditions are unified within the **single** relation

$$det \; \mathbf{C} \neq 0 \tag{43}$$

where C is the commutator of the (square of the) quark mass matrices

$$[mm^\dagger, \; m'm'^\dagger] \equiv i\mathbf{C}$$

Here the quantities m and m' are the 3 by 3 quark mass matrices which were discussed in some detail in the section on mass matrices. The theory does not tell us what these mass matrices are. In fact their "determination", from first principles, is one of the most outstanding problems in particle physics. Although we can not determine the mass matrices, we can measure their eigenvalues (the 6 quark masses) and the four independent measurables of the quark mixing matrix, e.g., three independent moduli and the quantity J. What is highly remarkable about the above commutator is that its determinant is given by [23]

$$det \; \mathbf{C} = -2J \; (m_t^2 - m_c^2)(m_c^2 - m_u^2)(m_u^2 - m_t^2)(m_b^2 - m_s^2)(m_s^2 - m_d^2)(m_d^2 - m_b^2) \tag{44}$$

where J is the invariant, related to CP violation, which we have already discussed in some detail. We showed how to construct it in any parametrization. In the KM parametrization it is given by $J = sin^2\theta_1 sin\theta_2 sin\theta_3 cos\theta_1 cos\theta_2 cos\theta_3 sin\delta$. Thus we see that the above 14 conditions are unified in the determinant. In other words **CP is violated if and only if** $det \; \mathbf{C} \neq 0$. This statement hardly needs a proof. If the determinant is nonzero then the quark mixing matrix can not possibly be made real with any rephasing of the quark fields, because J can not be zero. Thus, from Eq. (26), we conclude that CP is violated (see also below). It is easy to give a derivation of the Eq. (44). Before doing so, however, we note that if the quark mass matrices happen to be hermitian one may instead of the commutator take

$$[m, \; m'] \equiv iC$$

whose determinant is given by

$$det \; C = -2J \; (m_t - m_c)(m_c - m_u)(m_u - m_t)(m_b - m_s)(m_s - m_d)(m_d - m_b) \tag{45}$$

This relation is very useful, in model calculations, because one can show [30] that in the Standard Model the mass matrices may be taken to be hermitian, without loss of generality. As the derivations of Eqs. (44) and (45) follow the same pattern we give just one of them. The commutator C is, by definition, hermitian and traceless. Thus its eigenvalues are real. In fact they are measurables, even though C itself is not a measurable. The determinant of any traceless 3 by 3 matrix may be computed from the trace of the third power of the matrix, i.e.,

$$det\ C = \frac{1}{3}tr(C^3) = \frac{(-i)^3}{3}tr\{[m,m']^3\}$$

Spelling out the trace, we find 8 terms. However, using the identity $tr(AB) = tr(BA)$ we find the simple result

$$
\begin{aligned}
det\ C &= \frac{(-i)^3}{3}\ 3tr[m^2m'^2mm' - m'^2m^2m'm] \\
&= -2\ Imtr[m^2m'^2mm']
\end{aligned}
\tag{46}
$$

Using the relations $UmU^\dagger = D$ and its primed version (see Eq. (18)), which are valid for hermitian mass matrices, and the definition $V \equiv UU'^\dagger$, where we have dropped the subscripts L, we have

$$det\ C = -2\ Imtr[U^\dagger D^2 UU'^\dagger D'^2 U'U^\dagger DUU'^\dagger D'U']$$

$$= -2\ Imtr[D^2VD'^2V^\dagger DVD'V^\dagger]$$

where the D and D' are the diagonal mass matrices, see Eqs. (18) and (19). Spelling out the trace, in the above equation, we find

$$
\begin{aligned}
Imtr[...] &= Im \sum_{\alpha,\beta} \sum_{j,k} d_\alpha^2 V_{\alpha j} d_j'^2 V_{\beta j}^* d_\beta V_{\beta k} d_k' V_{\alpha k}^* \\
&= J \sum_{\gamma,l} \epsilon_{\alpha\beta\gamma}\ \epsilon_{jkl} d_\alpha^2 d_\beta d_j'^2 d_k'
\end{aligned}
$$

Here d_α (d_j') refer to the quark masses, e.g., $d_1 = m_u$, etc. and we have used the Eq. (40). The remaining sums, over γ and l give "Vandermonde determinants", i.e., the factor $(m_t - m_c)(m_c - m_u)(m_u - m_t)$ and the corresponding one for the down-kind quarks. Putting everything together, we find the result quoted in Eq. (45). In the calculation of the detC the only difference is that we obtain the square of the masses. The determinants above are the simplest ones in an infinite class of such quantities. We shall not discuss them here and refer the reader to the literature [24]. The det C is more fundamental than the $detC$ because it does not depend on the sign of the mass and we know that the sign of the mass in the Lagrangian is irrelevant. Nevertheless, in model calcuations $detC$ is very useful (see, for example [31]). It is instructive to look at the commutator C from a different angle [32]. Let us consider it in the "frame" where we have diagonalized one of the mass matrices, say m. Then $m \rightarrow D$ and $m' \rightarrow \hat{m}'$. We find

$$C = -i[D, \hat{m}'] = -i \begin{pmatrix} 0 & (m_u - m_c)\hat{m}'_{12} & (m_u - m_t)\hat{m}'_{13} \\ (m_c - m_u)\hat{m}'_{21} & 0 & (m_c - m_t)\hat{m}'_{23} \\ (m_t - m_u)\hat{m}'_{31} & (m_t - m_c)\hat{m}'_{32} & 0 \end{pmatrix}$$

Thus the determinant of C is proportional to $Im\{\hat{m}'_{12}\hat{m}'_{23}\hat{m}'_{31}\}$. Nonvanishing of the determinant would imply that the latter imaginary part must be nonvanishing. In other words, both mass matrices can not be real (simultaneously). This explicit calculation also shows that the above product is a measurable quantity.

14   Manifestations of CP Violation

Suppose that CP were a good symmetry in nature. Then the CP operator would commute with the Hamiltonian, $[H, CP] = 0$. Note, however, that CP does not commute with the charge operator Q. Therefore, even in the limit of CP conservation not all particles would have well defined CP eigenvalues. However a particle and its CP-image would be degenerate, as can be seen from the relations

$$CP \, |\alpha> = exp(i\theta)|\bar{\alpha}>$$

$$H \, CP|\alpha> = CP \, H|\alpha>, \qquad H|\alpha> = m|\alpha>$$

Here θ is a phase and $\bar{\alpha}$ denotes the CP-image of the particle α. For a charged particle then this image, with the same mass but opposite charge would be naturally identified with the antiparticle. Note that by charge we mean the collection of all the "internal charges" that the particle may have, such as the electric charge, baryon number, strangeness, etc. In the absence of weak interactions the CP image of K^0 is its antiparticle $\bar{K}^0$ both having the same masses but opposite strangeness "charges". In the presence of weak interactions, however, strangeness is not a good quantum number and, therefore, the neutral kaons K^0 and $\bar{K}^0$ are no longer "good" particles, i.e., they do not have well defined masses and lifetimes (this irrespectively of whether CP is a good symmetry or not). K^0 and $\bar{K}^0$ mix through weak interactions. If CP were a good symmetry, in order to find the physical states of the neutral kaons, we would have had to look for the CP eigenstates. Taking $\bar{K}^0> = CP|K^0>$ we could have constructed the eigenstates

$$K^0_\pm = \frac{1}{\sqrt{2}}[K^0> \pm \bar{K}^0>]$$

corresponding to CP eigenvalues ± 1 respectively. These would have been the physical states as they are just those combinations which could not mix, if CP were conserved. In a CP conserving world, there can be no transition from a state with $CP = 1$ to a state with $CP = -1$.

Now, in the real life, we know since 1964 that CP is not a good symmetry [5] and the theoretical analysis becomes more complicated. We still can define that $\bar{K}^0 >= CP|K^0 >$ and construct the above CP eigenstates but, in general, they would not be the physical states, with well defined masses and lifetimes (eigenstates of the Hamiltonian) because CP does not commute with the Hamiltonian. Since up to now CP violation has only been seen in the decays of the K_L^0 in the following we shall focus our attention on the neutral kaon system but the general analysis is also valid for other similar systems, for example, the neutral beauty systems.

In discussing CP violation one distinguishes between direct CP and indirect CP violation as follows.

• Indirect CP violation

In this scenario one assumes that all transitions in nature are mediated by an "S matrix" which is CP conserving. Nevertheless, CP is violated because some particles (states) are "CP impure", i.e. they are linear combinations of CP even and CP odd eigenstates. Thus in such a scenario one has

$$|particle >= a|CP = +1 > +b|CP = -1 >$$

where a and b are constants. In such a scenario, once these coefficients have been determined from some CP violating processes there is no more to be learned. We know that the long-lived neutral kaon, K_L, is predominantly a $CP = -1$ state as it does not like to decay into two pions (which by Bose-Einstein statistics has $CP = 1$) in spite of the fact that such decay channels are much favoured by phase space. However, it does have a small CP violating branching ratio into two pions. Thus, in the framework of indirect CP violation one would say that, for *particle* $= K_L$, the coefficient b is large and responsible for all CP conserving decays of the K_L and the coefficient a is small and determines all CP violating decays. The short-lived neutral kaon which is known to be predominantly CP even will then contain a small CP odd admixture. In the limit where CP is conserved, in such a description, one of the coefficients in the above relation is expected to go to zero. In field theory it is, in general, not possible to have such a scenario because of higher order corrections. However, the direct CP violation effects, although generally present, may turn out to be too small to be observable at least in the near future.

• Direct CP violation

CP violation is said to be direct if the states are CP pure but the S matrix, or the transition Hamiltonian is a linear combination of CP even and CP odd parts, $H = H_+ + H_-$, where the first term conserves CP and the second violates it by inducing transitons between states with opposite CP eigenvalues, such as

$$< odd|H_-|even > \neq 0$$

In such a picture CP violation is far from "universal". The matrix element of the CP violating Hamiltonian is expected to be different for various CP violating processes.

As mentioned above, charged particles, such as the charged kaons, are not CP eigenstates. Under a CP transformation a positive kaon goes into a negative kaon, etc. Thus the indirect CP violation is not relevant for them. CP violation for such systems could exhibit itself, for example, via the so called Okubo effect, if the probability for a transition and its CP image were found to be different (see the section on CP transformations). One can imagine many such processes involving decays of charged particles but no CP asymmetry has been seen. In fact, the only CP violating processes observed up to now are the decays $K_L \to \pi^+\pi^-$, $K_L \to \pi^0\pi^0$ (with small branching ratios of the order of 10^{-3}) together with the observation that the rate of $K_L \to \pi^-e^+\nu$ is slightly larger than that of $K_L \to \pi^+e^-\nu$. That is it! At present, all data on CP violation (except the results of the NA31 experiment at CERN) can be explained by assuming indirect CP violation and it is not clear whether the direct CP violation has been seen or not [33].

An important point to be kept in mind is that the states K^0 and $\bar{K}^0$ are orthogonal, even in the presence of CP violation. Why? Because they are eigenstates of strangeness with opposite eigenvalues, ± 1. This is easily checked by working out the matrix element of the strangeness between these two states, $< K^0|S|\bar{K}^0 >$. Similarly the states $K^0_\pm$ are orthogonal as they are eigenstates of CP with eigenvalues ± 1. However, the physical neutral kaons K_L and K_S are *not* orthogonal to each other because of the following

$$H|K_X >= (M_X - \frac{i}{2}\Gamma_X)|K_X >\equiv \lambda_X|K_X > \qquad (47)$$

where X stands for either L or S; M denotes the mass of the particle and Γ is the width. Since the effective Hamiltonian H is not hermitian its eigenvalues are not real and the usual proof in quantum mechanics of the orthogonality of the eigenvectors corresponding to different eigenvalues does not go through. This fact is of great conceptual importance. In quantum mechanics, we are used to decompositions of the kind

$$\psi = a_1\psi_1 + a_2\psi_2 + ...a_n\psi_n$$

where we identify the $|a_j|^2$ as the probability of finding the system in the state j, etc. Such an interpretation requires our states ψ_j to be orthogonal to one another. We know that, for example K^0 is a linear combination of K_L and K_S (see below) but we are not allowed to identify the coefficients in that expansion with probabilities because our states in that expansion are not orthogonal.

Writing K_L and K_S as linear combinations of K^0 and $\bar{K}^0$ we have

$$|K_L> = p_L|K^0> + q_L|\bar{K}^0> \tag{48}$$

$$|K_S> = p_S|K^0> + q_S|\bar{K}^0> \tag{49}$$

where $|p_X|^2 + |q_X|^2 = 1$, for $X = L, S$. Now using the Schrödinger equation

$$i \begin{pmatrix} \frac{d}{dt}K^0(t) \\ \frac{d}{dt}\bar{K}^0(t) \end{pmatrix} = \begin{pmatrix} M_{11} & M_{12} \\ M_{21} & M_{22} \end{pmatrix} \begin{pmatrix} K^0(t) \\ \bar{K}^0(t) \end{pmatrix} \tag{50}$$

we may derive the eigenvectors of the Hamiltonian (the matrix on the RHS) and thus determine the physical neutral kaons. Doing so we obtain the following relations

$$
\begin{aligned}
M_{11} &= (p_S q_L \lambda_S - q_S p_L \lambda_L)/\Delta \\
M_{22} &= (p_S q_L \lambda_L - q_S p_L \lambda_S)/\Delta \\
M_{12} &= p_S p_L(\lambda_L - \lambda_S)/\Delta \\
M_{11} &= q_S q_L(\lambda_S - \lambda_L)/\Delta
\end{aligned}
$$

where $\Delta = p_S q_L - q_S p_L$. CPT invariance requires that the amplitude $K^0 \leftrightarrow K^0$ be equal to that of $\bar{K}^0 \leftrightarrow \bar{K}^0$, i.e., $M_{11} = M_{22}$. Thus $p_S q_L = -p_L q_S$ if CPT is valid. Time-reversal invariance gives the restriction that $|M_{12}| = |M_{21}|$, i.e., $|p_S p_L| = |q_S q_L|$. Finally, the assumption of CP symmetry is equivalent to requiring both CPT and T symmetries. Invoking CPT symmetry and using the relation $|p_X|^2 + |q_X|^2 = 1$, see after Eq. (49), we find that the physical neutral kaons may be expressed as

$$|K_L> = N_\epsilon[(1 + \epsilon)|K^0> - (1 - \epsilon)|\bar{K}^0>] \tag{51}$$

$$|K_S> = N_\epsilon[(1 + \epsilon)|K^0> + (1 - \epsilon)|\bar{K}^0>] \tag{52}$$

where $1/N_\epsilon = \sqrt{2(1 + |\epsilon|^2)}$. This parameter ϵ characterizes the strength of the indirect CP violation in the neutral kaon system. Evidently, this analysis is also valid for other similar systems, such as the neutral charm mesons and the neutral beauty mesons. The relevant quantities when dealing with the CP violating two-pion decay modes of the K_L are

$$\eta_{+-} \equiv \frac{A(K_L \to \pi^+\pi^-)}{A(K_S \to \pi^+\pi^-)} \approx \epsilon + \epsilon' \tag{53}$$

$$\eta_{00} \equiv \frac{A(K_L \to \pi^0 \pi^0)}{A(K_S \to \pi^0 \pi^0)} \approx \epsilon - 2\epsilon' \tag{54}$$

Here, in addition to the indirect CP violation parametrized by ϵ the direct CP violation, if there, manifests itself through the parameter ϵ'. The latter parameter is expected to be small because of the empirical $\Delta I = 1/2$ rule. Some years ago, one was expecting $|\epsilon'/\epsilon|$ to be perhaps as large as one or two percents. Unfortunately due to several reasons, such as the large mass of the top quark the present theoretical expectations give a much smaller value. Anyhow, the experiments (see [33] and references therein) have shown that the above ratio is at most at the level of 10^{-3}, perhaps much smaller and one does not yet have a conclusive evidence for the presence of direct CP violation in the K_L decays. Thus, we must wait for the next round of these experiments which are expected to give results in a few years. Due to lack of time, I shall not discuss the matter any further here. The interested reader may consult the articles in Refs. [34] and [33].

15   Comments on CP Violation in the Beauty Sector

Generally, by neutral B-mesons, in the following, we shall mean $\bar{b}d \equiv B^0$ and $b\bar{d} \equiv \bar{B}^0$. The corresponding systems with strange and antistrange quarks do not concern us here. The most important item on the "shopping list" of the future B-factories is "search for CP violation". For a review see, for example Ref. [35]. Furthermore, the subject was discussed in detail at this School, in the lectures by Danilov [27]. Therefore, here I shall only make a few comments.

As mentioned in the last section, in the neutral kaon system, the bulk of the observed CP violation (perhaps all of it) can be explained by invoking indirect CP violation [33]. However, in the case of the neutral B system, in the Standard Model, one does not expect an observable CP violating effect to originate from the "indirect" mechanism alone. The reason being that, to a good approximation, the parameter ϵ, for the neutral B-mesons, is a pure phase. One can simply show that by introducing the analogs of the K_L and K_S for the B's, i.e.,

$$|B_L> = N_B[(1 + \epsilon_B)|B^0> - (1 - \epsilon_B)|\bar{B}^0>] \tag{55}$$

$$|B_S> = N_B[(1 + \epsilon_B)|B^0> + (1 - \epsilon_B)|\bar{B}^0>] \tag{56}$$

For a pure phase, i.e., $\epsilon_B^* = -\epsilon_B$ we see that, in the above equations, the coefficient of B^0 is simply the complex conjugate of the coefficient of the $\bar{B}^0$. Thus, we may write each such coefficient as a modulus times a phase (i.e., in the form $r \, exp(\pm i\phi)$) and remove the phase ϕ by redefining the B^0 and the $\bar{B}^0$ fields. The moduli are, from the normalization conditions, equal to $1/\sqrt{2}$. Therefore, we get that in the Standard Model, to a good approximation, the physical B's are eigenstates of CP with opposite eigenvalues. However, here we shall not remove the phase in order to follow the usual convention. This phase is coming from the box-diagram giving $B - \bar{B}$ mixing, where the vertices of the box contain the elements of the quark mixing matrix. We may, if we wish, redefine the phases here but then, for consistency, we must do so everywhere such as in the decay amplitudes of the B^0 and $\bar{B}^0$.

The above notation B_L, B_S, which we have chosen in analogy with the case of neutral kaons, is not so suitable for the B's. In the case of the neutral kaons the long-lived one has a lifetime of about $5x10^{-8}$ seconds and lives some 500 times longer than the short-lived one. But, the neutral B's are both short-lived, with lifetimes of the order of picoseconds and the ratio of their lifetimes is approximately one. Therefore, in the literature the physical neutral B's are usually referred to as B_H and B_L, where H stands for "heavy" and L for "light". This is because the difference of the masses of these objects, Δm, is expected to be more important than the difference of their widths, $\Delta\Gamma$. Going back to the parameter ϵ (we suppress the subscript B), we see that its being purely imaginary means that there is plenty of "mixing" in the B-system. In other words, the physical neutral B-mesons are "half" B^0 and "half" $\bar{B}^0$. The most promising mechanism for CP violation in the B-system is based on this maximal mixing where one looks for interference between final states which can be reached from both B^0 and $\bar{B}^0$, as I shall explain now (for a review see, for example [35])

Suppose now that at the time $t = 0$ we have a B^0. This assumption "costs" quite a bit because in order to make sure that we have a B^0 we must tag the associately produced $\bar{B}^0$ on the "opposite side", for example by looking for its semileptonic decays using the fact that $b \to l^-$ while $\bar{b} \to l^+$, where l can be a muon or an electron. Due to mixing, our tagged B^0 will evolve in time into a state $|B^0(t) >$ which is a linear combination of $|B^0 >$ and $|\bar{B}^0 >$

$$|B^0(t) > = e^{-i(M_B - i\Gamma/2)t} \{c(t)|B^0 > +i\omega s(t)|\bar{B}^0 >\} \tag{57}$$

Here M_B is the average mass of the neutral B meson eigenstates, their mass difference being ΔM; Γ is their common width (we neglect the difference of their widths) and $c(t) = cos(\frac{\Delta m}{2}t)$, $s(t) = sin(\frac{\Delta m}{2}t)$. The parameter ω defined by

$$\omega = \frac{V_{td}V_{tb}^*}{V_{td}^*V_{tb}} = e^{-2i\delta_m} \tag{58}$$

arises from the box diagram description of the B^0-$\bar{B}^0$ mixing with δ_m being the appropriate weak phase originating from the quark mixing matrix. The idea is then to look for the decay of our $B^0(t)$ into a final state f such that the amplitudes for $B^0 \to f$ and $\bar{B}^0 \to f$ are as large as possible and comparable with each other. Then one can get a large interference between these two amplitudes. That, by itself, does not demonstrate that there has been CP violation. In order to look for CP violation, we may compare the rate for the process we have been talking about with the rate of its CP-image. Nonequality of the two rates would be a signal of CP violation. In general, of course, we have four possible channels, viz., $B^0(t) \to f$, $\bar{B}^0(t) \to \bar{f}$, $\bar{B}^0(t) \to f$ and $B^0(t) \to \bar{f}$, where $\bar{f}$ denotes the CP conjugate of the state f. The corresponding amplitudes are

$$
\begin{aligned}
< f|T|B^0 > &= e^{i\delta} M_f e^{i\alpha_f} \\
< f|T|\bar{B}^0 > &= e^{-i\delta} M'_f e^{i\alpha'_f}
\end{aligned}
\tag{59}
$$

Here the modulii of the transition matrix elements have been denoted by M and M' respectively and δ is the single decay weak phase associated with $V_{cs}V_{cb}^*$. The quantities α_f and α'_f are the "strong" phases, those which do not change sign when going from a transition to its CP conjugate process. The amplitudes for the CP conjugate transitions are given by

$$
\begin{aligned}
< \bar{f}|T|\bar{B}^0 > &= e^{-i\delta} M_f e^{i\alpha_f} \\
< \bar{f}|T|B^0 > &= e^{i\delta} M'_f e^{i\alpha'_f}
\end{aligned}
\tag{60}
$$

Putting the decay amplitudes in (59) and (60) into Eq. (57) and the corresponding one for $\bar{B}^0(t)$ we finds the following four decay rates of the time-evolved $B^0(t)$ and $\bar{B}^0(t)$ into the mutually CP conjugate final state f and $\bar{f}$

$$
\begin{aligned}
\Gamma(B \to f) &= e^{-\Gamma t} R_f \{1 + \Delta_f cos(\Delta mt) + sin(\Delta mt)[A_f sin\phi + D_f cos\phi]\} \\
\Gamma(\bar{B} \to \bar{f}) &= e^{-\Gamma t} R_f \{1 + \Delta_f cos(\Delta mt) + sin(\Delta mt)[-A_f sin\phi + D_f cos\phi]\} \\
\Gamma(\bar{B} \to f) &= e^{-\Gamma t} R_f \{1 - \Delta_f cos(\Delta mt) + sin(\Delta mt)[-A_f sin\phi - D_f cos\phi]\} \\
\Gamma(B \to \bar{f}) &= e^{-\Gamma t} R_f \{1 - \Delta_f cos(\Delta mt) + sin(\Delta mt)[A_f sin\phi - D_f cos\phi]\}
\end{aligned}
\tag{61}
$$

where

$$
\begin{aligned}
R_f &= \frac{1}{2}(M_f^2 + M'^2_f)\Gamma_f^{(0)} \\
\Delta_f &= \frac{M_f^2 - M'^2_f}{M_f^2 + M'^2_f} \\
A_f &= \frac{2M_f M'_f cos(\alpha_f - \alpha'_f)}{M_f^2 + M'^2_f} \\
D_f &= \frac{2M_f M'_f sin(\alpha_f - \alpha'_f)}{M_f^2 + M'^2_f}
\end{aligned}
\tag{62}
$$

where $\Gamma_f^{(0)}$ is the phase space factor and we have that $\Delta_f^2 + A_f^2 + D_f^2 = 1$. Thus these measurables correspond to three independent quantities. Furthermore $\phi = 2(\delta_M + \delta)$ is the CP violating weak phase for this process. Its presence can be exhibited by showing that either $\Gamma(B \to f) \neq \Gamma(\bar{B} \to \bar{f})$ or $\Gamma(B \to \bar{f}) \neq \Gamma(\bar{B} \to f)$.

Up to now, in the literature, special emphasis has been put on two-body final states which are CP eigenstates, i.e., their own CP images such as $J/\psi K_S$ [36] or $\pi^+\pi^-$, see the lectures by Danilov [27]. Then, only two of the above four channels are independent. The resulting asymmetry provides then a particularly clean determination of the appropriate weak phase because only one (dominant) decay amplitude intervenes in the process and it cancels in the asymmetry. The problem is, however, the limitation to a few decay channels with small branching ratios. To circumvent this limitation, it has been proposed that one could study the two or three body decays of neutral B mesons to final states which are not CP eigenstates [37]. These proposals can help in reducing the number of B mesons required by a factor of two to four.

Recently, J. Bernabeu and I [38] have considered the possibility of using semi-inclusive decays of neutral B mesons which we hope may serve the double purpose of establishing CP violation in the B system and determining the appropriate weak phase. We advocate looking inte the semi-inclusive decays of the neutral B-mesons into the final states $K_S X$, with $X = \bar{c}c$, where the charms quarks emerge as "jets" rather than J/ψ, etc.

The b quark has two dominant decay modes, viz., $b \to c\bar{u}d$ and $b \to c\bar{c}s$. The relative strength of these transitions has been estimated in various models. In their recent update, Altarelli and Petrarca [39] find that the branching ratio for $b \to c\bar{c}s$ lies in the range 11% to 24%, by taking "heavy masses" or the favoured "light masses" for the quarks respectively. Our results indicate that about a 7% fraction of this transition should produce the semi-inclusive channel which we find to be of great interest for studying CP violation in the B system i.e., the transition $B, \bar{B} \to K_S X(c\bar{c})$. It is essential that this rate be measured to see whether the CP violation studied is experimentally feasible. A prerequisite for the kind of study we have in mind is the use of good vertex detectors to the channel $K_S + \bar{c}c$.

Experimentally all we know about the inclusive decay channels of b involving kaons is that $B \to K^\pm$ anything and $B \to K^0/\bar{K}^0$ anything have [40] the branching fractions $(85 \pm 11)\%$ and $(63 \pm 8)\%$ respectively. Since the sum of these two branching fractions is larger than unity we may conclude that a substantial fraction of B decays (approximately 50%) produce (at least) two kaons. This result, not by itself conclusive, looks promising. The mechanism $b \to c\bar{c}s$ is also responsible for J/ψ production in B decays. The inclusive branching ratio for $b \to J/\psi X$ is known to be [40] about 1%. Our estimate of the semi-inclusive branching ratio for $B \rightarrow K_S X(c\bar{c})$, in some sense complementary to it, is again about 1%.

16 Acknowledgements

I am indebted to Professor Tom Ferbel for asking me to give these lectures. This work has been supported by the Swedish National Research Council (NFR).

References

[1] T. D. Lee, Particle Physics and Introduction to Field Theory (harwood academic publishers, 1981)
W. M. Gibson and B. R. Pollard, "Symmetry Principles in Elementary Particle Physics" (Cambridge Univ. Press, 1976)
E. P. Wigner, "Symmetries and Reflections" (MIT Press, 1972)
J. J: Sakurai, "Invariance Principles and Elementary Particles" (Princeton Univ. Press, 1964)
Wigner Z. Phys. 43 (1927) 624, Gtt. Nach. Math. Naturv. Kl (1932) 546

[2] S. L. Glashow, Nucl. Phys. 22 (1961) 579
S. Weinberg, Phys. Rev. Lett. 19 (1967) 1264
A. Salam, in Elementary Particle Theory, Ed. N. Svartholm (Almqvist and Wiksell, 1968)

[3] C. S. Wu et al., Phys. Rev. 105 (1957) 1413
R. L. Garwin, L. M. Lederman and M. Weinrich, ibid, p. 1415
J. I. Friedman, V. L. Telegdi, ibid, p. 1681 and Phys. Rev. 106 (1957) 1290

[4] T. D. Lee and C. N. Yang, Phys. Rev. 104 (1956) 254

[5] J. H. Christenson, J. W. Cronin, V. L. Fitch and R. Turlay, Phys. Rev. Lett. 13, (1964) 138

[6] P. K. Kabir, The CP Puzzle (Academic Press, 1968)

[7] L. Wolfenstein, Phys. Rev. Lett. 13 (1964) 562

[8] K. Kleinknecht, Proc. of the 1974 Int. Conf. on High Energy Phys. (London) p.III-23

[9] M. Kobayashi and T. Maskawa, Prog. Theor. Phys. 49 (1973) 652

[10] T. D. Lee, Phys. Reports 9C (1973) 143

[11] S. Weinberg, Phys. Rev. Lett. 37 (1976) 657
P. Sikivie Phys. Lett. 65B (1976) 141

[12] R. N. Mohapatra in "CP Violation", Ed. C. Jarlskog (World Scientific, 1989) p. 384

[13] J. D. Bjorken and S. D. Drell, "Relativistic Quantum Field Theory" (McGraw-Hill, 1965)

[14] R. D. Peccei in "CP Violation", Ed. C. Jarlskog (World Scientific, 1989) p. 503

[15] See, for example S. Rudaz, Phys. Rev. D41 (1990) 2619

[16] See, for example, E.S. Abers and B. W. Lee, Phys. Rep. 9C (1973) 1
For reviews by the present author see
C. Jarlskog, CERN Yellow Reports 85-11, p. 260 and p. 277; 82-04, p. 63

[17] S. L. Glashow, J. Iliopoulos and L. Maiani, Phys. Rev. D2 (1970) 1285

[18] F.D. Murnaghan, "The Unitary and Rotation Groups" (D. C. Spartan, 1962)

[19] N. Cabibbo, Phys. Rev. Lett. 10 (1963) 531
M. Gell-Mann and M. Levy, Nuovo Cimento 16 (1960) 705

[20] F. Dydak, lectures given at this School

[21] Z. Maki, M. Nakagawa and S. Sakata, Proc. of 1962 Inter. Conf. (CERN, Geneva) p. 663; Prog. Theor. Phys. 28 (1962) 870

[22] C. Jarlskog, Phys. Lett. B241 (1990) 579

[23] C. Jarlskog, Phys. Rev. Lett. 55 (1985) 1039

[24] C. Jarlskog, Zeit. f. Phys. C29 (1985) 491

[25] D-D Wu, Phys. Rev. D33 (1986) 860;
O. W. Greenberg, Phys. Rev. D32 (1985) 1841

[26] L. Wolfenstein, Phys. Rev. Lett. 51 (1983) 1945

[27] M. Danilov, lectures given at this School

[28] C. Jarlskog and R. Stora, Phys. Lett. B208 (1988) 268
J. D. Bjorken (private communication) has also, independently, pointed out the importance of one of the unitarity triangles (the c-triangle)

[29] G. Källen, "Elementary Particle Physics" (Addison-Wesley, 1964)

[30] P. H. Frampton and C. Jarlskog, Phys. Lett. 154B (1985) 421

[31] C. H. Albright, C. Jarlskog and B.-Å Lindholm, Phys. Lett. 199B (1987) 553; Phys. Rev. D38 (1988) 872

[32] C. Jarlskog, Phys. Rev. D35 (1987) 1685 and references therein

[33] See for example
G. Barr (NA31 Collaboration), in Proc. Joint Int. Lepton-Photon Symp. and Euro-Phys. Conf. on High Energy Phys. (Geneva 1991) 179
B. Winstein (E731 Collaboration) ibid 186

[34] See, for example
E. A. Paschos and U. Türke, Phys. Rep. 178 (1989) 146
K. Kleinknecht *in* "CP Violation", Ed. C. Jarlskog (World Scientific, 1989) 41

[35] See, for example
"B Decays", Ed. S. Stone (World Scientific, 1992)
I. I. Bigi, V. A. Khoze, N. G. Uraltsev and A. I. Sanda *in* "CP Violation" Ed. C. Jarlskog (World Scientific, 1989) 175

[36] A. Carter and A. Sanda, Phys. Rev. Lett. 45 (1980) 952; Phys. Rev. D23 (1981) 1567
I. Bigi and A. Sanda, Nucl. Phys. B193 (1981) 85, B281 (1987) 41

[37] B. Kayser, M. Kuroda, R. Peccei and A. Sanda, Phys. Lett. B237 (1990) 508
T. Mannel, W. Roberts and Z. Ryzak, Phys. Lett. B248 (1990) 392
J. Rosner, Phys. Rev D42 (1990) 3732
I. Dunietz, H. Lipkin, H. Quinn, A. Snyder and W. Toki, Phys. Rev. D43 (1991) 2193
R. Aleksan, I. Dunietz, B. Kayser and F. Le Diberder, Nucl. Phys. B361 (1991) 141

[38] J. Bernabeu and C. Jarlskog, to be published

[39] G. Altarelli and S. Petrarca, Phys. Lett. B261 (1991) 303

[40] Particle Data Group, Phys. Rev D45 (1992) S1

PINNING DOWN THE STANDARD MODEL

F. Dydak

MPI of Physics, Munich, Germany, and CERN, Geneva, Switzerland

INTRODUCTION

Since the coming into operation of CERN's Large Electron-Positron storage ring (LEP), in 1989, the Electroweak Standard Model has been tested with an unprecedented precision, both concerning the number of independent physics quantities which have been measured and the accuracy of the measurements. In these lecture notes, the most significant experimental results from LEP and elsewhere, and their implications for the Electroweak Standard Model are reviewed.

As is well known, the Electroweak Standard Model in its minimal form (Minimal Standard Model, MSM) is very successful in describing a large variety of data. Yet there are many aspects of the model which are poorly tested, and there is a general consensus of opinion that the Electroweak MSM is not the ultimate theory. Both aspects stimulate experimentalists to measure as many quantities as precisely as possible – in the hope of better understanding the Electroweak MSM and of finding a clue to the wider theory behind.

Up to now LEP has not revealed a new particle or a new phenomenon, but it has produced a wealth of data well suited to high-precision tests of the Electroweak MSM. Why is LEP so exciting, taking into account that testing the Electroweak MSM has already quite a respectable history of 15 years?

In the leptonic sector of the Electroweak MSM, the lack of knowledge is concentrated in the neutrino area: masses and mixing angles are unknown; not even the particle type (Dirac or Majorana) is known. However, LEP added a very stringent measurement of the number of light neutrino families. As for charged leptons, LEP is an ideal machine for precisely measuring their weak neutral-current couplings.

In the quark sector, the initial hopes of discovering the t-quark were not fulfilled. The LEP studies have so far concentrated on testing the weak neutral-current couplings of quarks. Eventually, LEP will contribute a lot to the knowledge on weak heavy-quark decays; however such studies need large statistics and have only begun in earnest in 1992, with the successful operation of Silicon microvertex-detectors.

As for gauge bosons, LEP delivered a measurement of the Z-mass with very high precision. The W-mass which is currently measured by the UA2, CDF and D0 Collaborations, will only be accessible when LEP is upgraded to LEP200. LEP's measurement precision (± 100 MeV) will, however, not be better than what is likely to be achieved at the FNAL $p\bar{p}$-collider.

In the Higgs sector, LEP cleared up the experimental situation in an impressive way: as will be discussed below, there is no 'light' Higgs-boson.

Finally, thanks to the high-precision measurements which became possible with the advent of LEP, electroweak radiative effects became directly measurable or, equivalently, have to be taken into account in order to obtain consistent results. This aspect confirms the nature of the Electroweak MSM as that of a renormalizable gauge theory, and permits insight into the properties of particles which are too heavy to be directly produced at LEP but can be sensed through loop effects.

Altogether, LEP made a significant impact on testing the Electroweak MSM and will continue to do so. However, as will also be discussed below, LEP is complemented by other experiments which are perhaps less striking from the point of view of precision, but address different aspects of the Electroweak MSM.

This review is not complete in itself but is rather to be seen in the context of the overall programme of this School. Important aspects of the Electroweak MSM are covered in the lectures on B-physics by M. Danilov, on the CKM-matrix and CP-violation by C. Jarlskog, on the t-quark by P. Tipton, and last but not least in J. Siegrist's lectures on the opportunities for electroweak physics at the future hadron colliders LHC and SSC. Very important aspects of the Electroweak MSM are dealt with in the lectures of M. Spiro on non-accelerator particle physics. The interested reader is invited to consult the respective lecture notes.

THE PERFORMANCE OF LEP AND OF THE SLC

LEP was in operation for three months in 1989, for six months in 1990, and for seven months each in 1991 and 1992. The total score of hadronic and leptonic Z-decays recorded on tape up to the end of 1992 is given in Table 1, as well as the breakdown for the four large LEP detectors, ALEPH, DELPHI, L3, and OPAL. The results presented in these lecture notes are based on the data obtained up to a point between the end of 1991 and the end of 1992, i.e. on a total of up to 5 million Z-decays. The combined operation efficiency of LEP and of its associated injector complex was about 45%, as compared with the theoretical maximum of about 85%. The number of recorded events fell short by a factor of three with respect to earlier hopes. The main reason for the shortfall were the relatively low operation efficiency and the so-called beam-beam effect, which blows up the transverse beam size and thus limits the maximum attainable luminosity already at lower currents than anticipated. Nonetheless, the performance of LEP is generally considered satisfactory.

Table 1. Number of hadronic and leptonic Z-decays recorded by the LEP experiments (in units of 10^3)

	hadronic	leptonic
ALEPH	1140	140
DELPHI	1050	100
L3	1100	100
OPAL	1170	150
Total	4460	490

The detectors ALEPH, DELPHI, L3, and OPAL were all the time taking data when LEP was in operation. Existing differences in the design of the four detectors have not yet shown up in the quality of the results which have been presented. The most important features of the detectors are their good coverage of the solid angle which is close to 4π, and their redundancy in the triggering systems; both properties are common to all four detectors, and constitute important assets for the quality of the data and their analysis.

So far, LEP has been operated at the Z-peak energy, and at the centre-of-mass energy settings of approximately $m_Z \pm 1$, $m_Z \pm 2$, and $m_Z \pm 3$ GeV. In 1992, however, all data were taken solely at the peak position, with no scanning across the resonance.

Figure 1 shows the integrated luminosity delivered by LEP to each of the four LEP experiments in the years 1990–92. Altogether, each experiment has logged data on tape which correspond to an effective integrated luminosity of 50 pb^{-1}, out of which 8 pb^{-1} correspond to data taking off the Z-peak. The steady increase of performance over the years of operation is quite apparent from Fig. 1.

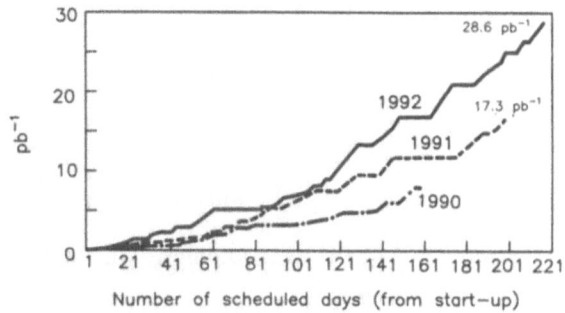

Figure 1. Integrated LEP luminosity delivered in 1990–92

Figure 2 shows the peak luminosity per machine fill, in 1992, as a function of the total beam current, for the ALEPH experiment. In the 8-bunch operation mode which was implemented later in 1992, the peak luminosity was typically $8 \times 10^{30} cm^{-2} s^{-1}$.

An important machine parameter is the absolute energy which enters directly into the measurement of the Z-mass. The absolute energy scale used to be based on a measurement of the velocities of positrons and protons circulating with the same momentum of 20 GeV in the machine, resulting in a calibration of the magnetic field at this momentum. This calibration was then scaled up to 45 GeV with the help of flux-loop measurements. A careful assessment of the measurement errors resulted in an estimate of the absolute scale error at the Z-pole of 2.4×10^{-4}, i.e. 22 MeV. This figure suggested, however, a greater precision than warranted: as the dominant contribution to the scale error arose from the distortion of the magnetic field by a few μm thick Ni layer in the beam pipe, the effective thickness of which is not well known, a round number of 20 MeV was quoted for quite some time, as the best guess for the uncertainty of the machine's centre-of-mass energy at the Z-pole.

A new method became available after the observation, in 1991, of a transverse polarization of the LEP beam of 10%, by means of a Compton-scattering laser polarimeter [1]. Under the influence of a weak oscillating horizontal magnetic field, the spins of the orbiting electrons are rotated away from their nominal vertical axis, and a depolarizing resonance is observed if the depolarizing magnetic field is in phase with the spin precession.

As the number of spin precessions per revolution is linearly related to the beam energy, with only the precisely measured mass m_e and the anomalous magnetic moment $(g_e - 2)/2$ of the electron as further ingredients, the beam energy is in essence measured via a frequency which traditionally can be measured with very good precision. The intrinsic precision of the method is very high, however inherent machine instabilities limit the attainable precision at present to 8×10^{-5}.

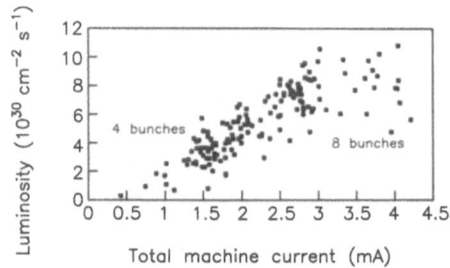

Figure 2. ALEPH peak luminosity in 1992

Instabilities of the beam energy between measurements are due to temperature effects and the changes in bending power when corrector dipoles are activated with a view to tuning the horizontal beam orbit. Surprisingly, it also turned out that the 'tidal' forces of the sun and the moon on the earth's crust cause a variation of the 27 km circumference of LEP by 1 mm, causing a peak-to-peak variation of the beam energy of 10 MeV (see Fig. 3), which is non-negligible for a precision determination of the Z-mass. This insight is the more interesting because it is the first incident in particle physics measurements that the graviational interaction had to be taken into account in the analysis of a particle physics experiment.

SLAC's Linear Collider, the SLC, had its first physics run before LEP, in the spring of 1989, albeit with poor luminosity. Eventually the running of this machine which exploits a novel technique in accelerator construction, became better understood and the luminosity went up considerably, so as to produce some ten Z's per hour. With the newly built SLD detector replacing the old war-horse MARK II, and a polarized electron gun successfully commissioned, the stage was set, in 1992, for a first measurement of the left-right asymmetry A^{LR} in electron-positron collisions at the Z-pole. Since then, the luminosity was further improved, and so was the longitudinal electron polarization, which went up from initially 20% to nowadays 60%.

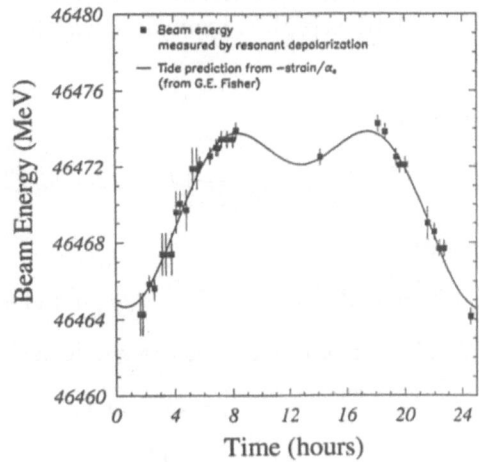

Figure 3. Variation of LEP's beam energy due to tidal forces

EXPERIMENTAL OBSERVABLES

One of the striking aspects of the experimental tests of the Electroweak MSM is the great variety of the experimental observables. They permit measurements at very different scales of momentum transfer, ranging from 10^{-10} GeV^2 up to 10^4 GeV^2. They not only provide independent tests of the Electroweak MSM but offer also different sensitivity on possible new phenomena beyond the Electroweak MSM.

In these lecture notes, we restrict ourselves in view of the wealth of available experimental data to a discussion of the following processes and experimental observables:

- Line-shape, asymmetries and polarization at the Z-pole

- W-production in $p\bar{p}$-collisions: W-mass

- Neutrino-quark scattering: neutral- to charged-current ratio

- Neutrino-electron scattering: cross-sections

RADIATIVE CORRECTIONS

One of the salient features of a renormalizable gauge theory is that radiative corrections can be calculated in principle to any order of perturbation, with finite results which can be tested by experiment. This holds for any experimental observable, however the attainable experimental precision renders some observables more attractive than others. To illustrate the point, we discuss below the radiative corrections of some observables at the Z-pole in more detail.

The precise measurement of the Z-line-shape and the Z-decay characteristics has attracted a lot of attention. We recall here the basic formulae, also with a view to introducing our nomenclature. In Born approximation, the $e^+e^- \to f\bar{f}$ matrix element reads as follows:

$$\mathcal{M} = Q_e Q_f \frac{4\pi\alpha}{s} J_e^{em} J_f^{em}$$
$$+ \sqrt{2}\rho G_F m_Z^2 \frac{1}{s - m_Z^2 + im_Z\Gamma_Z}(J_e^3 - 2\sin^2\theta_w J_e^{em})(J_f^3 - 2\sin^2\theta_w J_f^{em}),$$

with the fermion current

$$J_f^3 - 2\sin^2\theta_w J_f^{em} = \bar{u}_f[\gamma_\mu(I_f^3 - 2Q_f\sin^2\theta_w) + \gamma_\mu\gamma_5 I_f^3]v_f,$$

and the vector and axial-vector coupling constants, defined as

$$g_{Vf} = I_f^3 - 2Q_f\sin^2\theta_w$$

$$g_{Af} = I_f^3.$$

The Veltman parameter

$$\rho = \frac{G_F^{NC}}{G_F^{CC}} = \frac{m_W^2}{m_Z^2\cos^2\theta_w}$$

is equal to unity in the Electroweak MSM, and the definitions of the electroweak mixing parameter via the W- and Z-masses, and via the electromagnetic and weak coupling constants, coincide in Born approximation:

$$\sin^2\theta_w = 1 - \frac{m_W^2}{m_Z^2} = \frac{e^2}{g^2}.$$

Electroweak radiative corrections are large compared with the experimental errors.

They are customarily divided into two categories:

- *Photonic corrections*

  They involve in first order an extra photon which is added to the Born diagrams, either in the form of a real bremsstrahlung photon, or as a virtual photon loop. The corrections are very large – of the order of 100% – and depend on the experimental cuts. The dominant contribution is due to initial-state bremsstrahlung, and is customarily represented by a radiator fuction $\mathcal{H}(z, s)$ which is folded with the Born cross-section. For example, the Z-line-shape then takes the form

$$\sigma^f(s) = \int_0^1 \sigma_f^0(zs)\mathcal{H}(z, s)dz.$$

  Figure 4 shows a sample of diagrams of photonic corrections.

- *Non-photonic corrections*

  They involve all other corrections: notably the self-energy corrections of the vector bosons, and virtual W- and Z-loops (see Fig. 5). These corrections are still large – of the order of 10% – but independent of experimental cuts. The bulk of these non-photonic corrections can be absorbed into 'running' (i.e. Q^2-dependent) coupling constants, while preserving the Born approximation formulae. This concept, which has been pioneered by Kennedy and Lynn [2], has become known as the 'improved Born approximation' [3].

Figure 4. Diagrams of photonic corrections

Figure 5. Diagrams of non-photonic corrections

Throughout this report, the improved Born approximation and the notion of running coupling constants are utilized for presenting and interpreting the data. All formulae given below are understood to be valid only in the improved Born approximation: in each case a small correction is being ignored. The present level of the experimental accuracy still permits this approach.

The Z-line-shape for the final-state fermion f reads as follows:

$$\sigma_f(s) = \frac{s}{(s-m_Z^2)^2 + \frac{s^2\Gamma_Z^2}{m_Z^2}} \left(\frac{12\pi}{m_Z^2}\Gamma_e\Gamma_f + I_f N_c \frac{s-m_Z^2}{s} \right) + \frac{4\pi}{3} N_c Q_f^2 \frac{\bar\alpha^2}{s} \,,$$

where I_f denotes the $Z-\gamma$ interference term, and $\bar\alpha$ the fine structure constant at $Q^2 = m_Z^2$. The Z-mass and widths are understood as the physical line-shape parameters. The decay width of the fermion f is

$$\Gamma_f = \frac{G_F m_Z^3}{6\sqrt{2}\pi}(\bar{g}_{Vf}^2 + \bar{g}_{Af}^2)\,, \tag{1}$$

with

$$\bar{g}_{Vf} = \sqrt{\bar\rho}(I_f^3 - 2Q_f \sin^2\bar\theta_w)$$

and

$$\bar{g}_{Af} = \sqrt{\bar\rho}I_f^3\,,$$

where the Fermi constant G_F, unlike other coupling constants, is not running but used with its numerical value $G_F = 1.16637\times10^{-5}$ GeV^{-2}.

The forward-backward asymmetry of the fermion f is given by

$$A_f^{FB} = \frac{3}{4}\mathcal{A}_e\mathcal{A}_f\,, \tag{2}$$

with

$$\mathcal{A}_e = 2\frac{\bar{g}_{Ve}\bar{g}_{Ae}}{\bar{g}_{Ve}^2 + \bar{g}_{Ae}^2}$$

and

$$\mathcal{A}_f = 2\frac{\bar{g}_{Vf}\bar{g}_{Af}}{\bar{g}_{Vf}^2 + \bar{g}_{Af}^2}\,.$$

The W- and Z-masses are, in terms of running coupling constants, given by

$$m_W^2 = \frac{\pi \bar{\alpha}}{\sqrt{2} G_{\mathrm{F}} \sin^2 \bar{\theta}_{\mathrm{w}}}$$

and

$$m_Z^2 = \frac{\pi \bar{\alpha}}{\sqrt{2} \bar{\rho} G_{\mathrm{F}} \sin^2 \bar{\theta}_{\mathrm{w}} \cos^2 \bar{\theta}_{\mathrm{w}}} \, .$$

The running electroweak mixing parameter, $\sin^2 \bar{\theta}_{\mathrm{w}}$, is related to the W- and Z-masses by

$$\sin^2 \bar{\theta}_{\mathrm{w}} = \frac{\bar{e}^2}{\bar{g}^2} = 1 - \frac{m_W^2}{\bar{\rho} m_Z^2} \, . \tag{3}$$

Other definitions of the running weak mixing parameter found in the literature, such as $\sin^2 \theta_{MS}$, $\sin^2 \theta_{\mathrm{w}}(m_Z^2)$, and $\sin^2 \theta_{\mathrm{w}}^*$, are, apart from very small corrections, equivalent. The running Veltman parameter is given by

$$\bar{\rho} = 1 + \Delta \bar{\rho} = 1 + \frac{3\sqrt{2}}{16\pi^2} G_{\mathrm{F}} m_t^2 \, , \tag{4}$$

exhibiting a quadratic dependence on the mass of the t-quark which permits conclusions on the range of allowed values of m_t.

The procedure which is applied in the data analysis is depicted in Fig. 6. The raw data are corrected for the limited apparatus acceptance and resolution, and other mundane effects such as dead electronic channels. The result are 'perfect' data, which can then be confronted with theoretical expectations. The latter start from Born cross-sections, and are modified by applying photonic and non-photonic radiative corrections leading to physical cross-sections which an ideal apparatus would measure. A fit of the perfect data to the theoretically expected physical cross-sections leads to the best values of free input parameters such as $m_Z, \Gamma_Z, \sigma_h^0, \Gamma_h, \Gamma_e, \Gamma_\mu, \Gamma_\tau$, and Γ_{inv}. The point is that up to this stage the only assumption is a Breit-Wigner form of the Z-line-shape and the validity of QED for the calculation of the radiator function. Thus, apart from numerically unimportant contributions from the $Z-\gamma$ interference, and from non-photonic radiative corrections, the procedure leads to determinations of the parameters of the Z-line shape which are independent of the Electroweak MSM.

Before we go into a discussion of the precise measurement of observables at the Z-pole, we address the question of the particle spectrum which is at work, and which has to be taken into account when calculating radiative corrections. The primary question there concerns the Higgs-boson H^0, the existence of which is assumed in the Electroweak MSM. Therefore, in the subsequent section, we review the experimental information on the existence of the Higgs-boson H^0.

SEARCH FOR THE MSM HIGGS-BOSON H^0

A considerable effort has gone into the experimental search for the neutral Higgs-boson which is required by the Electroweak MSM, as the agent responsible for the fermions and gauge bosons acquiring their masses. Before LEP, quite a number of searches were performed, with results that were hardly conclusive. At LEP, unambiguous limits on the mass of the Higgs-boson could be set, because of its well-defined coupling to fermions and the high luminosity delivered by the machine.

The search was performed through the decay $Z \to H^0 Z^*$, with $Z^* \to e^+ e^-, \mu^+ \mu^-$, and $\nu \bar{\nu}$. The search for a heavy H^0 is limited by the integrated luminosity, since

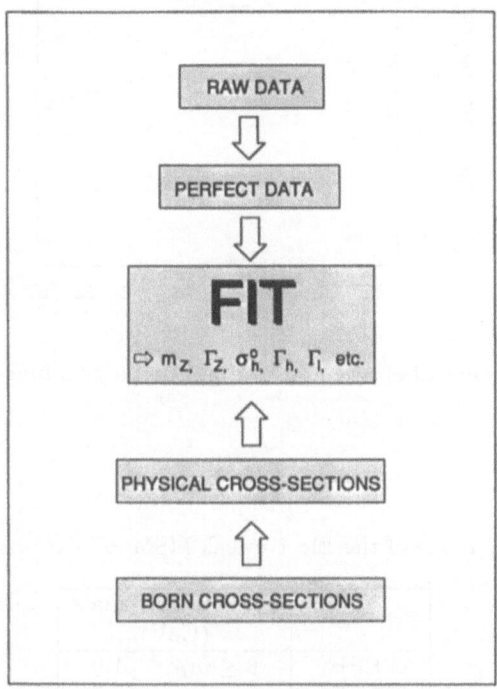

Figure 6. Procedure of the data analysis

the cross-section for $Z \rightarrow H^0 Z^*$ is relatively steeply falling with increasing m_H. The signature is an $e^+e^-, \mu^+\mu^-$, or $\nu\bar{\nu}$ pair from Z^* decay, together with the decay products of the heavy H^0, mostly $b\bar{b}$. The cuts are carefully optimized so as to extend the sensitivity to the highest masses while keeping the background negligible.

The global efficiency of observing a Higgs-boson is given in Fig. 7 as a function of its mass [4], for each of the four LEP experiments. The efficiency is hovering around 50 %, with relatively little difference between the experiments.

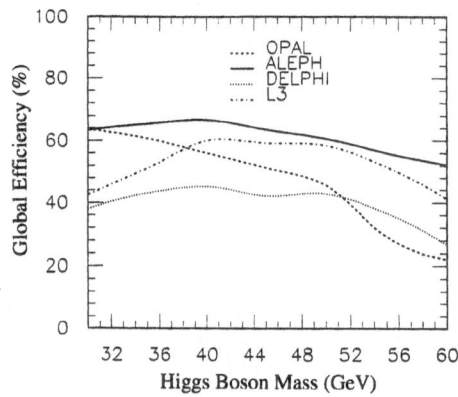

Figure 7. Efficiency of observing the Higgs-boson as a function of its mass

Table 2. Mass range of the Electroweak MSM H^0 excluded at 95% CL

	Excluded H^0 mass (GeV)
ALEPH	$0 \leq m_H \leq 51.0$
DELPHI	$0 \leq m_H \leq 42.0$
L3	$0 \leq m_H \leq 47.6$
OPAL	$0 \leq m_H \leq 47.3$
Combined	$0 \leq m_H \leq 57$

The mass range excluded by the LEP experiments is given in Table 2 [5]. A particular property of the search at LEP is that the range of excluded masses extends to zero at its lower end. Because of the limitation coming from the available integrated luminosity it is logical to combine the results from the four experiments and thus obtain an even better lower limit on m_H. The combined limit on the mass of the Higgs-boson is 60 GeV at 95% CL. This procedure, however, makes use of the non-observation of candidate events near the combined upper mass limit. This non-obsersation permits the intersection of the 95 % CL straight horizontal line of three events with the curve of the number of expected events which falls with increasing mass of the Higgs-boson.

As a matter of fact, a few Higgs candidates have been observed which are interpreted as either poorly measured events, because of apparatus imperfections, or else four-fermion events $e^+e^- \rightarrow l^+l^-q\bar{q}$. Table 3 gives an account of the so far observed candidate events (ALEPH has not observed a candidate event).

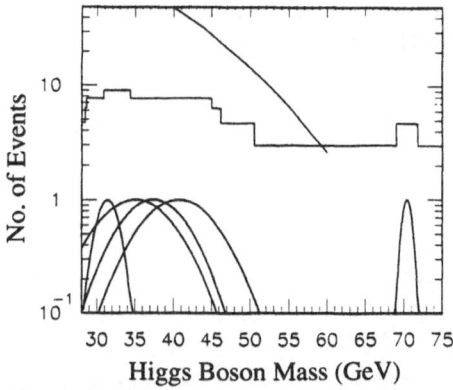

Figure 8. Number of Higgs events expected. The Gaussian curves correspond to the observed candidate events

Table 3. Higgs candidates seen by LEP experiments

Experiment	Decay channel	Observed mass
DELPHI	$e^+e^-q\bar{q}$	35.0 ± 5.0
L3	$e^+e^-q\bar{q}$	31.4 ± 1.5
L3	$\mu^+\mu^-q\bar{q}$	70.4 ± 0.7
OPAL	$e^+e^-q\bar{q}$	37.4 ± 4.4
OPAL	$\nu\bar{\nu}q\bar{q}$	40.7 ± 4.9

The candidate events loosen the 95 % CL but only in the vicinity of the observed mass. The number of Higgs events corresponding to a 95 % CL is then calculated by adding for each event a Gaussian distribution on top of the three-event line (see Fig. 8, taken from Ref. [4].

As there happens to be no candidate event near the 60 GeV combined limit, the result is not altered by this more detailed consideration.

THE Z-LINE-SHAPE

Below, the results on the Z-line-shape parameters as determined by the four LEP Collaborations are presented [6]. The errors quoted for the individual experiments are obtained after unfolding all systematic errors that we consider common to all four experiments, and then adding quadratically the remaining, experiment-specific, systematic and statistical errors. This allows a judgement of the compatibility of the results. After averaging, we add our own estimate of the common systematic error, thus obtaining the overall experimental result, which can then be compared with the expectation from the Electroweak MSM. The latter is calculated for $m_Z = 91.187 \pm 0.007$ GeV, for $m_H = 60$ and 1000 GeV, and $\alpha_s = 0.120 \pm 0.006$. The expectations of the Electroweak MSM is shown on the subsequent plots, with the m_t-dependence in the range $80 \leq m_t \leq 250$ GeV emphasized.

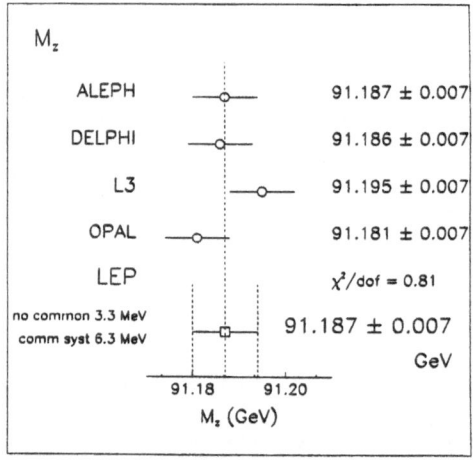

Figure 9. Measurements of the Z-mass

In the fit of the Z-line-shape, the Z-mass is numerically stable, being virtually unaffected by other fit parameters. The statistical precision is dominated by the abundant hadronic decays. Figure 9 shows the Z-mass measurements performed by the LEP experiments, from an energy scan across the resonance. The overall error is dominated by the uncertainty of the absolute energy calibration of LEP.

The measurements of the Z-mass shown in Fig. 9 are in good agreement. The average is 91.187 ± 0.007 GeV.

A common fit to the Z-line-shape as measured with $Z \rightarrow q\bar{q}, e^+e^-, \mu^+\mu^-$, and $\tau^+\tau^-$ events permits the simultaneous determination of six resonance parameters which can, for example, be chosen as $m_Z, \Gamma_Z, \sigma_h^0, \Gamma_e/\Gamma_h, \Gamma_\mu/\Gamma_h,$ and Γ_τ/Γ_h, a set of reasonably uncorrelated parameters.

As for the total width Γ_Z, the measurements are in good agreement (see Fig. 10). The common systematic error is estimated to be ± 4.5 MeV, partly due to point-to-point errors in the LEP energy settings. As can be seen from Fig. 10 the common systematic error is much smaller than the experimental error, so there is quite some room for improvement in the measurement of Γ_Z. The current average, $\Gamma_Z = 2488 \pm 7$ MeV, is in good agreement with the Electroweak MSM prediction. The ultimate precision on Γ_Z will be about 5 MeV.

The hadronic peak cross-section (in Born approximation, before radiative corrections),

$$\sigma_h^0 = \frac{12\pi}{m_Z^2} \frac{\Gamma_e \Gamma_h}{\Gamma_Z^2},$$
(5)

is statistically well determined. A systematic error of $\pm 0.3\%$ (± 0.12 nb) arises from theoretical uncertainties in the Bhabha cross-section, which is utilized to determine the luminosity from the observed $e^+ e^- \rightarrow e^+ e^-$ events at small polar angles. The measurements of σ_h^0 are shown in Fig. 11. They are well compatible, and the average is in good agreement with the Electroweak MSM prediction. A common systematic error of $\pm 0.15\%$ (± 2.6 MeV) owing to the theoretical uncertainty of the Bhabha cross-section has been assumed. The measurements are in good agreement, as they are with the Electroweak MSM prediction.

The measurement of the electronic Z-width is complicated by the fact that Bhabha scattering proceeds both through s- and t-channel diagrams. Therefore, an additional, even dominant, systematic uncertainty arises from the theoretical uncertainty of the t-channel subtraction, which is estimated to be $\pm 0.3\%$ (± 0.25 MeV) of the electronic width. The measurements are in good agreement, and are quite consistent with the MSM prediction.

For the measurement of the muonic Z-width, the common systematic error is smaller than for Γ_e, i.e. $\pm 0.15\%$ (± 0.13 MeV), arising from the uncertainty of the Bhabha cross-section. The common systematic error is small compared with the overall error, so that quite some improvement is possible by further running. The measurements are in good agreement as they are with the Electroweak MSM prediction.

For the measurement of the tauonic Z-width, the common systematic error is the same as for Γ_μ, i.e. $\pm 0.15\%$ (± 0.13 MeV) from the uncertainty of the Bhabha cross-section. This error is small compared with the experimental error, and leaves much room for improvement. The measurements are in good agreement as they are with the Electroweak MSM prediction.

Figure 12 displays the measurements of the leptonic width, assuming e-μ-τ universality. The common systematic error is estimated to be $\pm 0.2\%$ (± 0.17 MeV) owing to the uncertainty of the Bhabha cross-section and to the t-channel subtraction in large-angle Bhabha scattering. The measurements agree well with each other, and with the Electroweak MSM prediction. A significant improvement of the measurement is possible.

The Z-partial widths exhibit a significant dependence on m_t through non-photonic radiative corrections: varying m_t from 80 to 250 GeV increases the partial widths by about 1.5%. As this m_t-dependence tends to cancel in ratios of partial widths, quantities such as σ_h^0 (Eq. 5) or $R_{hl} = \Gamma_h/\Gamma_l$ provide a particularly stable testing ground of the Electroweak MSM. Figure 13 shows the experimental results for R_{hl}. The common systematic uncertainty is estimated to be $\pm 0.1\%$ of R_{hl}, owing to the uncertainty of the t-channel subtraction, which is very small compared with the overall experimental

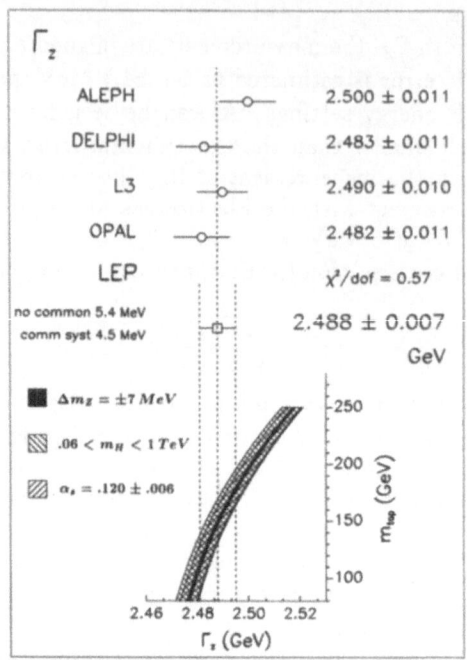

Figure 10. Measurements of the total Z-width

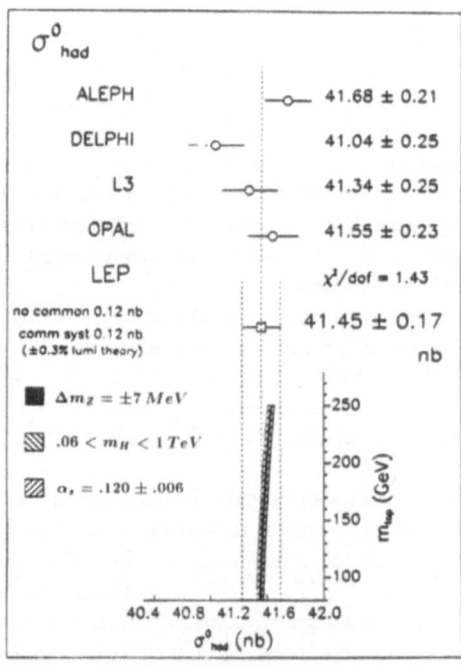

Figure 11. Measurements of the hadronic peak cross-section of the Z

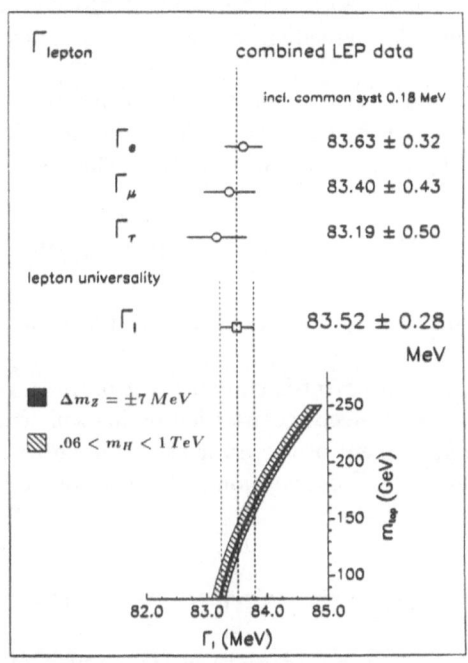

Figure 12. Measurements of the leptonic Z-width

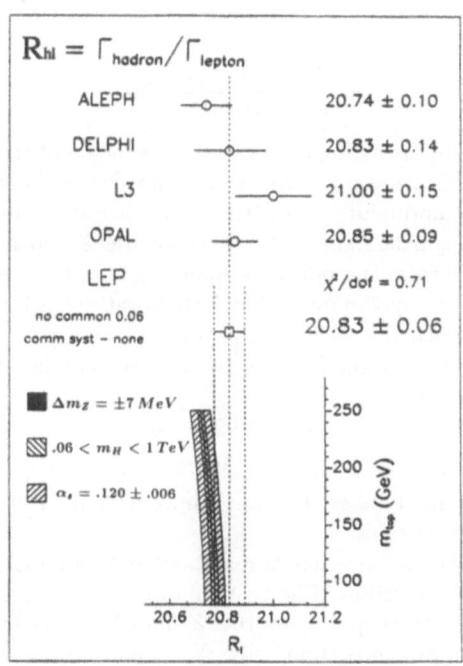

Figure 13. Measurements of $R_{hl} = \Gamma_h/\Gamma_l$

error. There is good agreement with the Electroweak MSM prediction. It will be most interesting to see whether this agreement persists when the error is reduced by further data taking.

The ratio R_{hl} permits a measurement of the strong coupling constant α_s, evaluated at $Q^2 = m_Z^2$. This measurement is of particular interest since it has a well-defined Q^2, and is not plagued by theoretical uncertainties. The determination of α_s uses the theoretical prediction [7]

$$R_{hl} = R^0(1 + 1.05\frac{\alpha_s}{\pi} + 0.9(\frac{\alpha_s}{\pi})^2 - 13(\frac{\alpha_s}{\pi})^3) ,$$

where $R^0 = 19.95 \pm 0.03$ is the prediction of the Electroweak MSM, before QCD corrections.

The result of this measurement is $\alpha_s = 0.131 \pm 0.010$, a value which tends to be larger than α_s as measured in deep-inelastic lepton-hadron scattering, but not (yet?) significantly so. An improvement of the error of this quantity which is possible from increased statistics of leptonic Z-decays would therefore be very interesting.

THE NUMBER OF LIGHT NEUTRINO FAMILIES

As is well known, the Z-width is a measure of the number of neutrino families, N_ν, provided the neutrino mass is well below $m_Z/2$. There are several ways of determining N_ν. As statistical precision is not the main concern, we adopt here a method that is nearly model independent. From

$$N_\nu = \frac{\Gamma_{\text{inv}}}{\Gamma_\nu} = \frac{1}{\Gamma_\nu}(\Gamma_Z - \Gamma_h - 3\Gamma_l) ,$$

we obtain

$$N_\nu = \frac{\Gamma_l}{\Gamma_\nu}(\sqrt{\frac{12\pi R_{hl}}{\sigma_h^0 m_Z^2}} - R_{hl} - 3) . \tag{6}$$

In this formula we utilize the average of the measurements of R_{hl}, σ_h^0, and m_Z. The only information from the Electroweak MSM that is needed is the ratio – not the absolute values – of the leptonic and neutrino widths, which is independent of m_t. The common systematic error in the measurement of the invisible Z-width is estimated to be ± 3 MeV, stemming mainly from the point-to-point errors in the LEP energy settings. The measurements performed by the four LEP Collaborations are in good agreement with each other, as well as with the Electroweak MSM prediction.

Utilizing Eq. 6, we obtain for the number of neutrino families, from the LEP measurements, the average

$$N_\nu = 2.99 \pm 0.03 .$$

This value is quite consistent with the belief in three families of quarks and leptons with next to massless neutrinos. Further improvement of N_ν is expected from the accumulation of more statistics.

Figure 14 shows the experimental results of the four LEP Collaborations on the number of light neutrino families. Figure 15 shows the so far most prominent achievement at LEP: the precise mapping of the Z-line-shape as measured with hadronic events. The data are quite consistent with $N_\nu = 3$, whereas $N_\nu = 2$ and $N_\nu = 4$ are clearly ruled out. The data shown in Fig. 15 are from the ALEPH experiment. The analogous data from the DELPHI, L3, and OPAL experiments look equally convincing.

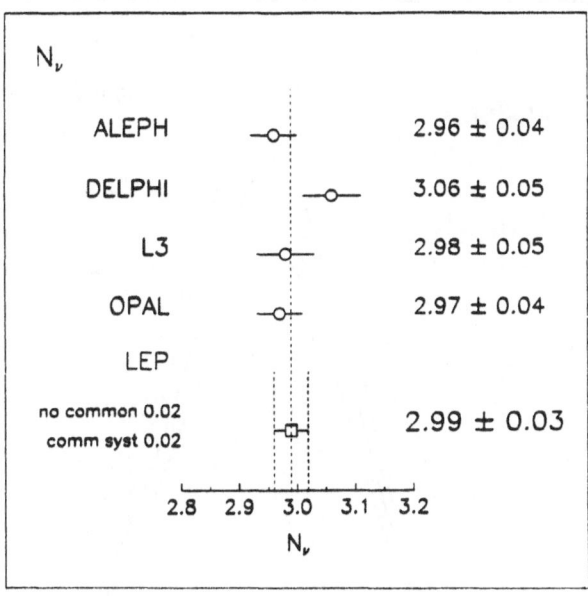

Figure 14. Measurements of the number of light neutrino families

ASYMMETRIES AT THE Z-POLE

On the peak, the leptonic forward-backward asymmetry is given in the improved Born approximation by

$$A_l^{\mathrm{FB}} = \frac{3}{4}\mathcal{A}_e\mathcal{A}_l \,,$$

with

$$\mathcal{A}_l = 2\frac{\bar{g}_{Vl}\bar{g}_{Al}}{\bar{g}_{Vl}^2 + \bar{g}_{Al}^2} \,.$$

The forward-backward asymmetry is determined by a fit to the distribution in the polar angle Θ:

$$\frac{d\sigma}{d\cos\Theta} = const \times [1 + \cos^2\Theta + \frac{8}{3}A^{\mathrm{FB}}\cos\Theta] \,.$$

From the measurement of the forward-backward asymmetry of leptons, together with the measurement of the leptonic width (Eq. 1), the effective leptonic coupling constants $\bar{g}_{Vl}$ and $\bar{g}_{Al}$ can be determined (see the section on the coupling constants $\bar{g}_{Vl}$ and $\bar{g}_{Al}$).

All data presented on the leptonic forward-backward asymmetry are well compatible with lepton universality and with the Electroweak MSM prediction.

The leptonic forward-backward asymmetry at the Z-pole is an important quantity to be measured, however this asymmetry is proportional to the square of the leptonic vector coupling constant which renders a high-precision measurement difficult to achieve as the leptonic vector coupling constant tends to be close to zero.

Another important asymmetry at the Z-pole is the left-right asymmetry

$$A^{\mathrm{LR}} = \frac{\sigma_{\mathrm{L}} - \sigma_{\mathrm{R}}}{\sigma_{\mathrm{L}} + \sigma_{\mathrm{R}}} = 2\frac{\bar{g}_{Ve}\bar{g}_{Ae}}{\bar{g}_{Ve}^2 + \bar{g}_{Ae}^2} \,,$$

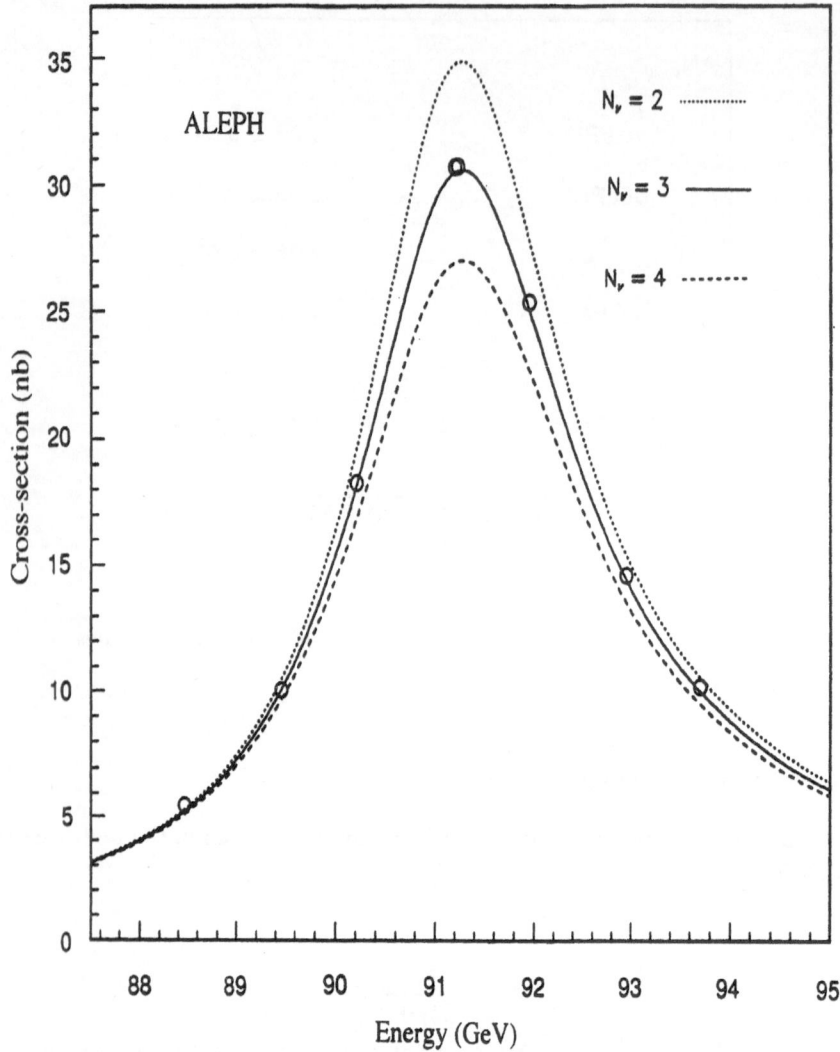

Figure 15. Measurement of the Z-line-shape, and comparison with the Electroweak MSM expectation for $N_\nu = 2$, 3, and 4 (ALEPH data)

which is the relative difference of the e^+e^--annihilation cross-section for longitudinally polarized electrons with left and right helicity. As this measurement involves just the sum and the difference of total cross-sections, it has a small statistical and systematic error since there is no ambiguity in the event selection, and little influence from acceptance and resolution effects.

At SLAC, in 1992 the first measurement of A^{LR} was performed by the SLD Collaboration [8] working at the Stanford Linear Collider, SLD. This machine has the unique feature that longitudinally polarized electrons can be brought into collision, at the Z-pole, with unpolarized positrons. The polarized electrons are emitted from a photocathode which is illuminated with circularly polarized laser light, the helicity of which is randomized on a pulse-to-pulse basis. After acceleration, the electron polarization is measured at the intersect with a Compton polarimeter.

Unlike the leptonic forward-backward asymmetry, A^{LR} depends linearly on the leptonic vector coupling constant. A^{LR} is very sensitive to $\sin^2 \bar{\theta}_{\text{w}}$ ($\Delta A^{\text{LR}} \sim 8\Delta \sin^2 \bar{\theta}_{\text{w}}$), and is primarily sensitive to the Higgs-boson mass m_H when electroweak radiative corrections are taken into account. In this respect, A^{LR} complements another sensitive quantity, the W-mass which is primarily sensitive to m_t.

Based on a sample of nearly 5000 Z-decays, the result for the left-right asymmetry was

$$A^{\text{LR}} = 0.02 \pm 0.07 ,$$

corresponding to $\sin^2 \bar{\theta}_{\text{w}} = 0.247 \pm 0.009$. This precision is not yet competitive but it gives a clear sign of the potential of this measurement once more data will have been collected.

τ-POLARIZATION AND LIFETIME

The τ-polarization is defined as

$$\mathcal{P}_\tau = \frac{\sigma_{\text{L}} - \sigma_{\text{R}}}{\sigma_{\text{L}} + \sigma_{\text{R}}} = 2\frac{\bar{g}_{\text{V}\tau}\bar{g}_{\text{A}\tau}}{\bar{g}_{\text{V}\tau}^2 + \bar{g}_{\text{A}\tau}^2} ,$$

where σ_{L} and σ_{R} denote the cross-sections for the production of left- and right-handed τ's, respectively. It measures the same combination of coupling constants for the τ as A^{LR} measures for the electron.

Experimentally, the τ-polarization is determined from the momentum spectrum of the charged particles emitted in the decay channels $\tau \rightarrow e\nu\bar{\nu}, \tau \rightarrow \mu\nu\bar{\nu}, \tau \rightarrow \pi\nu, \tau \rightarrow \rho\nu$, and $\tau \rightarrow a_1\nu$. Figure 16 shows the results for $\mathcal{P}_\tau$ as obtained by the four LEP experiments. The combined result, $\mathcal{P}_\tau = 0.142 \pm 0.017$, agrees well with the prediction of the Electroweak MSM.

The lifetime of the τ-lepton is a good testing ground for $\mu - \tau$ universality. The τ-lifetime is expected to be related to the μ-lifetime as follows:

$$\tau_\tau = \tau_\mu \left(\frac{G_F^\mu}{G_F^\tau}\right)^2 \left(\frac{m_\mu}{m_\tau}\right)^5 BR(\tau \rightarrow e\nu\bar{\nu}) .$$

With the masses and lifetimes of the τ- and μ-leptons measured, and with the leptonic branching ratio of the τ-lepton known, the Fermi constants G_F^μ and G_F^τ can be compared, a question which has aroused quite some interest in the recent past [9].

With the help of Silicon microvertex-detectors, the ALEPH, DELPHI, and OPAL Collaborations were able to measure precisely the τ-lifetime. Figure 17 taken from Ref. [10] shows the results of the LEP experiments together with the average from

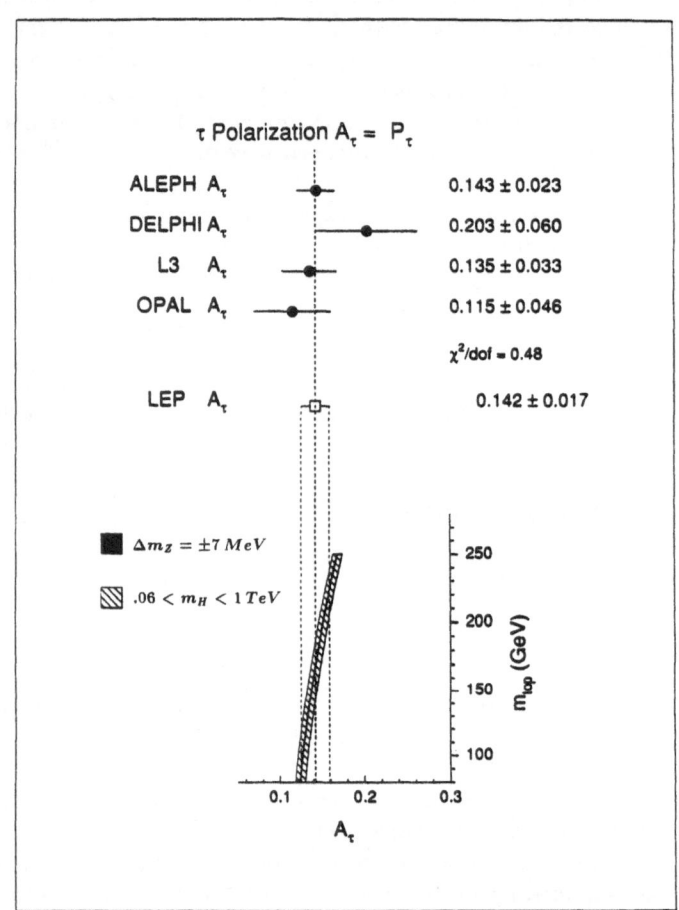

Figure 16. Measurements of the τ-polarization at LEP

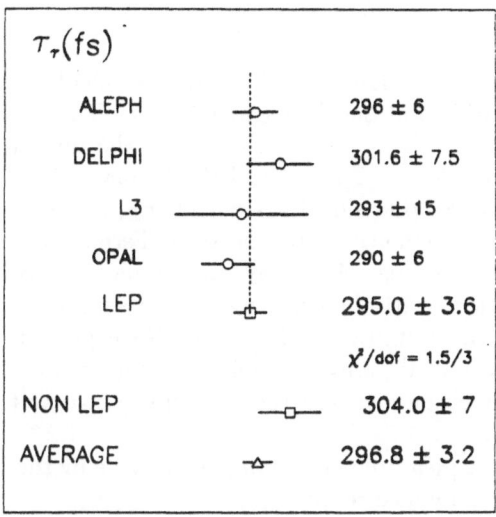

Figure 17. Measurements of the τ-lifetime

earlier experiments, exhibiting clearly the progress due to the LEP machine. The new world average of the τ-lifetime is

$$\tau_\tau = 296.8 \pm 3.2 \text{ fs} .$$

Also the leptonic branching ratio $BR(\tau \to e\nu\bar{\nu})$ has been measured recently with greater precision, chiefly by the CLEO Collaboration [11] but also by the LEP Collaborations [10]. The new world average is

$$BR = 17.71 \pm 0.12 \% .$$

Finally, and most importantly, the mass of the τ-lepton has been measured recently with much greater precision by the BES Collaboration [12] from a scan near the $\tau^+\tau^-$ production threshold. Their result is

$$m_\tau = 1776.9^{+0.4}_{-0.5} \pm 0.2 \text{ MeV} .$$

From all these results, the ratio of the Fermi coupling constants is

$$\frac{G_F^\tau}{G_F^\mu} = 0.987 \pm 0.006 ,$$

where unity is expected. The experimentally determined ratio is 2.2σ away from unity, just enough to attract quite some attention in a time when all measured quantities agree typically better than within 1σ with the expectation from the Electroweak MSM. The deviation from unity in terms of σ is nearly the same as before although the measurement error has decreased by a factor of two. Further progress would need even better measurements of the τ-lifetime and of the leptonic branching ratio of the τ-lepton.

W-MASS

The procedure of testing the Electroweak MSM is to calculate observables on the basis of three input quantities, and then to compare the predictions with precise experimental measurements of the observables. In principle, the three input quantities are the fine-structure constant α, the Z-mass, and the W-mass. Unfortunately, the W-mass is known with much less precision than α and m_Z, so by convention it is replaced by the closely related but much better measured Fermi coupling constant G_F^μ as derived from the muon lifetime. This way m_W becomes an observable which can be calculated and compared with the experimental measurement. This explains the large interest in a measurement of m_W which is as precise as possible. In fact m_W is besides the left-right asymmetry A^{LR} the most important observable which has prospects of high experimental precision and is at the same time sensitive to electroweak radiative corrections.

The most precise measurements of m_W were made by the UA2 [15] and CDF [16] Collaborations. The combined result is [17]

$$m_W = 80.22 \pm 0.26 \text{ GeV}.$$

This result is expected to be further improved by the CDF and D0 Collaborations which continue to take data at the Fermilab $p\bar{p}$-collider. If the measurement is viewed not as an absolute measurement of m_W but rather of the *difference* to the well-known Z-mass, the $p\bar{p}$-collider is a quite powerful instrument for a precise determination of the W-mass. An error below 100 MeV can be reasonably expected, rendering a further improvement by the LEP collaborations highly non-trivial when LEP 200 will become available in 1995.

The W-mass predicted by the Electroweak MSM is

$$m_W = 80.23 \pm 0.12 \text{ GeV},$$

where the error is as usual due to the standard range of variation of m_t and m_H. The agreement with the measured mass is very good.

R_ν AND $R_{\bar{\nu}}$

The measurement of the neutral- to charged-current ratios in neutrino-nucleon (R_ν) and antineutrino-nucleon ($R_{\bar{\nu}}$) scattering is closely equivalent to the direct measurement of the W-mass. This has been first pointed out by Stuart [19], and is due to a fortuitous numerical cancellation which occurs for $\sin^2 \bar{\theta}_{\mathrm{w}} \sim 0.23$.

On an isoscalar target, the ratios R_ν and $R_{\bar{\nu}}$ are given by the following expressions:

$$R_\nu = \bar{\rho}^2(\tfrac{1}{2} - \sin^2 \bar{\theta}_{\mathrm{w}} + \tfrac{5}{9}(1 + r)\sin^4 \bar{\theta}_{\mathrm{w}})$$
$$R_{\bar{\nu}} = \bar{\rho}^2(\tfrac{1}{2} - \sin^2 \bar{\theta}_{\mathrm{w}} + \tfrac{5}{9}(1 + \tfrac{1}{r})\sin^4 \bar{\theta}_{\mathrm{w}}),$$

where $r = (CC)_{\bar{\nu}}/(CC)_\nu$ is the ratio of charged-current events in antineutrino- to neutrino-scattering.

Around $\sin^2 \bar{\theta}_{\mathrm{w}} \sim 0.23$, R_ν is very sensitive to $\sin^2 \bar{\theta}_{\mathrm{w}}$, whereas $R_{\bar{\nu}}$ is insensitive to $\sin^2 \bar{\theta}_{\mathrm{w}}$ but measures in essence the Veltman parameter $\bar{\rho}$ (Eq. 4), albeit with less precision since the antineutrino-nucleon cross-section is lower and the antineutrino flux is more scarce than the neutrino flux.

The above formulae can easily be verified by employing the Quark-Parton-Model of the nucleons which serve as target for neutrino scattering, and the neutral-current

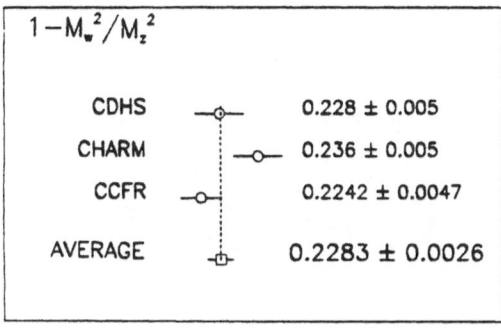

Figure 18. $sin^2\theta_{\rm w} = 1 - (\frac{m_W}{m_Z})^2$ from neutrino-nucleon scattering

coupling constants of u- and d-quarks but as has been shown by Llewellyn-Smith [18] their validity extends well beyond the Quark-Parton-Model of the nucleons.

Changing the parameters $\bar{\rho}$ and $\sin^2\bar{\theta}_{\rm w}$ to the new parameters m_W/m_Z and $\sin^2\bar{\theta}_{\rm w}$, keeping in mind the relation (3), one obtains

$$R_\nu = (m_W/m_Z)^4(\tfrac{1}{2} - \sin^2\bar{\theta}_{\rm w} + \tfrac{5}{9}(1+r)\sin^4\bar{\theta}_{\rm w})/(1 - \sin^2\bar{\theta}_{\rm w})^2$$
$$= (m_W/m_Z)^4\tfrac{1}{2}(1 + 0.15\sin^2\bar{\theta}_{\rm w}) .$$

That means that with $\sin^2\bar{\theta}_{\rm w} \sim 0.23$, the ratio R_ν is only very weakly dependent on $\sin^2\bar{\theta}_{\rm w}$, but rather measures m_W/m_Z.

After the pioneering high-precision experiments of the CDHS [20] and CHARM [21] collaborations, also the CCFR Collaboration [22] which works at the Fermilab high-intensity quad-triplet neutrino beam, has achieved a result with comparable precision. The experimental results for m_W/m_Z as derived from the measured neutral- to charged-current ratios is shown in Fig. 18 taken from Ref. [10]. The average of $1 - (m_W/m_Z)^2$ is 0.2283 ± 0.0026, onto which a common systematic error of ± 0.0045 is to be added which arises from the uncertainty of the effective mass of the c-quark which is excited in about 10% of the charged-current neutrino scatterings, and thereby affects the neutral- to charged-current cross-section ratio.

This result, together with the precisely known Z-mass, gives a measurement of the W-mass:

$$m_W = 80.10 \pm 0.27 \text{ GeV} ,$$

which has the same precision as the direct W-mass measurement, and is also in good agreement with the expectation from the Electroweak MSM.

Whilst this progress in experimental precision in neutrino-nucleon scattering is impressive, a further improvement of the measurement of the neutral- to charged-current ratio seems not worthwhile without a significant improvement in the understanding of the behaviour of c-quark excitation in neutrino-nucleon scattering.

NEUTRINO-ELECTRON SCATTERING

Neutrino-electron scattering started 20 years ago as an extremely daring enterprise, with hardly a handful of events observed after strenuous efforts. With time, it devel-

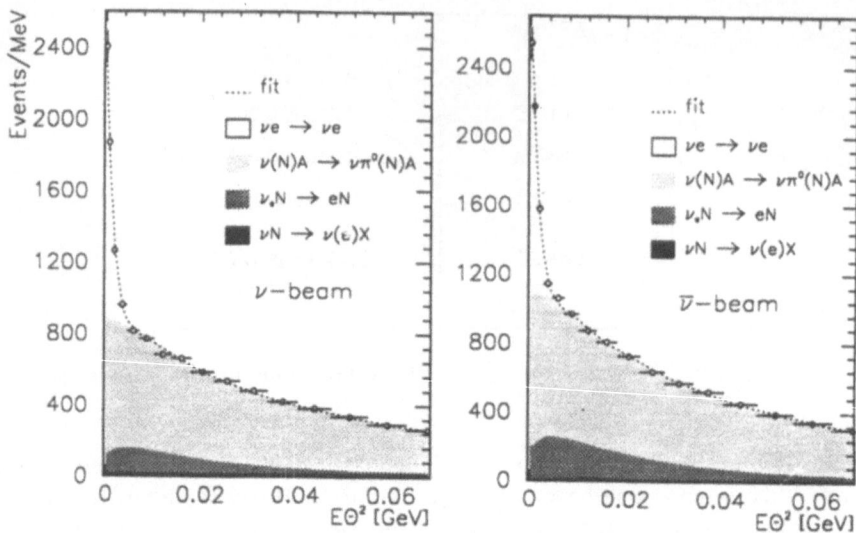

Figure 19. Transverse energy distribution of neutrino-electron scattering (CHARM II data)

oped into a big business, and became almost a routine job. Today, the standard is set at the level of many thousand events per experiment.

The cross-section of neutrino-electron scattering is given by

$$\sigma(\nu_\mu e) = \frac{2G_F^2 m_e}{\pi} E_\nu \left((\bar{g}_{Ve} + \bar{g}_{Ae})^2 + \frac{1}{3}(\bar{g}_{Ve} - \bar{g}_{Ae})^2 \right) \ .$$

As the cross-section is proportional to the electron mass, in contrast to neutrino-nucleon scattering which is proportional to the much larger mass of the nucleon, the event rate is low and the background is high, rendering the experimental observation very difficult.

The experimental signature is a very narrow forward peak in the differential cross-section which is a direct consequence of the small electron mass. Figure 19 shows recent results from the CHARM II Collaboration which worked at a high-flux wide-band neutrino beam at CERN. The data clearly exhibit the forward peak in the relevant variable $E\Theta^2$ – the product of the energy of the outgoing muon and its scattering angle squared – on top of a large but much broader background which is due to neutrino-nucleon scattering.

Pioneered by the CHARM Collaboration (the precursor of the CHARM II Collaboration), the standard way of extracting $\sin^2 \bar{\theta}_w$ from neutrino-electron scattering is from the ratio of neutrino- to antineutrino cross-sections:

$$\frac{\sigma(\nu_\mu e)}{\sigma(\bar{\nu}_\mu e)} = 3 \frac{1 - 4\sin^2 \bar{\theta}_w + \frac{16}{3}\sin^4 \bar{\theta}_w}{1 - 4\sin^2 \bar{\theta}_w + 16\sin^4 \bar{\theta}_w} \ ,$$

which, again for fortuitous numerical reasons, is very sensitive to $\sin^2 \bar{\theta}_w$ at $\sin^2 \bar{\theta}_w \sim$ 0.23. At the same time, systematic errors arising e.g. from the determination of the neutrino fluxes, are much reduced as only the ratio between neutrino and antineutrino fluxes is involved.

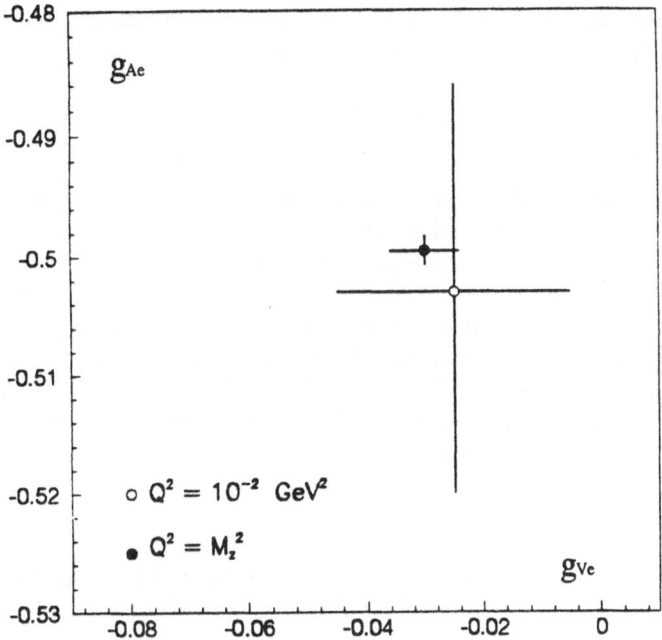

Figure 20. Comparison of results on the electron coupling constants from neutrino-electron scattering and from LEP

From about 3000 neutrino and 3000 antineutrino events observed, CHARM II's result for $\sin^2 \bar{\theta}_{\mathrm{w}}$ is [23]

$$\sin^2 \bar{\theta}_{\mathrm{w}} = 0.232 \pm 0.009 \ .$$

At the same time, by analyzing also the energy spectrum of the outgoing electrons, the vector- and axial-vector coupling constants of the electron have been determined:

$$\bar{g}_{Ve} = -0.025 \pm 0.020$$
$$\bar{g}_{Ae} = -0.503 \pm 0.017 \ .$$

In Figure 20 taken from Ref. [10], these coupling constants are compared with those which have been measured at LEP at a Q^2 which is six orders of magnitude larger. The Q^2-dependence expected from within the Electroweak MSM being very small, the agreement between the experimental results from neutrino-electron scattering and from LEP is impressive. Further progress on the precision in particular of the neutrino-electron scattering results is in principle possible but would be a major undertaking.

THE COUPLING CONSTANTS $\bar{g}_{Vl}$ AND $\bar{g}_{Al}$

The coupling constants $\bar{g}_{Vl}$ and $\bar{g}_{Al}$ are essentially determined from the leptonic Z-widths, the leptonic forward-backward asymmetries, and the τ-polarization. The leptonic forward-backward asymmetries (Eq. 2) determine essentially the ratio $\bar{g}_{Vl}/\bar{g}_{Al}$, whereas the leptonic Z-width (Eq. 1) determines essentially the axial-vector coupling constant $\bar{g}_{Al}$.

The results from the LEP experiments for the coupling constants are in good agreement with each other. The overall results are [6]

$$\bar{g}_{Vl} = -0.0372 \pm 0.0024$$
$$\bar{g}_{Al} = -0.4999 \pm 0.0009 \,,$$

and agree well with the MSM prediction. Note that the vector coupling constant, although small, is significantly different from zero.

The precision with which the leptonic vector and axial-vector coupling constants are known today is remarkable. Figure 21 from Ref. [14] recalls the experimental knowledge back in 1986. The results then stemmed from neutrino-electron scattering, polarized electron-deuteron scattering, polarized muon-carbon scattering, and $e^+e^- \to l^+l^-$ annihilation as measured at PEP and PETRA. With the advent of the LEP data, the errors have shrunk to such an extent that quite some magnification is needed in order to display the status as of today.

$\sin^2 \bar{\theta}_w$ AND m_t

The same measurements which served for determining the leptonic vector and axial-vector coupling constants, also serve for determining the electroweak mixing parameter,

$$\sin^2 \bar{\theta}_w = 1 - \frac{m_W^2}{\bar{\rho} m_Z^2} \,.$$

For example, the leptonic Z-width reads in terms of $\sin^2 \bar{\theta}_w$ as

$$\Gamma_l = \frac{\bar{\alpha} m_Z}{12 \sin^2 \bar{\theta}_w \cos^2 \bar{\theta}_w} \left\{ \left(\frac{1}{2} - 2\sin^2 \bar{\theta}_w \right)^2 + \left(\frac{1}{2} \right)^2 \right\} \,.$$

Note that the electroweak mixing parameter, if determined this way, has the knowledge about m_t and m_H incorporated. Because of this, the numerical value of $\sin^2 \bar{\theta}_w$ will stand whatever values m_t and m_H eventually take.

Figure 22 displays the measurements of $\sin^2 \bar{\theta}_w$ from the LEP experiments. The common systematic error is estimated to be ± 0.0003, arising chiefly from a propagation of the theoretical uncertainty of the Bhabha cross-section.

The value of the effective electroweak mixing parameter from LEP,

$$\sin^2 \bar{\theta}_w = 0.2319 \pm 0.0007 \,,$$

constitutes the most precise determination so far, surpassing now clearly the precision from the neutral- to charged-current ratio as measured in neutrino-nucleon scattering, and from the W- and Z-masses.

As pointed out above, the value of $\sin^2 \bar{\theta}_w$ determined from Γ_l is independent of m_t, in contrast with determinations of $\sin^2 \bar{\theta}_w$ from the ratio m_W/m_Z in the $p\bar{p}$ collider experiments [15, 16] and, equivalently, the neutral- to charged-current ratio in neutrino-nucleon scattering experiments [20, 21], and from the Z-mass. The question of which value of m_t makes all these measurements consistent is answered in Fig. 23. A fit to the experimental data yields the result [24]

$$m_t = 160^{+16+18}_{-17-20} \text{ GeV} \,,$$

where the first error is experimental. and the second is due to a variation of m_H in the range from 60 to 1000 GeV. The fit result is insensitive to whether the strong

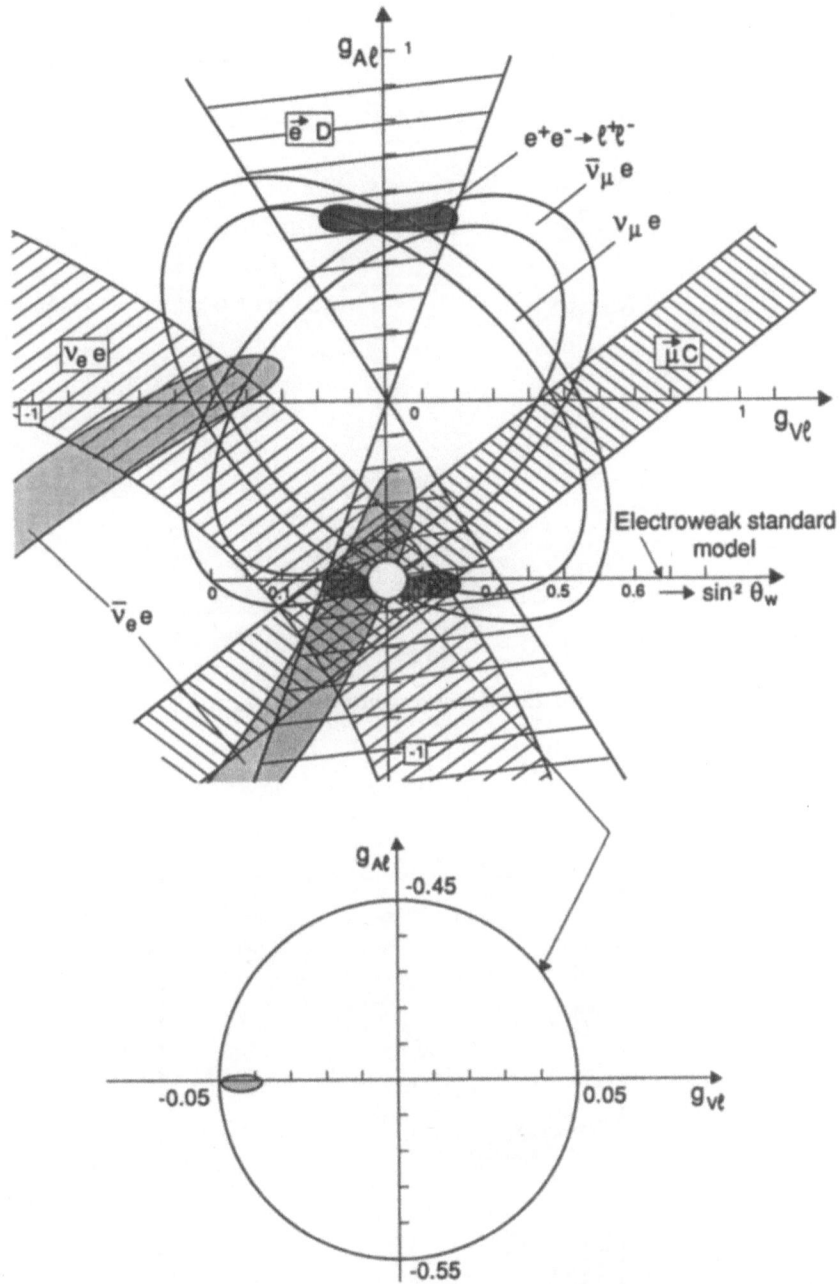

Figure 21. The knowledge of $\bar{g}_{Vl}$ and $\bar{g}_{Al}$ from various experiments, in 1986 and today

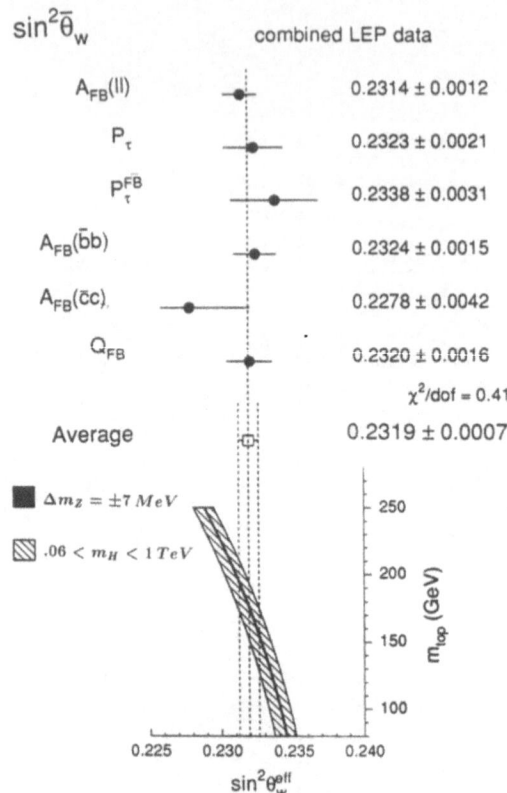

Figure 22. Measurements of $\sin^2 \bar{\theta}_w$

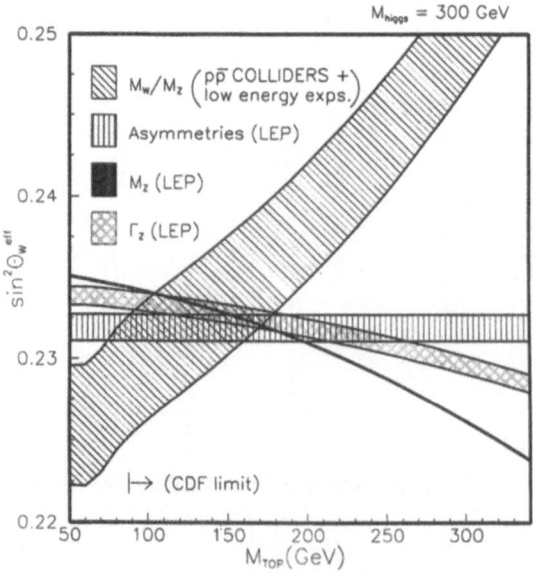

Figure 23. $\sin^2 \bar{\theta}_{w^*}$ as a function of m_t, for $m_H = 300$ GeV, from various experiments

coupling constant α_s is fixed to the value determined from the event shapes of hadronic Z-decays, or is left as a free fit parameter.

This global fit to all relevant electroweak data constitutes essentially the best guess which we have at present about the mass of the t-quark, of course together with the lower limit of $m_t > 91$ GeV as determined by the CDF Collaboration [25] (see also P. Tipton's lectures at this School).

The global fit to the relevant electroweak data favours a small value of the mass of the Higgs-boson but not significantly so. A more definitive bound on m_H will be obtained once m_t will be known and no longer be a free parameter.

SUMMARY

The amount of high-quality data, which has been accumulated in particular at LEP, in such a short time, is impressive. No less impressive is how well the Electroweak Minimal Standard Model has been doing in defending itself. ALL data from LEP are in good agreement with the predictions of the Electroweak Minimal Standard Model.

The highlights of these first results on high-precision Z-physics may be summarized as follows:

- $N_\nu = 2.99 \pm 0.03$

- The mass of the minimal standard Higgs-boson is larger than 60 GeV at 95% CL

- $\sin^2 \bar\theta_{\rm w} = 0.2319 \pm 0.0007$

- $m_t = 160^{+16+18}_{-17-20}$ GeV

Acknowledgement

I wish to express my sincere thanks to the organizer of this superb Physics School, Tom Ferbel, for his invitation to give these lectures and for his warm hospitality at St. Croix, and for his gentle but steady pressure on me to finally get these lecture notes written up.

References

[1] L. Knudsen et al., Phys. Lett. **B270** (1991) 97.

[2] D.C. Kennedy and B.W. Lynn, Nucl.Phys. **B322** (1989) 1.

[3] G. Altarelli, Proc. Int. Symposium on Lepton and Photon Interactions at High Energies, Stanford, 1989, ed. M. Riordan (World Scientific, Singapore, 1990), p. 286.

[4] E. Gross, The search for the SM Higgs-boson at LEP 1 – A Comparative Study, preprint CERN-PPE/92-91.

[5] T. Mori, Searches for the Standard Model Higgs-boson at LEP, Talk given at the International Conference on High Energy Physics, Dallas, USA, 1992.

[6] M. Pepe-Altarelli, Tests of the electroweak theory at LEP, Talk given at the Moriond Conference on Electroweak Interactions and Unified Theories, Les Arcs, France, 1993; Frascati preprint LNF-93/019.

[7] S. G. Gorishny, A. L. Kataev and S. A. Larin, Phys. Lett. **B259** (1991) 144; L. R. Surguladze and M. A. Samuel, Phys. Rev. Lett. **66** (1991) 560.

[8] P. C. Rowson, First results from polarized beam running at the SLC/SLD, Talk given at the International Conference on High Energy Physics, Dallas, USA, 1992; B. A. Schumm et al. (SLD Collaboration), preprint SLAC-PUB-5995.

[9] See, for example, J. Ellis, Proc. Joint International Lepton-Photon Symposium and Europhysics Conference on High Energy Physics, World Scientific, Singapore (eds. S. Hegarty et al.), Vol. 2, p. 27.

[10] L. Rolandi, Precision tests of the electroweak interaction, Talk given at the International Conference on High Energy Physics, Dallas, USA, 1992; preprint CERN-PPE/91-175.

[11] K. Gan, τ-physics at CLEO, Talk given at the International Conference on High Energy Physics, Dallas, USA, 1992.

[12] F. Porter, Measurement of the mass of the τ-lepton, Talk given at the International Conference on High Energy Physics, Dallas, USA, 1992.

[13] V. Hatton et al., LEP absolute energy in 1990, LEP Performance Note 12 (CERN, December 1990).

[14] F. Dydak, Proc. Les Houches School, Session XLIV, 1985, eds. P. Ramond and R. Stora (Elsevier Science Publishers B.V., Amsterdam, 1987), p. 127.

[15] J. Alitti et al. (UA2 Collaboration), Phys. Lett. **B276** (1992) 354.

[16] F. Abe et al. (CDF Collaboration), Phys. Rev. Lett. **65** (1990) 2243.

[17] K. Hikasa et al. (Particle Data Group), Phys. Rev. **D45** (1992) 1.

[18] C. H. Llewellyn-Smith, Nucl. Phys. **B228** (1983) 205.

[19] R. G. Stuart, Z. Phys. **C34** (1987) 445.

[20] H. Abramowicz et al. (CDHS Collaboration), Phys. Rev. Lett. **57** (1986) 298; A. Blondel et al. (CDHS Collaboration), Z. Phys. **C45** (1990) 361.

[21] J.V. Allaby et al. (CHARM Collaboration), Phys. Lett. **B177** (1986) 446; Z. Phys. **C36** (1987) 611.

[22] T. Bolton, Results from the CCFR experiment, Talk given at the International Conference on High Energy Physics, Dallas, USA, 1992.

[23] A. Staude, Neutral-current coupling constants from neutrino-electron scattering, Talk given at the International Conference on High Energy Physics, Dallas, USA, 1992; D. Geiregat et al. (CHARM II Collaboration), Phys. Lett. **B259** (1991) 499; P. Vilain et al. (CHARM II Collaboration), Phys. Lett. **B281** (1992) 159.

[24] M. Martinez, private communication.

[25] F. Abe et al. (CDF Collaboration), Phys. Rev. **D43** (1991) 664.

ISSUES IN B PHYSICS

Michael V. Danilov

Institute of Theoretical and
Experimental Physics
117259 Moscow, Russia

INTRODUCTION

Beauty particles are actively studied now at all major accelerators. Beauty physics is a part of the scientific program for the future supercolliders SSC and LHC. There are many proposals to build a B factory - a specialized accelerator for studies of B mesons. The main motivation behind this large interest is the hope of discovering new phenomena beyond the so-called Standard Model (SM) - a theory which describes at present all known effects in particle physics. Beauty decays can be sensitive to phenomena at a larger mass scale than presently available at accelerators, for instance to charged Higgs bosons or supersymmetric particles. Further studies of the b quark are expected to play a crucial role in solving the puzzle of CP violation.

B mesons provide information on three out of the four independent parameters of the Cabibbo-Kobayashi-Maskawa (CKM) matrix, V, which rotates the quark mass eigenstates (d, s, b) into the weak eigenstates (d', s', b'):

$$\begin{pmatrix} d' \\ s' \\ b' \end{pmatrix} = \begin{pmatrix} V_{ud} & V_{us} & V_{ub} \\ V_{cd} & V_{cs} & V_{cb} \\ V_{td} & V_{ts} & V_{tb} \end{pmatrix} \begin{pmatrix} d \\ s \\ b \end{pmatrix}$$

It is possible that a more fundamental theory will eventually specify these parameters. Meanwhile experimental measurements are extremely important in giving some indication of what a more complete theory should look like.

Finally, beauty hadrons provide unique possibilities for studying strong interactions. The large masses of the b and c quarks considerably simplify theoretical calculations and allow, in many cases, quantitative tests of QCD. The field witnessed recently a rapid theoretical development. There are hopes that the so-called Heavy Quark Effective Theory (HQET) will provide a model independent description of beauty particles.

In this paper we review the present knowledge of beauty hadron properties with emphasis on recent results relevant to future CP violation studies: B lifetime, hadronic decays, semileptonic $b \rightarrow c$ and $b \rightarrow u$ transitions, $B\overline{B}$ mixing and searches for $b \rightarrow s\gamma$ transitions. The available information on the CKM matrix elements is then summarized in terms of the unitarity triangle.

Techniques and Concepts of High Energy Physics VII
Edited by T. Ferbel, Plenum Press, New York, 1994

1. B Hadron Lifetimes

In 1983, the MAC [1] and MARK II [2] collaborations measured for the first time the lifetime of B-hadrons which was found to be about 1 ps. This came as a big surprise. Theoretical expectations based on a naive assumption that $|V_{cb}| \sim |V_{us}|$ were an order of magnitude smaller. This demonstrated once again that quark couplings are not fixed in the SM and should be measured experimentally. Large progress in the determination of B-hadron lifetimes has been achieved recently mainly due to results from LEP. The average lifetime for a mixture of B-hadrons produced at the Z^0 mass is known now to an accuracy of 5% and information on individual lifetimes of B^+, B^0, B_s, and Λ_b has started to become available.

1.1. Average B-Lifetime

The efficiency for B vertex reconstruction is quite low. Therefore, practically all experiments use partial B reconstruction. Generally, a single lepton from a B decay is required. Leptons with large total and transverse momenta are usually selected (typically above 4 GeV/c and 1-2 GeV/c respectively). Such a selection has a reasonable efficiency of about 1/2 and suppresses background from charm decays and misidentified hadrons. It provides a sample of leptons from B decays with more than 80% purity.

At e^+e^- colliders the luminous region is small in the plane perpendicular to the beams. Therefore a displacement of lepton trajectories from the beam spot (the so-called impact parameter) can be translated into a B lifetime. B momenta are not known on an event by event basis. However, they can be estimated using the b quark fragmentation function. The LEP results obtained using the lepton impact parameter technique are summarized in Table 1 [3, 4] .

Table 1. Average B hadron lifetime .

Experiment	B Lifetime [ps]
ALEPH	1.49 ± 0.07
DELPHI	1.38 ± 0.05
L3	1.36 ± 0.07
OPAL	1.37 ± 0.09

The impact parameter method can be employed for hadrons as well. The most accurate measurement using this method was made by DELPHI [3] : $\tau_B = 1.41 \pm 0.04 \pm 0.6$ This agrees nicely with the values obtained using leptons. The average B lifetime is known with a high accuracy limited by systematic errors. In this paper we will use the average b lifetime of [4]

$$\langle \tau_b \rangle = 1.32 \pm 0.06 \; ps.$$

Several other methods have been used for the determination of τ_b (for a detailed discussion see [4]). These give results consistent with the impact parameter method with no improvement in accuracy.

1.2 Specific B Hadron Lifetime Measurements

By requiring a D^{*+} together with a high-p_T ℓ^- the MARK II Collaboration enriched their sample with $\overline{B}^0$ decays *, allowing them to perform the first direct measurement of the B^0 lifetime [5]

$$\tau_{B^0} = 1.20^{+0.52+0.16}_{-0.36-0.14} \, ps.$$

ALEPH [6] and DELPHI [6] extended this method to $D^0\ell^-$ combinations which are enriched with B^- decays and measured both τ_{B^0} and τ_{B^+} . Moreover, by studying $D_s^+\ell^-$ and $\Lambda_c^+\ell^-$ combinations they obtained evidence for B_s^0 and Λ_b^0 respectively. The lifetimes are summarized in Table 2 [3, 7] . They do not differ much in accordance

Table 2. Specific B hadron lifetime .

B Hadron	Lifetime [ps]
B^0	1.46 ± 0.18
B^+	1.33 ± 0.19
B_s	$1.05 \pm 0.31 \pm 0.10$
Λ_b	$0.98^{+0.23}_{-0.18} \pm 0.08$

with the theoretical expectation of drastically smaller nonspectator effects in beauty hadrons in comparison with charm particles.

2. Nonleptonic Decays of B mesons

The masses of B mesons have been measured using reconstructed hadronic B decays. Recent CLEO results are shown in Fig.1. CLEO reconstructed 601.5 ± 57.9 and $492.7\pm$

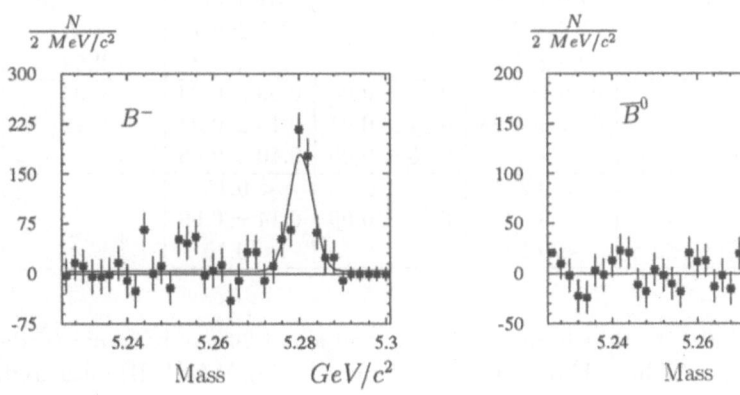

Figure 1. Beam constrained masses for B mesons after background subtraction.

43.6 charged and neutral B mesons respectively. This is an order of magnitude increase in comparison with their previous analysis. The difference between the $\overline{B}^0$ and B^- masses is consistent with zero: $M_{B^0} - M_{B^+} = 0.26 \pm 0.38 \, MeV/c^2$ [8].

The mass splitting has important implications for theoretical estimates of the ratio, f_+/f_0, where f_0 and f_+ are the branching ratios for $\Upsilon(4S) \to B^0\overline{B}^0$ and $\Upsilon(4S) \to B^+B^-$

---

*References in this paper to a specific charged state also imply the charge conjugate state.

decays respectively. Knowledge of f_+ and f_0 is important for the determination of B meson branching ratios and hence for the extraction of the fundamental parameters of the SM. Recent theoretical estimates [9, 10, 11], which include the momentum dependence of the $\Upsilon(4S) \to B\bar{B}$ vertex function and the effects of the B^- and B^0 form factors, indicate that $f_+/f_0 = 1.00 \pm 0.05$, where the error reflects the theoretical uncertainty in the momentum dependence of the vertex function. Throughout this paper the ratio f_+/f_0 is fixed to one and the systematic error is not included in the branching ratios of B mesons.

Tables 3 and 4 provide the B^0 and B^- hadronic branching ratios measured by ARGUS [12, 13] and CLEO [8, 14, 15].

Table 3. Branching ratios for nonleptonic $\overline{B}^0$ decays in % .

Mode	ARGUS	CLEO	average	BSW model
$\bar{B}^0 \to D^+\pi^-$	0.48 ± 0.16	0.26 ± 0.07	0.29 ± 0.06	0.39
$\bar{B}^0 \to D^+\rho^-$	0.9 ± 0.6	0.71 ± 0.23	0.73 ± 0.21	1.03
$\bar{B}^0 \to D^0\rho^0$	< 0.3	< 0.08	< 0.08	0.00
$\bar{B}^0 \to D^+\pi^-\pi^-\pi^+$	—	0.8 ± 0.3	0.8 ± 0.3	
$\bar{B}^0 \to D^+D_s^-$	1.4 ± 1.2	0.9 ± 0.5	1.0 ± 0.5	1.56
$\bar{B}^0 \to D^+D_s^{*-}$	2.3 ± 1.9	—	2.3 ± 1.9	0.92
$\bar{B}^0 \to D^{*+}\pi^-$	0.28 ± 0.11	0.27 ± 0.07	0.27 ± 0.06	0.31
$\bar{B}^0 \to D^{*+}\pi^-\pi^0$	1.8 ± 0.6	—	1.8 ± 0.6	
$\bar{B}^0 \to D^{*+}\rho^-$	0.7 ± 0.4	0.73 ± 0.19	0.72 ± 0.17	0.97
$\bar{B}^0 \to D^{*+}\pi^-\pi^-\pi^+$	1.2 ± 0.6	1.6 ± 0.5	1.4 ± 0.4	
$\bar{B}^0 \to D^{*+}a_1^-$	—	1.8 ± 0.8	1.8 ± 0.8	1.34
$\bar{B}^0 \to D^{*+}\pi^-\pi^-\pi^+\pi^0$	4.1 ± 2.2	—	4.1 ± 2.2	
$\bar{B}^0 \to D^{*+}D_s^-$	1.2 ± 0.9	1.8 ± 1.0	1.5 ± 0.7	0.7
$\bar{B}^0 \to D^{*+}D_s^{*-}$	2.3 ± 1.3	—	2.3 ± 1.3	2.54
$\bar{B}^0 \to J/\psi K^0$	0.08 ± 0.06	0.10 ± 0.04	0.09 ± 0.03	0.03
$\bar{B}^0 \to J/\psi \bar{K}^{*0}$	0.11 ± 0.05	0.14 ± 0.04	0.13 ± 0.03	0.12
$\bar{B}^0 \to J/\psi K^-\pi^+$	—	0.10 ± 0.05	0.10 ± 0.05	
$\bar{B}^0 \to \psi' K^0$	< 0.28	< 0.15	< 0.15	
$\bar{B}^0 \to \psi' \bar{K}^{*0}$	< 0.23	0.14 ± 0.09	0.14 ± 0.09	
$\bar{B}^0 \to \psi' K^-\pi^+$	< 0.10	—	< 0.10	

At present about 15% of hadronic B^- decays and about 20% of hadronic $\overline{B}^0$ decays have been observed. All branching ratios are calculated using MARK III measurements for the decays of D mesons [16]. For the decays $D^{*+} \to \pi^+ D^0$ and $D^{*0} \to \pi^0 D^0$ we use the recent CLEO values of $68.1 \pm 1.0 \pm 1.3\%$ and $63.5 \pm 4.2\%$, respectively, for the branching ratios [17]. D_s^+ branching ratios are reliably known only relative to $BR(D_s^+ \to \phi\pi^+)$. The latter has been measured using indirect methods and it is hard to estimate the relevant systematic errors. Therefore a fixed value of 2.7% is assumed throughout this paper for $BR(D_s^+ \to \phi\pi^+)$ [19]. In order to compare different experiments, results have been rescaled where necessary to these common values of charm branching ratios.

The last column of Tables 3 and 4 shows the predictions of the BSW model [20] with two free parameters of the model fixed by a fit to the measured branching ratios [14].

134

Table 4. Branching ratios for nonleptonic B^- decays in % .

Mode	ARGUS	CLEO	average	BSW model
$B^- \to D^0 \pi^-$	0.20 ± 0.10	0.40 ± 0.10	0.30 ± 0.07	0.29
$B^- \to D^0 \rho^-$	1.3 ± 0.6	1.02 ± 0.30	1.08 ± 0.27	0.91
$B^- \to D^0 \pi^- \pi^- \pi^+$	—	1.2 ± 0.4	1.2 ± 0.4	
$B^- \to D^0 D_s^-$	2.1 ± 1.1	2.2 ± 0.9	2.2 ± 0.7	1.56
$B^- \to D^0 D_s^{*-}$	1.4 ± 1.0	-	1.4 ± 1.0	0.92
$B^- \to D^{*0} \pi^-$	0.40 ± 0.18	0.35 ± 0.13	0.37 ± 0.10	0.20
$B^- \to D^{*0} \rho^-$	1.0 ± 0.7	1.14 ± 0.39	1.10 ± 0.34	0.71
$B^- \to D^{*0} D_s^-$	1.1 ± 0.8	—	1.1 ± 0.8	0.70
$B^- \to D^{*0} D_s^{*-}$	2.7 ± 1.5	—	2.7 ± 1.5	2.52
$B^- \to D^{*+} \pi^- \pi^-$	0.26 ± 0.16	< 0.4	0.26 ± 0.16	
$B^- \to D^{*+} \pi^- \pi^- \pi^0$	1.8 ± 0.9	—	1.8 ± 0.9	
$B^- \to D^{*+} \pi^- \pi^- \pi^- \pi^+$	< 1.0	—	< 1.0	
$B^- \to J/\psi K^-$	0.07 ± 0.03	0.08 ± 0.03	0.08 ± 0.02	0.03
$B^- \to J/\psi K^{*-}$	0.16 ± 0.11	0.13 ± 0.09	0.14 ± 0.07	0.12
$B^- \to J/\psi K^- \pi^- \pi^+$	< 0.16	0.12 ± 0.07	0.12 ± 0.07	
$B^- \to \psi' K^-$	0.18 ± 0.09	< 0.05	—	
$B^- \to \psi' K^{*-}$	< 0.49	< 0.35	< 0.35	
$B^- \to \psi' K^- \pi^- \pi^+$	0.19 ± 0.12	—	0.19 ± 0.12	
$B^- \to \chi_{c_1} K^-$	0.19 ± 0.14	0.10 ± 0.06	0.11 ± 0.06	

2.1 B meson Decays into Charmonium States

B meson decays to charmonium states like $B^0 \to J/\psi K_s^0$ or $B^0 \to J/\psi K^{*0}$ are considered to be among the best candidates for future studies of CP violation. Decays to χ_c and η_c mesons could also be useful, provided the corresponding branching ratios are large enough.

ARGUS [21] obtained the first evidence for inclusive and exclusive B decays into χ_{c_1} mesons with

$$BR(B \to \chi_{c_1} X) = (1.05 \pm 0.35 \pm 0.25)\%$$

and

$$BR(B^- \to \chi_{c_1} K^-) = (0.19 \pm 0.13 \pm 0.06)\%.$$

CLEO [22] has confirmed the existence of χ_{c_1} production. Their preliminary analysis led to

$$BR(B \to \chi_{c_1} X) = (0.54 \pm 0.15 \pm 0.14)\%$$

and

$$BR(B^- \to \chi_{c_1} K^-) = (0.10 \pm 0.06)\%.$$

Averaging the ARGUS and CLEO results one obtains

$$BR(B \to \chi_{c_1} X) = (0.64 \pm 0.19)\%,$$
$$BR(B^- \to \chi_{c_1} K^-) = (0.11 \pm 0.055)\%.$$

The inclusive branching ratio is twice as large as the theoretical predictions [23]. After subtracting cascade contributions, one concludes that direct yields of J/ψ, ψ'

and χ_{c_1} are comparable, being approximately 0.70%, 0.46%, and 0.60%, respectively. This again does not agree well with the theoretical predictions [23], although the errors are large. Such a large branching ratio implies that the analogous decay $B^0 \to \chi_{c_1} K_s^0$ will be useful in future searches for CP violation in B decays, particularly if the background allows the use of direct χ_{c_1} decays into $3(\pi^+\pi^-)$, $2(\pi^+\pi^-)$, $K^+K^-\pi^+\pi^-$ or other hadronic final states.

The KEK Monte Carlo study [24] demonstrates that this is indeed the case. Fig.2 shows clear χ_{c_1}, J/ψ and η_c signals in four-prong final states which together with K_S^0 form B meson. There is also a χ_{c0} signal but the branching ratio for this decay is expected to be much smaller than assumed in the simulation . The KEK group concluded that the decay $B \to \chi_{c_1} K_S^0$ with $\chi_{c_1} \to \gamma J/\psi$ will have one half of the sensitivity of the gold plated mode $B \to J/\psi K_S^0$. Similar sensitivity is expected for a combination of B decays into K_S^0 and χ_{c_1} , J/ψ or η_c when charmonium states are reconstructed in their four-prong decay modes.

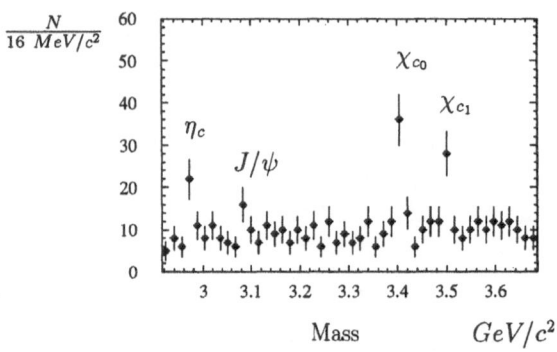

Figure 2. Invariant mass distribution for four-prong final states.

B decays into $J/\psi K^{*0}$ with $K^{*0} \to K_s^0\pi^0$ can also be useful for CP violation studies [25], depending on the relative helicity population of the K^{*0}. In particular, the (0,0) final state in this decay is a pure CP eigenstate. The relative fraction of the (0,0) state can be determined by measuring J/ψ or K^{*0} polarization.

The first evidence for J/ψ polarization was obtained by ARGUS [26] from an analysis of the decay angle Θ of the lepton from the J/ψ decay in the cm-system of the J/ψ meson with respect to the direction of the J/ψ in the B meson cm-system. For fast J/ψ mesons with momenta above 1.4 GeV/c the angular distribution exhibits a $\sin^2 \Theta$ distribution (see Fig.3a) as expected for a (0,0) state (in contrast to the $1 + \cos^2 \Theta$ distribution for $(\pm 1, \pm 1)$ states). A fit to the measured angular distribution of the form

$$\frac{d\sigma}{d\Omega} \propto 1 + \alpha \times \cos^2 \Theta$$

yields $\alpha = -1.17 \pm 0.17$. This implies that J/ψ mesons from the decay $B \to J/\psi K^*$, which dominate the rate in the high-momentum region, are predominantly produced with helicity zero. Recently CLEO confirmed this result [27] $\alpha = -0.78 \pm 0.09 \pm 0.05$ (see Fig.3b).

As a further demonstration that the J/ψ mesons produced in $B \to J/\psi K^*$ decays are longitudinally polarized, ARGUS has studied the distributions of the helicity angles for the J/ψ and K^* decays in 12 events containing B mesons reconstructed in these two-body channels. The corresponding distributions are shown in Fig.4. Both are in good agreement with the expectation for a $J/\psi K^*$ system in a (0,0) state. The best fit corresponds to a pure (0,0) helicity state, a CP eigenstate, with a lower limit of $\Gamma_{00}/\Gamma_{tot} > 0.78$ at the 95% CL.

Corresponding distributions for 39 $J/\psi K^*$ events reconstructed by CLEO [27] are shown in Fig.5. The fit to these distributions gives $\Gamma_{00}/\Gamma_{tot} = 0.78 \pm 0.10 \pm 0.10$.

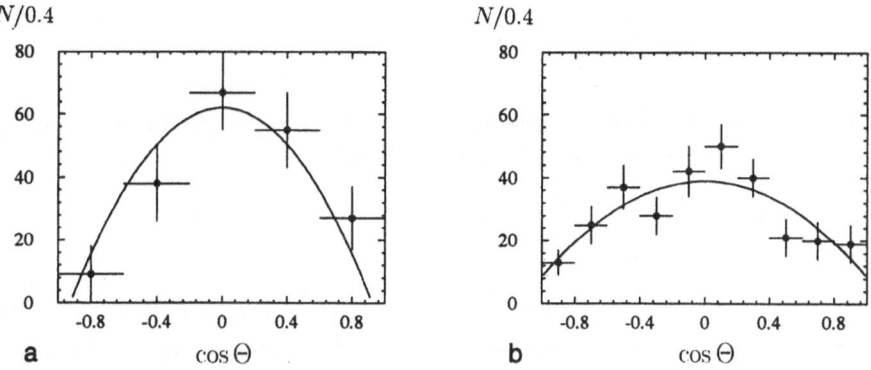

Figure 3. ARGUS(a) and CLEO(b) decay angular distributions of leptons from J/ψ mesons produced in $\Upsilon(4S)$ decays with $p(J/\psi) > 1.4\ GeV/c$.

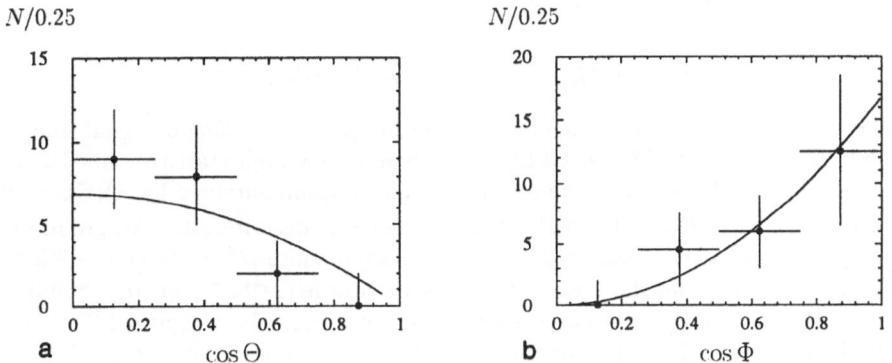

Figure 4. Acceptance corrected distributions of the helicity angles for decays of a) J/ψ and b) K^* mesons from $B \to J/\psi K^*$ decay (ARGUS).

These results show that the dilution of CP asymmetry is small, and the sensitivity of the decay $B^0 \to J/\psi K^{*0}$ in searches for CP violation will be comparable to that for $B^0 \to J/\psi K_S^0$. Observation of the decay $B^- \to \chi_{c_1} K^-$ and dominance of the helicity (0,0) state in the decay $\overline{B}^0 \to J/\psi K^{*0}$ suggest that the luminosity required to observe CP violation could be substantially smaller than the conservative estimates used in B factory design studies [25].

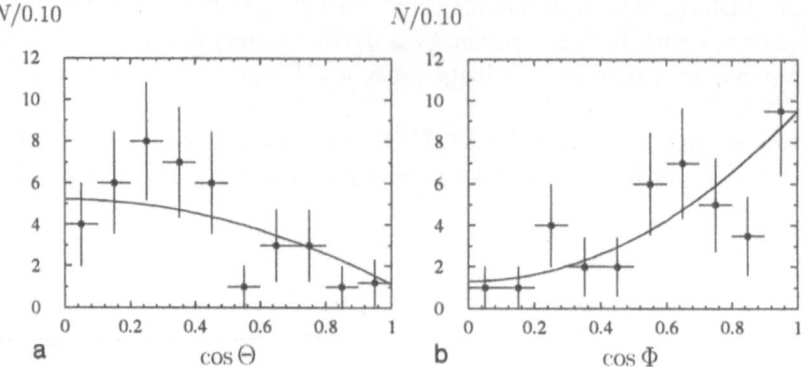

Figure 5. Acceptance corrected distributions of the helicity angles for decays of $a)$ J/ψ and $b)$ K^* mesons from $B \to J/\psi K^*$ decay (CLEO).

3. Semileptonic B Decays

The semileptonic modes for B decays are the best understood both theoretically and experimentally. Their study provides at present the leading method for determining the values of CKM matrix elements $|V_{cb}|$ and $|V_{ub}|$.

3.1 The Decay $B \to D^*\ell\nu$

Exclusive semileptonic B decays into D^* and D mesons have been measured using the missing mass technique [28]. The recoil mass squared against a $D^*\ell$ or $D\ell$ combination, M_{recoil}^2, is calculated from the $D^{(*)}$ and ℓ momenta, neglecting the small momentum of the B meson in the laboratory frame:

$$M_{recoil}^2 = (E_{beam} - E_{D^{(*)}} - E_\ell)^2 - (\vec{p}_{D^{(*)}} + \vec{p}_\ell)^2.$$

The resolution on M_{recoil}^2 is sufficient to achieve a good separation of signal from background. Recently ARGUS and CLEO performed a new high statistics analysis of the decay $B \to D^*\ell\nu$. Fig.6 shows the recoil mass distribution obtained by ARGUS [29] for the decay $\overline{B}^0 \to D^{*+}\ell^-\overline{\nu}$. The shaded area in the figure describes the background which can be due to higher mass excited D^{**} mesons decaying into D^{*+} or to nonresonant $D^*\pi$ production. Fig.7 shows the invariant mass spectrum $M(D^{*+}\pi^-)$ for $M_{rec}^2 > 0$ indicating an enhancement at masses of P-wave states. From the fit, a signal of 30 ± 10 D^{**0} events is determined for the sum of contributions from both resonances. If the background in Fig.6 were to result solely from the decays of P-level D^{**} mesons, one would expect to reconstruct 35 ± 9 events. The agreement between the two numbers demonstrates that D^{**} production dominates the background for the decay $\overline{B}^0 \to D^{*+}\ell^-\overline{\nu}$. Thus ARGUS obtained the first direct evidence for semileptonic B decays into D^{**} mesons. The branching ratios were found to be

$$BR(\overline{B}^0 \to D^{*+}\ell^-\overline{\nu}) = 4.5 \pm 0.4 \pm 0.5\%,$$

$$BR(\overline{B}^0 \to D^{**+}\ell^-\overline{\nu}) = 2.3 \pm 0.4 \pm 0.4\%$$

where D^{**+} denotes a mixture of $1^3P_{1,2}$, 1^1P_1, 2^1S_0 and 2^3S_1 states with the weights predicted by the GISW model [30]. Using the BHKT model [31] one arrives at a somewhat smaller branching ratio

$$BR(\overline{B}^0 \to D^{**+}\ell^-\overline{\nu}) = 2.0 \pm 0.5 \pm 0.3\%$$

which coincides within systematic errors with the value obtained using the GISW model.

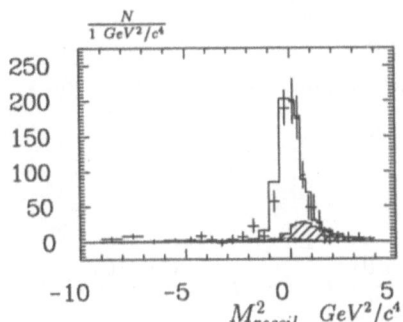

Figure 6. Distribution on recoil mass against $D^{*+}\ell^-$.

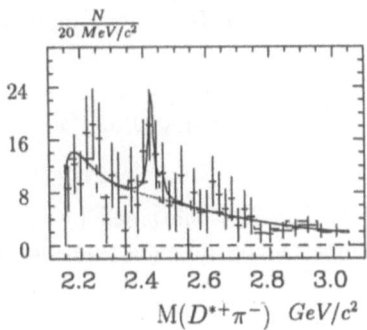

Figure 7. $D^{*+}\pi^-$ invariant mass distribution [29]

The decay $\overline{B}^0 \to D^{*+}\ell^-\overline{\nu}$ followed by $D^{*+} \to D^0\pi^+$ is completely described by specifying q^2 and the three angular degrees of freedom defined in Fig.8.

Neglecting the masses of the e^- and μ^-, the differential decay width can be written as:

$$\frac{d^4\Gamma(q^2, \cos\theta, \cos\theta^*, \chi)}{dq^2\, d\cos\theta\, d\cos\theta^*\, d\chi} = BR(D^{*+} \to D^0\pi^+)\frac{G_F^2}{(2\pi)^3}\,|V_{cb}|^2\,\frac{q^2 p}{12M_{\overline{B}^0}^2} \times$$
$$\left| \sum_\lambda d^1_{0,\lambda}(\theta^*) \cdot d^1_{\lambda,1}(\theta) \cdot e^{-i(\lambda+1)\chi} \cdot H_\lambda(q^2) \right|^2$$

where p is the D^{*+}-momentum in the $\overline{B}^0$ rest frame and $H_\lambda(q^2)$ are the helicity form factors. One qualitatively expects $|H_-|^2 > |H_+|^2$ since the c-quark in the D^{*+}-meson is produced with predominantly negative helicity, and both helicities occur with equal amplitude for the spectator $\overline{d}$-quark. Hence, parity violation in the weak interaction should manifest itself as a forward-backward asymmetry A_{FB} in the distribution of $\cos\theta$ [32]:

$$\frac{d\Gamma(\cos\theta)}{d\cos\theta} \propto \alpha \cdot \sin^2\theta - \frac{4}{3}A_{FB}(3+\alpha) \cdot \cos\theta + 2$$

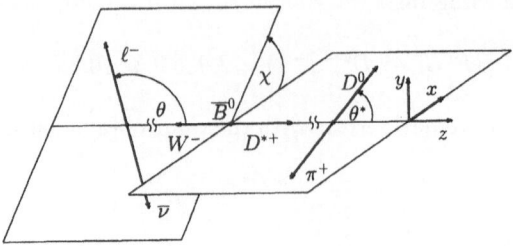

Figure 8. Definition of polar angles θ and θ^* and azimutal angle χ.

where

$$A_{FB} = \frac{\int_{-1}^{0} \frac{d\Gamma(\cos\theta)}{d\cos\theta} d\cos\theta - \int_{0}^{+1} \frac{d\Gamma(\cos\theta)}{d\cos\theta} d\cos\theta}{\int_{-1}^{+1} \frac{d\Gamma(\cos\theta)}{d\cos\theta} d\cos\theta} = \frac{3}{4} \cdot \frac{\Gamma_- - \Gamma_+}{\Gamma}.$$

The helicity alignment of the W^- is given by $\alpha = 2\Gamma^0/(\Gamma^+ + \Gamma^-) - 1$, which describes the D^{*+} polarization extracted from the D^{*+} decay angle distribution via:

$$\frac{d\Gamma(\cos\theta^*)}{d\cos\theta^*} \propto 1 + \alpha \cdot \cos^2\theta^*.$$

The distributions on $\cos\theta^*$ and $\cos\theta$ obtained by the ARGUS collaboration are shown in Fig.9. The simultaneous fit to these distributions and distributions on q^2 and M^2_{recoil} leads to

$$A_{FB} = \frac{3}{4} \cdot \frac{\Gamma^- - \Gamma^+}{\Gamma} = 0.20 \pm 0.08 \pm 0.06$$

and

$$\alpha = 2 \cdot \frac{\Gamma^L}{\Gamma^T} - 1 = 1.1 \pm 0.4 \pm 0.2.$$

The value for A_{FB} is in agreement with the predictions from the various models and shows that $b \to c$ transitions are left chiral. The ARGUS update on α is in agreement with previous measurements [33].

A similar analysis was performed by the CLEO collaboration [34]. It also demonstrated a forward-backward asymmetry in accordance with the SM predictions (see Fig.10a and 10b).

3.2 Measurement of the decay $\overline{B}^0 \to D^{*+}\ell^-\overline{\nu}$ with partial D^{*+} reconstruction

The missing mass technique can be used for the decay $\overline{B}^0 \to D^{*+}\ell^-\overline{\nu}$ even in the case where only the π^+ from the $D^{*+} \to \pi^+ D^0$ decay is detected [35]. The energy

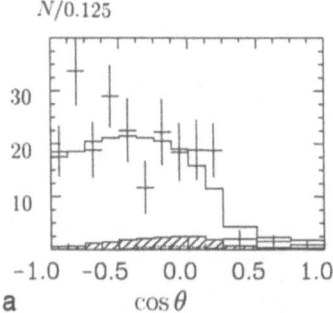
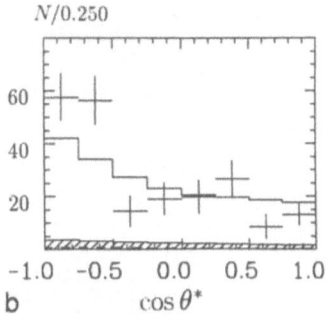

Figure 9. The measured $\cos\theta$ (a) and $\cos\theta^*$ (b) distributions, uncorrected for efficiency. Histogram is a theoretical predictions for $V - A$ coupling. The shaded histograms show the amount and the shape of the background.

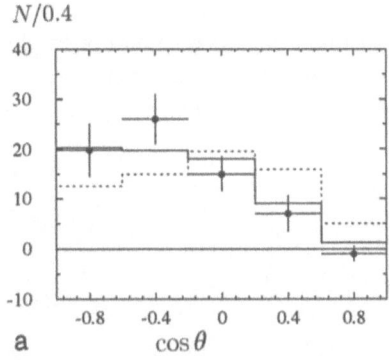
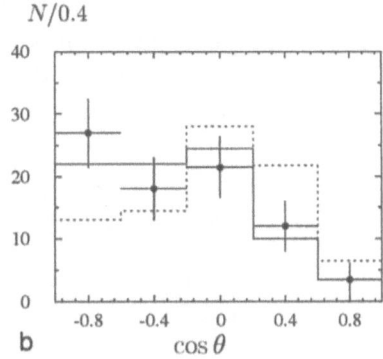

Figure 10. Distributions on $\cos\theta$ for semileptonic B decays into a) D^{*+} and b) D^{*0}. Theoretical predictions for $V - A$ and $V + A$ couplings are shown by solid and dotted lines respectively.

release in the latter decay is only 6 MeV. Therefore, the D^{*+} momentum can be well estimated using the momentum of the slow pion:

$$\vec{p}_{D^{*+}} = \alpha\vec{p}_\pi + \beta\vec{p}_\pi / \mid \vec{p}_\pi \mid$$

with parameters $\alpha = 8.23$ and $\beta = 0.41$ GeV/c fixed by Monte Carlo studies. The kinematic limit for the π^+ momentum in the studied process is about 200 MeV/c. Therefore only pions with momenta below this limit were used in the analysis.

The momentum of the lepton was required to be more than 1.4 GeV/c to suppress cascade charm decays as well as decays of $B \rightarrow D^{**}\ell\nu$, where the lepton momentum is expected to be softer.

The recoil mass distribution after subtraction of the contribution from continuum and misidentified hadrons is shown in Fig.11. A prominent peak near zero recoil mass in the right-sign combinations (points with errors) corresponds to the studied decay, while the wrong-sign combination distribution behaves smoothly in this region (histogram).

Monte Carlo simulation predicts similar shapes for the wrong- and right-sign background distributions, therefore the wrong-sign combinations were used to describe the background under the signal.

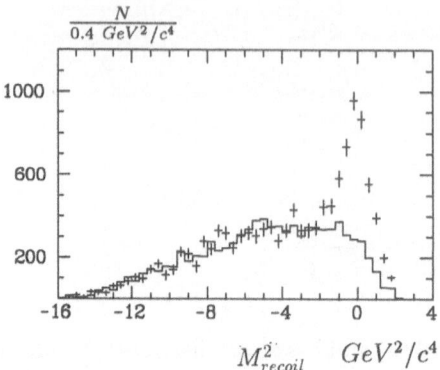

Figure 11. Recoil mass distribution for right- and wrong-sign π–ℓ combinations.

After wrong-sign combination subtraction, two functions corresponding to the decays $\overline{B}^0 \to D^{*+}\ell^-\overline{\nu}$ and $B \to D^{**}\ell\nu$ were fitted to the recoil mass distribution (see Fig.12). The fit found the number of the events from the studied decay to be equal

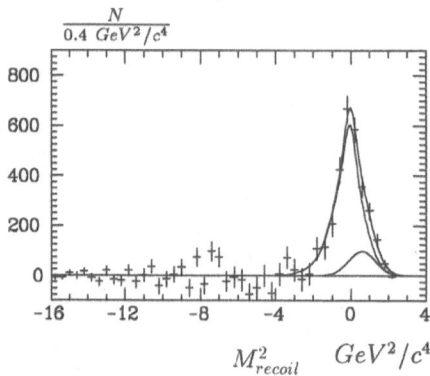

Figure 12. Recoil mass distribution for $\pi^+\ell^-$ combinations after background subtraction.

to $N(\overline{B}^0 \to D^{*+}\ell^-\overline{\nu}) = 2693 \pm 183 \pm 105$, which corresponds to

$$BR(\overline{B}^0 \to D^{*+}\ell^-\overline{\nu}) = (4.5 \pm 0.3 \pm 0.4)\%$$

if one uses $BR(D^{*+} \to D^0\pi^+) = (68.1 \pm 1.0 \pm 1.3)\%$ [17]. The GISW model [30] was used for efficiency calculations and extrapolation to small lepton momenta. This result is in good agreement with measurements using full D^{*+} reconstruction technique and has the added advantage of being independent of D^0 branching ratios.

Partially reconstructed decays $\overline{B}^0 \to D^{*+}\ell^-\overline{\nu}$ provide a large sample of tagged B^0 mesons. This sample can be used to measure the inclusive semileptonic branching ratio for neutral B mesons. Background in the region $|M^2_{recoil}| < 1\ GeV^2/c^4$ is relatively small (see Fig.11) and only partly due to charged B mesons.

The obtained semileptonic branching ratio is equal to

$$BR(\overline{B}^0 \to X\ell^-\overline{\nu}) = (9.7 \pm 1.3 \pm 1.2)\%.$$

Comparing this value with the branching ratio for a mixture of charged and neutral B mesons [36] one gets

$$\tau(B^+)/\tau(B^0) = 1.03 \pm 0.27 \pm 0.26.$$

Using the large samples of reconstructed B mesons CLEO obtained [8]

$$BR(B^0 \to X\ell^+\nu) = (10.8 \pm 2.1 \pm 1.4)\%,$$
$$BR(B^+ \to X\ell^+\nu) = (12.0 \pm 2.2 \pm 1.3)\%.$$

This corresponds to

$$\tau(B^+)/\tau(B^0) = 1.1 \pm 0.3.$$

Direct measurements at LEP give [3]

$$\tau(B^+)/\tau(B^0) = 0.92 \pm 0.21.$$

Finally, a comparison of semileptonic B decays to D^{*+} and D^{*0} leads to

$$\tau(B^+)/\tau(B^0) = 1.06 \pm 0.25.$$

Combining all these results we get

$$\tau(B^+)/\tau(B^0) = 1.01 \pm 0.13.$$

This agrees well with theoretical predictions of approximately equal B^0 and B^+ lifetimes.

The presently available experimental data on $BR(B \to D^*l\nu)$ and $BR(B \to Dl\nu)$ are summarized in table 5.

Table 5. Branching ratios for $B \to Dl\nu$ and $B \to D^*l\nu$ decays.

	ARGUS	CLEO
BR $(B^- \to D^0\ell^-\bar{\nu})$	$1.5 \pm 0.5 \pm 0.3\%$	$1.6 \pm 0.6 \pm 0.3\%$
BR $(\overline{B}^0 \to D^+\ell^-\bar{\nu})$	$1.5 \pm 0.5 \pm 0.3\%$	$1.8 \pm 0.8 \pm 0.3\%$
BR $(B^- \to D^{*0}\ell^-\bar{\nu})$	$5.1 \pm 1.2 \pm 1.0\%$	$4.1 \pm 0.8^{+0.8}_{-0.9}\%$
BR $(\overline{B}^0 \to D^{*+}\ell^-\bar{\nu})$	$4.5 \pm 0.4 \pm 0.5\%$ $4.5 \pm 0.3 \pm 0.4\%$	$3.8 \pm 0.4 \pm 0.6\%$

Theoretical models predict partial widths of exclusive semileptonic B decays in terms of CKM matrix elements. To transform measured branching ratios into decay rates we assume $\tau_{B^0}=\tau_{B^+}=(1.32 \pm 0.06) \times 10^{-12}$ sec [4]. In order to minimize effects due to possible differences between τ_{B^0} and τ_{B^+}, we use in the calculations average branching ratios for neutral and charged B mesons. Tables 6 and 7 list the values of $|V_{cb}|$ derived from branching ratios for exclusive semileptonic B decays using different theoretical models.

Unfortunately the resulting values of $|V_{cb}|$ are model dependent and it is difficult to estimate the theoretical uncertainty in this fundamental parameter of the SM.

Table 6. Determination of $|V_{cb}|$ from $B \to D^* \ell \nu$ decay.

Model	$|V_{cb}|$
BSW [37]	0.040 ± 0.004
SP [38]	0.037 ± 0.003
KS [39]	0.038 ± 0.004
GISW [30]	0.034 ± 0.003

Table 7. Determination of $|V_{cb}|$ from $B \to D \ell \nu$ decay.

Model	$|V_{cb}|$
BSW [37]	0.041 ± 0.006
SP [38]	0.046 ± 0.006
KS [39]	0.039 ± 0.005
GISW [30]	0.034 ± 0.005

3.3 Estimation of $|V_{cb}|$ Using HQET

Heavy quark effective theory (HQET) provides in principle a model independent method for $|V_{cb}|$ determination with well controlled theoretical uncertainties. In practice, however, it is still necessary to use models in order to extrapolate data to the region of phase space where the predictions of HQET are applicable. Fortunately, this model dependence will be reduced in future as more experimental data becomes available.

The possibility of $|V_{cb}|$ extraction from the decay $B \to D^* l \nu$ in the infinite quark mass limit was first discussed by Voloshin and Shifman [41]. They have shown that the decay rate for $B \to D^* l \nu$ when the momentum of the D^* is small can be found in a model independent way. Corrections to this prediction are of the order of $\mu^2/m_c^2 \sim 5\%$, where μ is a characteristic momentum of quarks in the meson. HQET generalizes this result. In this approach, the decay width over the full Dalitz plot depends only on the $|V_{cb}|$ matrix element and a single universal function, the Isgur-Wise function $\xi(y)$, where $y = v \cdot v'$ is a product of the four-velocities of the B and D^* mesons [42]. The product $|V_{cb}| \xi(1)$ can be determined by extrapolation from the full momentum interval [43].

For the determination of $|V_{cb}|$, the y distributions of D^{*+} and D^{*0} mesons from $\overline{B}^0 \to D^{*+} \ell^- \overline{\nu}$ and $B^- \to D^{*0} \ell^- \overline{\nu}$ decays have been investigated. The latter channel is particularly suitable since the reconstruction efficiency for the D^{*0} is practically independent of momentum, in contrast to the case for the D^{*+} in the former decay. Fig.13 shows $|V_{cb}| \xi(y)$ extracted from the ARGUS data [44, 29]. The shape of the Isgur-Wise function is not fixed by theory, but it is equal to one at the zero recoil point $y = 1$. The fit with a linear parameterization (dotted lines) for the Isgur-Wise function finds $|V_{cb}| = 0.043 \pm 0.004$ and $|V_{cb}| = 0.043 \pm 0.007$ for B^0 and B^+ respectively. Similar results are obtained for $|V_{cb}|$ under different parameterizations for the Isgur-Wise function, such as a single-pole model (solid lines) or an exponential dependence.

Recently de Rafael and Taron [45] obtained limits on the slope and the second derivative of the Isgur-Wise function at $y = 1$. If we use their limits in the whole y range then the value of $|V_{cb}|$ drops considerably (the fit results are shown in Fig.13 by

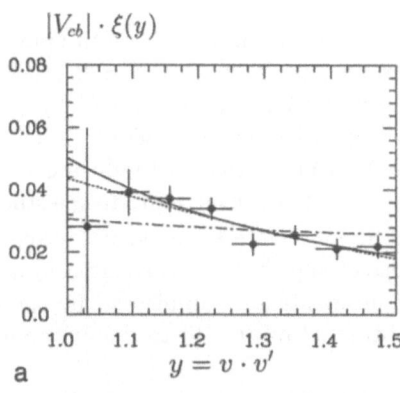
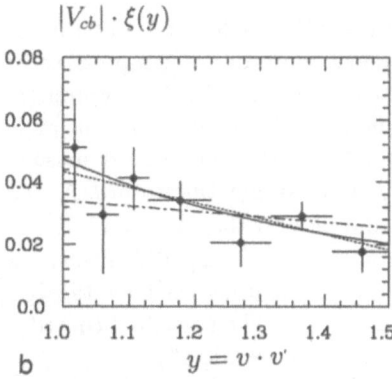

Figure 13. Distributions of $|V_{cb}|\xi(y)$ for (a) $\overline{B}^0 \rightarrow D^{*+}\ell^-\overline{\nu}$ and (b) $B^- \rightarrow D^{*0}\ell^-\overline{\nu}$ decays. The lines show the fit results (see text).

dashdotted lines). The quality of the fit is poor (probabilities are at a few percent level) but such a parametrization of $\xi(y)$ can not be completely discarded on this basis. This exercise demonstrates the need for a better theoretical understanding of the shape of the Isgur Wise function or enough of an increase in the data sample size to eliminate the extrapolation procedure problem.

3.4 Inclusive Lepton Spectrum

The inclusive semileptonic branching ratio of the B meson has been determined quite precisely. The latest results obtained at $\Upsilon(4S)$ energies are summarized in table 8.

Table 8. Recent measurements of B meson semileptonic branching ratio at the $\Upsilon(4S)$ energy.

	$BR(B \rightarrow X\ell^-\bar{\nu})$ (%)
ARGUS	$10.2 \pm 0.4 \pm 0.2$
CLEO	$10.3 \pm 0.2 \pm 0.4$
CLEO $(\ell\ell)$	$10.6 \pm 0.5 \pm 0.6$
CUSB	$10.0 \pm 0.4 \pm 0.3$
X Ball	$12.0 \pm 0.5 \pm 0.7$
Average	10.4 ± 0.2

The average branching ratio $BR(B \rightarrow \ell X) = (10.4 \pm 0.2)\%$ is too small to be understood in a straight-forward way in the spectator model where a branching ratio of 12-14% is predicted [46]. Measurements at the Z^0 peak, which actually represent average semileptonic branching ratios for an unknown mixture of b-flavoured hadrons, yield similar values of less than 11% [47] (except for L3).

Since the low momentum part of the lepton spectrum is dominated by leptons from charm decays, only the energetic part is commonly used for the determination of the B meson semileptonic branching ratio. Extrapolation to the whole momentum interval is usually performed using the theoretical models. One possible explanation of the discrepancy between experiment and theory could be an underestimation of the contribution from soft leptons. Recently, ARGUS performed the first measurement of the low energy part of the primary electron spectrum in B meson decays [35] .

In the ARGUS detector electrons are well identified over almost the entire momentum range. In order to suppress secondary electrons, the flavour of one B meson is tagged by the sign of a fast lepton ($tag^+ = e^+$ or μ^+, with $p_{tag^+} > 1.4\ GeV/c$) and then used to study the momentum spectrum of opposite-sign electrons. Secondary electrons from the untagged B meson have opposite sign to that of the primary ones (except those originating from D_s^-, J/ψ or τ^- decays). These contribute to the studied spectrum only in the case of B^0–$\overline{B}^0$ mixing and are therefore suppressed. Secondary electrons from the tagged B meson have the correct sign but are correlated in angle with the fast leptons used for tagging and can be drastically suppressed by the cut $\cos\theta_{tag^+e^-} > 0$. Their residual contribution is subtracted using a fit to the distribution of $\cos\theta_{tag^+e^-}$ (see Fig.14).

The background from secondary electrons originating from the decays of D_s^-, ψ and τ^- was estimated by Monte Carlo simulation. Backgrounds from continuum, hadrons misidentified as leptons, as well as electrons from photons converted in matter, were subtracted using the data.

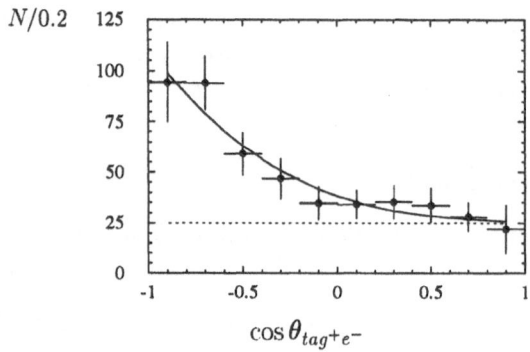

Figure 14. Angular distribution between electron and tagging lepton for the electron momenta $0.6 - 0.8\ GeV/c$.

The spectrum obtained after complete background subtraction is shown in Fig.15.

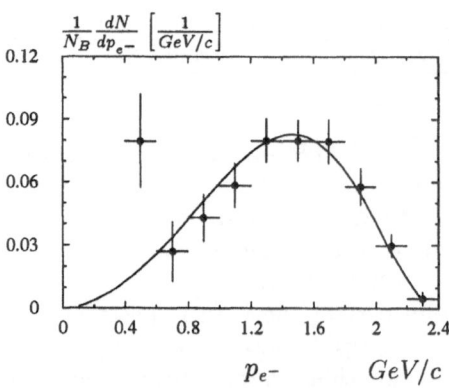

Figure 15. Momentum spectrum of primary electrons.

It agrees with the GISW model [30] using the parameters found in the previous ARGUS analysis [48], where only the hard part of the lepton spectrum was studied.

Thus the "lepton deficit" can not be explained by an unexpectedly large fraction of soft leptons.

3.5 Semileptonic $b \to u$ Decays

The existence of charmless semileptonic decays is reliably established. ARGUS [49] and CLEO [50] have observed leptons above the kinematic limit for the $b \to c\ell\nu$ transition. Both collaborations have checked that the observed excess cannot be explained by other processes including hypothetical non-$B\overline{B}$ decays of $\Upsilon(4S)$.

ARGUS and CLEO used quite different approaches for measuring the lepton spectrum above the kinematic limit for $b \to c\ell\nu$ transitions.

CLEO performed an almost inclusive analysis with a modest reduction of the continuum background through rejection of two jet events. The remaining continuum background was subtracted using measurements just below the $\Upsilon(4S)$ resonance. This method relies on extremely well known relative luminosities at the $\Upsilon(4S)$ and in the continuum (an uncertainty in the ratio of only 0.5%) and on stable operation of the detector. In order to achieve this CLEO switched from $\Upsilon(4S)$ to continuum data taking every three days.

CLEO presented results in two lepton momentum intervals $2.2 - 2.6 \ GeV/c$ and $2.4 - 2.6 \ GeV/c$ but only in the latter was the result statistically significant. The semileptonic branching ratio in this interval was found to be $(1.8 \pm 0.4 \pm 0.3) \cdot 10^{-4}$. In order to extrapolate the result in this tiny momentum region to the full momentum range one has to invoke a theoretical model, leading to considerable uncertainty. For example the ACCMM spectator model [51] yields to $|V_{ub}|/|V_{cb}| = 0.12 \pm 0.02$ while the GISW model [30] gives $|V_{ub}|/|V_{cb}| = 0.19 \pm 0.03$. One should remember that the errors on $|V_{ub}|/|V_{cb}|$ are not gaussian and do not reflect a large model dependence.

The ARGUS detector is hermetic enough to measure the neutrino momentum (approximated by missing momentum) in an event. Using this feature and the kinematics of B decays into ρ, ω, a_1, which dominate the endpoint region, the ARGUS collaboration suppressed the continuum contribution to a negligible level keeping the efficiency for $b \to u\ell\nu$ transitions at the 50% level.

For normalization, a lepton momentum interval of the same width below the endpoint for $b \to c\ell\nu$ transitions was used yielding

$$BR_{sl}(2.3 - 2.6 \ GeV/c)/BR_{sl}(2.0 - 2.3 \ GeV/c) = 5.4 \pm 0.9 \pm 0.8\%$$

This corresponds to $|V_{ub}|/|V_{cb}| = 0.11 \pm 0.01$ in the case of the ACCMM model.

The new CLEO analysis [3] is very similar to the ARGUS one. A missing momentum larger than $1 \ GeV/c$ in the hemisphere opposite to the lepton direction is required. Their selection criteria reduced the continuum contribution by a factor of 66 keeping the efficiency for the $b \to u\ell\nu$ transitions at 40 % . The lepton momentum spectrum obtained by CLEO after background subtraction is shown in Fig.16 .

There is a clear excess of leptons above the endpoint for $b \to c\ell\nu$ transitions. There are $37.8 \pm 9.3 \pm 3.3$ signal events in the momentum interval $2.4 - 2.6 \ GeV/c$. The efficiency in this momentum interval depends on compositions of exclusive final states in semileptonic $b \to u$ transitions. This leads to a large systematic (theoretical) uncertainty in determination of the semileptonic branching ratio

$$BR_{b \to u\ell\nu}(2.4 - 2.6 \ GeV/c) = (0.48 \pm 0.12 \pm 0.25) \cdot 10^{-4}.$$

This value is more than three times smaller than the previous CLEO results [50] and corresponds to $|V_{ub}|/|V_{cb}| = 0.062 \pm 0.010$ in the case of the ACCMM model. Thus

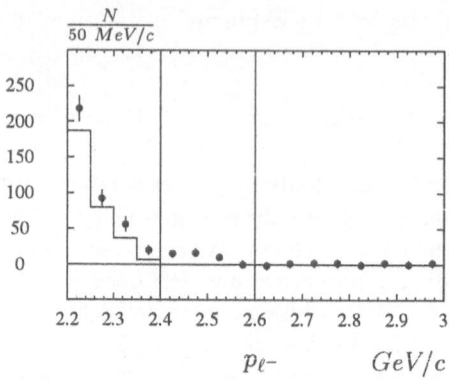

Figure 16. Lepton momentum spectrum for $\Upsilon(4S)$ decays. Histohgram shows the expected shape of the $b \to c$ contribution.

the $|V_{ub}|\,/\,|V_{cb}|$ ratio is smaller than previous ARGUS and CLEO values and still very uncertain due the extrapolation procedure and the model dependence of the event selection efficiency .

4. $B^0\overline{B}^0$ Oscillations

$B^0\overline{B}^0$ mixing is a very spectacular quantum mechanical phenomenon. Beauty is not a strictly conserved quantum number. Hence, B^0 and $\overline{B}^0$ mesons mix and the mass matrix eigenstates are linear combinations of B^0 and $\overline{B}^0$. Neglecting CP violation one obtains:

$$|B_1\rangle = \tfrac{1}{\sqrt{2}}(|B^0\rangle + |\overline{B}^0\rangle)$$
$$|B_2\rangle = \tfrac{1}{\sqrt{2}}(|B^0\rangle - |\overline{B}^0\rangle)$$

The two eigenstates are expected to have approximately equal decay widths [52] $\Gamma_1 \approx \Gamma_2 \approx \Gamma$ and this approximation will be used in this paper for the sake of simplicity. The time evolution of an initially pure B^0 state is then given by

$$w(t) \approx \tfrac{1}{2}\exp(-\Gamma t)\,[1 + \cos(\Delta M t)]$$
$$\overline{w}(t) \approx \tfrac{1}{2}\exp(-\Gamma t)\,[1 - \cos(\Delta M t)]$$

where $\Delta M = M_2 - M_1$ is the mass difference between the two eigenstates and $w(t)$ ($\overline{w}(t)$) is the probability for finding B^0 ($\overline{B}^0$) at a time t.

Analogous time dependences for neutral kaons were observed a long time ago. Unfortunately, measurements of time dependences are so far not possible for B mesons because of their short lifetime.

However, the existence of oscillations can be inferred from the time integrated rates $N(B^0)$ and $N(\overline{B}^0)$

$$N(B^0) = \int_0^\infty w(t)dt = \frac{1}{2\Gamma}(1 + \frac{1}{1 + (\frac{\Delta M}{\Gamma})^2})$$
$$N(\overline{B}^0) = \int_0^\infty \overline{w}(t)dt = \frac{1}{2\Gamma}(1 - \frac{1}{1 + (\frac{\Delta M}{\Gamma})^2})$$
(1)

The fraction of $\overline{B}^0$ meson decays in the initially pure B^0 beam depends on the ratio

$x = \frac{\Delta M}{\Gamma}$ of the oscillation frequency and decay width Γ :

$$\chi = \frac{N(\overline{B}^0)}{N(B^0) + N(\overline{B}^0)} \approx \frac{x^2}{2 + 2x^2}$$

The second equality can be easily obtained from eq.(1). For no $B^0\overline{B}^0$ mixing one has $\chi = 0$ while for complete mixing $\chi = 0.5$. Very often another mixing parameter r is used instead of χ :

$$r = \frac{N(\overline{B}^0)}{N(B^0)} = \frac{x^2}{2 + x^2}$$

There are simple relations between the mixing parameters r and χ:

$$r = \frac{\chi}{1 - \chi}$$
$$\chi = \frac{r}{1 + r}$$

In the Standard Model $B^0\overline{B}^0$ transitions are described by the box diagram shown in Fig.17 .

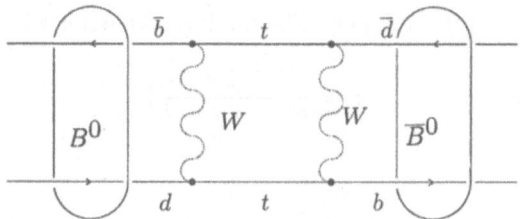

Figure 17. One of the diagrams for $B^0\overline{B}^0$ transitions.

The t quark exchange plays a dominant role in these diagrams. Thus the mixing parameter for B_d^0 mesons is given by:

$$x \approx \frac{G_F^2}{6\pi^2} B_B f_B^2 m_b \left| V_{tb}^* V_{td} \right|^2 m_t^2 F(\frac{m_t^2}{M_W^2}) \eta_{QCD} \tau_B \tag{2}$$

where $\eta_{QCD} \approx 0.55$ represents QCD correction to the box diagrams [53], V_{td} and $V_{tb} \approx 1$ are the CKM matrix elements which describe couplings between quarks of different generations and $F(\frac{m_t^2}{M_W^2})$ is a slowly decreasing function ($F(0) = 1$, $F(1) = 0.75$) which describes W propagator effects [54] . The factor $B_B f_B^2$ is a parametrization of the hadronic matrix element

$$\langle B^0 | \overline{d}\gamma_\mu(1 + \gamma_5)b\overline{d}\gamma_\mu(1 + \gamma_5)b | \overline{B}^0 \rangle = \frac{4}{3} B_B f_B^2 m_b$$

It accounts for the probability that the $\overline{b}d(b\overline{d})$ quarks inside the $B^0(\overline{B}^0)$ meson are close together, so that W exchange between them is possible. The value of $B_B f_B^2$ is a major source of uncertainty in eq.2.

So all but two parameters in eq.(2) can be estimated and the measurement of $B^0\overline{B}^0$ mixing provides information about V_{td} and m_t and thus put severe constraints on the free parameters of the SM. Moreover, hypothetical new particles like charged Higgs bosons, W_R, and supersymmetric particles can contribute to box diagrams and change the predictions of the SM.

That is why the observation of unexpectedly large $B^0\overline{B}^0$ mixing by the ARGUS collaboration [55] and confirmation by the CLEO group [56] attracted so much attention.

$B^0\overline{B}^0$ mixing manifests itself in the production of B^0B^0 or $\overline{B}^0\overline{B}^0$ pairs. Due to a low efficiency for the reconstruction of B mesons a partial reconstruction is usually used and is sufficient to tag the flavour of both B mesons.

At present the most accurate method for measuring the mixing rate is by tagging B^0 and $\overline{B}^0$ mesons with the charge of the lepton from semileptonic decays. B^0 mesons decay only to ℓ^+ while $\overline{B}^0$ mesons produce only ℓ^-. $B^0\overline{B}^0$ mixing leads, therefore, to like-sign lepton pairs. The ARGUS [57] and new CLEO [58] dilepton rates after background subtraction are listed in Table 9. In this method the number of like-sign lepton pairs from charged B decays must be subtracted. In order to estimate this number we take

$$\frac{f_+ \tau_{B+}}{f_0 \tau_{B^0}} = 1.0 \pm 0.14$$

and assume $\tau_{B^0}/\tau_{B+} = 1.0 \pm 0.1$ [59]. The mixing parameter r is equal to

$$r = \frac{N^{\pm\pm}(1+\lambda)}{N^{+-} - \lambda N^{\pm\pm}},$$

where

$$\lambda = \frac{f_+}{f_0}\left(\frac{\tau_{B+}}{\tau_{B^0}}\right)^2$$

and $N^{\pm\pm}$ and N^{+-} - number of like and opposite sign dileptons. This leads to

$$r = 0.194 \pm 0.038$$

Table 9. Dilepton rates and a mixing parameter r.

	ARGUS	CLEO
$N_{\ell^\pm\ell^\pm}$	48.1 ± 14.1	184.5 ± 23.7
$N_{\ell^+\ell^-}$	505 ± 27	2169.1 ± 52.4
r	0.21 ± 0.07	0.187 ± 0.045

In a second method one B^0 meson is reconstructed using the decay $B^0 \to D^{*-}l^+\nu$ while the second B is again tagged with the lepton charge. Table 10 gives the number of such events observed by ARGUS [57] and CLEO [60]. This method yields $r = 0.18 \pm 0.10$. Since in this method one B^0 meson is reconstructed the result does not depend on λ. Combining the two results on r one finds:

$$r = 0.192 \pm 0.036,$$
$$\chi = 0.161 \pm 0.030,$$
$$x = \Delta M \cdot \tau_b = 0.69 \pm 0.09$$

Table 10. The number of $B^0\ell^+$ and $B^0\ell^-$ events and the mixing parameter r.

	ARGUS	CLEO
$N_{B^0\ell^+}$	6.7 ± 3.4	1.9 ± 3.1
$N_{B^0\ell^-}$	29.1 ± 5.8	20.2 ± 6.3
r	0.23 ± 0.12	0.09 ± 0.16

As was discussed in the section 3 the decay $B^0 \to D^{*+}\ell^-\bar{\nu}$ can be reconstructed even if one measures only the ℓ^- and the slow pion from the D^{*+}. Using this partial reconstruction CLEO [58] obtained the momentum spectra for leptons from the second B decay shown in Fig.18 . The fit to these spectra gives 210 ± 27 and 1213 ± 54 direct leptons for like and unlike-sign cases. This corresponds to

$$r = 0.200 \pm 0.033 \pm 0.024 \pm 0.015$$

where the first error is statistical, the second error is systematic without λ, and the final error is due to λ. Using a similar method, ARGUS obtained a value for the mixing parameter of

$$r = 0.225 \pm 0.079 \pm 0.028,$$

which is independent of λ.

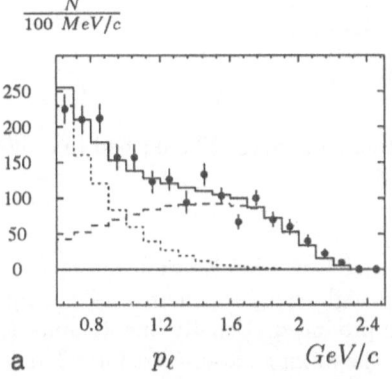
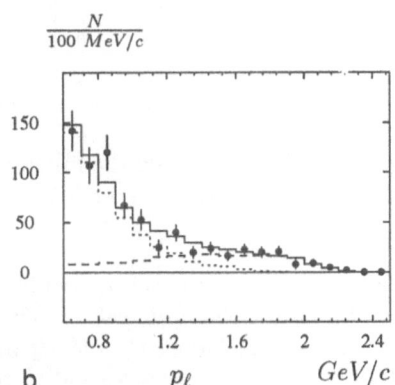

Figure 18. Lepton spectra for a) unlike-sign and b) like-sign leptons. Direct and cascade contributions are shown by the dashed and dotted histograms respectively.

The values of the mixing parameter r obtained with the method employing partial D^{*+} reconstruction agree nicely with the values obtained from dilepton studies. Unfortunately, the results from these two methods can not be averaged in a straight-forward way because of large correlations. Nevertheless, the new method provides important information because it has different systematic uncertainties.

Using this result, together with the estimate $B_B^{\frac{1}{2}}f_B = (220 \pm 40)\ MeV$ and $m_t = 145 \pm 26\ GeV/c^2$ from the analysis of electroweak radiative corrections, one obtains

$$|V_{td}| = 0.011 \pm 0.005.$$

The SM with three generations predicts a large rate for $B_s^0\overline{B}_s^0$ mixing:

$$\frac{x_s}{x_d} \approx \frac{|V_{ts}|^2}{|V_{td}|^2} > 5.$$

Table 11. $B^0\overline{B}^0$ mixing at high energies.

	$\chi = f_d\chi_d + f_s\chi_s$
ALEPH	$0.137 \pm 0.015 \pm 0.007$
DELPHI	$0.121 \pm 0.042 \pm 0.007$
L3	$0.121 \pm 0.017 \pm 0.006$
OPAL	$0.125 \pm 0.017 \pm 0.015$
LEP (average)	0.128 ± 0.010
UA1	0.145 ± 0.038
CDF	0.176 ± 0.050
Average	0.131 ± 0.010

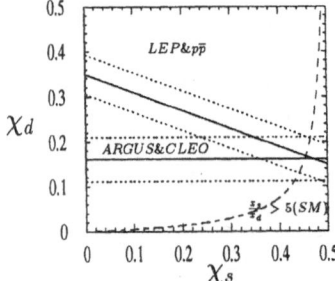

Figure 19. 90% CL limits on the mixing parameters χ_d and χ_s. The dashed line shows the allowed region $x_s/x_d > 5$ predicted by the SM.

Thus a determination of x_s would be a crucial test. Information on x_s can be obtained from the measurements of $B\bar{B}$ mixing at high energies, where a mixture of all beauty hadrons, including B_d^0 and B_s^0 mesons, is produced. Usually one assumes that $\chi = 0.375\chi_d + 0.15\chi_s$, where χ_d and χ_s represent the mixing parameters for B_d^0 and B_s^0 mesons respectively.

The latest measurements of $B\bar{B}$ mixing at high energies [3] are summarized in Table 11. Combining these results with those from ARGUS and CLEO on χ_d, one obtains a small allowed region in the χ_s-χ_d plane (see Figure 19), which is consistent with the practically complete $B_s^0\overline{B}_s^0$ mixing predicted by the SM. Using the usual assumptions on B_d^0 and B_s^0 production fractions one gets [3]:

$$\chi_s = 0.47 \pm 0.11.$$

5. A Search for $b \rightarrow s$ Transitions

The transitions $b \rightarrow s\gamma$ and $b \rightarrow sg$ can also provide information on the $|V_{ts}|$ matrix element, as can be seen in Fig.20. Recently CLEO [61] obtained new limits on decays induced by such transitions. These are collected in Table 12. Theoretical estimates for exclusive channels are model dependent. For example, predictions of $BR(B^0 \rightarrow$

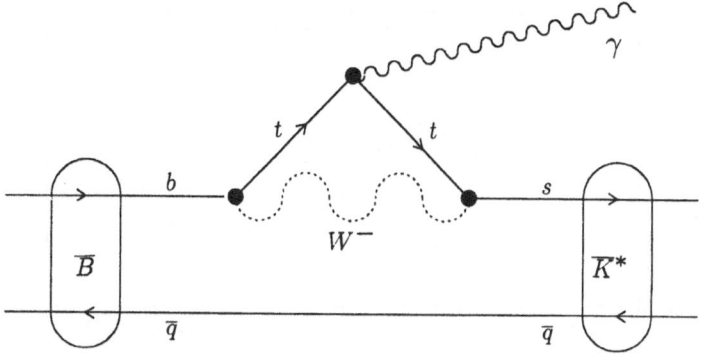

Figure 20. One of the diagrams for $\overline{B} \to \overline{K}^*\gamma$ decay.

Table 12. Upper limits on two-body charmless B decays .

Final state	BR $\cdot 10^4$ (90% CL)
$K^{*0}(892)\gamma$	0.7
$K^{*+}(892)\gamma$	2.8
$K^+\pi^-$	0.67
$K^+\pi^0$	0.49
$\pi^+\pi^-$	0.41
$\pi^+\pi^0$	0.54
$\pi^+\rho^0$	0.71
$K^+\phi^0$	0.67
$b \to s\gamma$	8.4

$K^{*0}(892)\gamma$) vary from 4.5% to 40% for $BR(b \to s\gamma)$ [62]. The inclusive branching ratio is estimated with less ambiguity to be about 4×10^{-4} for $m_t = 120$ GeV/c^2 [63, 62]. Thus CLEO may be on the brink of discovering $b \to s$ transitions.

6. Unitarity Triangle

Unitarity of the CKM matrix requires

$$V_{ub}^* + V_{td} - \sin\theta_C V_{cb} \approx 0.$$

Constraints on the sides of the unitarity triangle come from measurements of $B^0\overline{B}^0$ mixing (V_{td}/V_{cb}), charmless semileptonic B decays (V_{ub}/V_{cb}), and CP violation in K^0 decays. These constraints depend on the t quark mass, the decay constant of the B meson, and the bag parameter B_K. The shape of the unitarity triangle would be practically fixed, or even over-constrained, if these parameters were known precisely. Unfortunately this is not the case. Taking $B_K = 0.8 \pm 0.1$, $|V_{ub}| / |V_{cb}| = 0.062 \pm 0.018$ (derived using ACCMM model [51]), and $m_t = 137 \pm 39$ GeV/c^2 from a radiative correction analysis, Schmidtler and Schubert [64] performed a fit for different fixed values of f_B. The results of this fit are shown in Fig.21 where ρ and η are the Wolfenstein

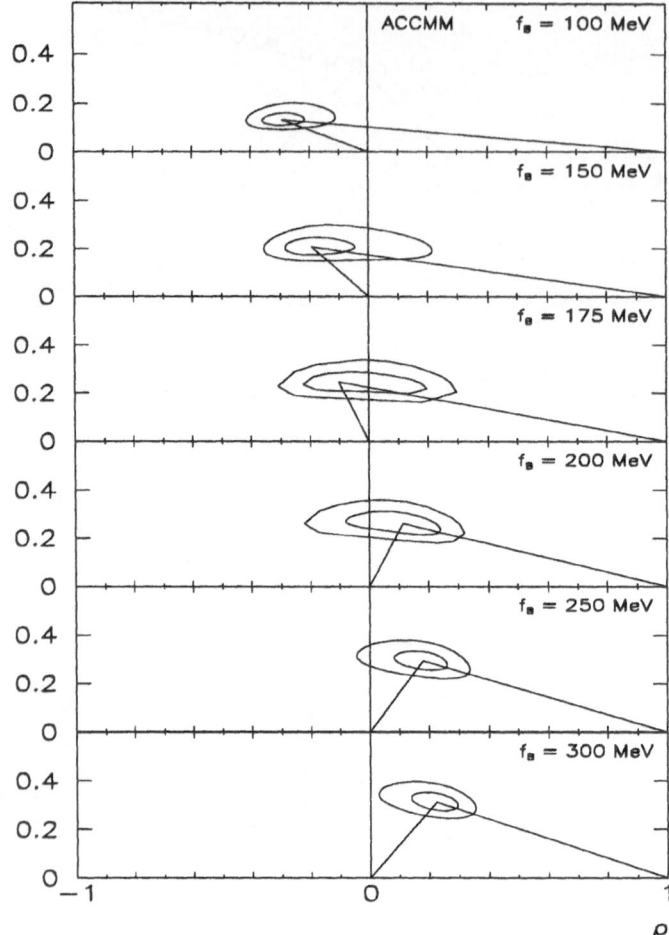

Figure 21. Unitarity triangle.

parameters [65] defined using the PDG parameterization of the CKM matrix [19] as

$$\rho = s_{13} \cos \delta_{13}/s_{12}s_{23},$$

$$\eta = s_{13} \sin \delta_{13}/s_{12}s_{23}.$$

The vertex of the triangle lies in the second quadrant for the small values of f_B long preferred by theorists. However, for the larger values of f_B obtained from recent lattice calculations [66] the vertex moves into the first quadrant. This would simplify a study of CP violation at B factories [25]. Values of $|V_{ub}|/|V_{cb}|$ derived using other models lead to somewhat different allowed regions in the $\rho - \eta$ plane. However, the general pattern remains unchanged.

Our knowledge of the fundamental parameters of the SM describing quark mixing is limited not by experimental errors but by theoretical uncertainties in the estimation of hadronic interactions and the unknown top quark mass. Recent developments in HQET, lattice calculations and other theoretical approaches offer some future prospect for a model independent extraction of these parameters from the experimental data.

Up to now all results on beauty particles are well described by the SM. Measurements of B_s mixing, $b \to s\gamma$ transitions and more precise determinations of $|V_{ub}|$, $|V_{cb}|$ and $|V_{td}|$ will provide important tests of it. Measurements of CP violation in B meson decays will test the internal consistency of the present theory.

Acknowledgements

I am especially indebted to I.Belyaev for giving me the most valuable help in preparation of this paper. I would like to acknowledge helpful discussions with P.Pakhlov on the determination of $|V_{cb}|$, and to thank P.Saull for help with the English. Finally, I wish to express my deep appreciation to the organizers of this wonderful school, especially to T.Ferbel.

References

[1] E.Fernandez *et al.*,(MAC Collaboration), *Phys.Rev.Lett.* **51** (1983) 1022.

[2] N.S.Lockyer *et al.*,(MARK II Collaboration), Collaboration,*Phys.Rev.Lett. 51* (1983) 1316.

[3] P.Drell,*Plenary Talk at the XXVI International Conference on High Energy Physics, Dallas*, 1992.

[4] W.B.Atwood,J.A.Jones,1992, in *S.Stone (editor) "B decays"*,261, World Scientific Publishing Co.Pte.Ltd. (Singapore, New Jersey, London, Hong Kong).

[5] S.R.Wagner *et al.*, (MARK II Collaboration), *Phys.Rev.Lett.* **64** (1990) 1095.

[6] V.Sharma, *Talk at the XXVI International Conference on High Energy Physics, Dallas*, 1992.

[7] M.Feindt, *Talk at the XXVI International Conference on High Energy Physics, Dallas*, 1992.

[8] D.S.Akerib *et al.*, (CLEO Collaboration), *Contributed paper to XXVI International Conference on High Energy Physics, Dallas*, 1992.

[9] D.Atwood and W.J.Marciano, *Phys.Rev.* **D41** (1990) 1736.

[10] G.P.Lepage, *Phys.Rev.* **D42** (1990) 3251.

[11] N.Byers and E.Eichten, *Phys.Rev.* **D42** (1990) 3885.

[12] H.Albrecht *et al.*,(ARGUS Collaboration), *Z.Phys.* **C48** (1990) 543.

[13] H.Albrecht *et al.*, (ARGUS Collaboration), *Preprint DESY* 91-121 (1991).

[14] K.Berkelman and S.Stone, *Preprint CLNS* 91/1089, 1991.

[15] D.Bortoletto *et al.*, (CLEO Collaboration) *Preprint CLNS* 91/1102 (1991).

[16] J.Adler et al., (MARK III Collaboration) *Phys.Rev.Lett.* **60** (1988) 89;

J.Adler et al., (MARK III Collaboration) *Phys.Lett.* **B208** (1988) 152.

[17] F.Butler et al., (CLEO Collaboration), *Phys.Rev.Lett.* **69** (1992) 201.

[18] D.Coffman et al., (MARK III Collaboration), *Phys.Rev.Lett.* **68** (1992) 282.

[19] Particle Data Group, *Phys. Lett.* **B239** (1990) .

[20] M.Bauer, B.Stech and M.Wirbel, *Z.Phys.* **C34** (1987) 103.

[21] H.Albrecht et al., (ARGUS Collaboration), *Phys.Lett.* **B277**, (1992) 209.

[22] CLEO Collaboration,*Contributed paper to the LP-HEP Conference*, Geneva, 1991.

[23] P.H.Cox, S.Hovater, and S.T.Jones, *Phys.Rev.* **D32** (1985) 1157.
J.H.Kühn and R.Rückl, *Phys.Lett.* **135B** (1984) 477.
N.G.Deshpande and J.Trampetic, *Phys.Rev.* **D41** (1990) 986.
J.H.Kühn, S.Nussinov and R.Rückl, *Z.Phys.* **C5** (1980) 117.

[24] K.Abe et al., KEK Report 92-3,1992.

[25] See, e.g., *SLAC Workshop on B factory, Preprint* SLAC-373(1991).

[26] H.Schröder, *Int. Conf. on High Energy Physics*, Singapore, 1990.

[27] A.Bean et al., (CLEO Collaboration), *Contributed paper to XXVI International Conference on High Energy Physics, Dallas*, 1992.

[28] H.Albrecht et al. (ARGUS Collaboration), *Phys.Lett* **B197** (1987) 452.

[29] H.Albrecht et al. (ARGUS Collaboration), Preprint DESY-92-146

[30] B.Grinstein, N.Isgur, D.Scora, M.B.Wise, *Phys.Rev.* **D39** (1989) 799.

[31] S.Balk, F.Hussain, J.G.Körner, G.Thompson, *Preprint MZ-TH-92-22* (1992).

[32] J.G.Körner, G.A.Schuler, *Z.Phys.* **C38** (1988) 511.

[33] H.Albrecht et al. (ARGUS Collaboration), *Phys.Lett* **B219** (1989) 121.

[34] H.Yamamoto,*Talk at the San Miniato Conference*, 1992.

[35] M.Danilov,*Talk at the 1992 Recontre de Physique de la Vallée d'Aoste at La Thuile.*
Yu.Zaitsev,*Talk at the XXVI International Conference on High Energy Physics, Dallas*, 1992.

[36] H.Albrecht et al.,(ARGUS collaboration), Phys.Lett. **B249** (1990) 359.
S.Henderson et al., (CLEO collaboration), Phys.Rev. **D45** (1992) 2212.

[37] M.Wirbel, B.Shtech and M.Bauer, *Z.Phys.* **C29** (1985) 637.

[38] F.Schöbert and H.Pietschmann, *Europhys. Lett.* **2**, 583 (1986) .

[39] J.S.Korner and G.A.Shuler, *Z.Phys.* **C38** (1988) 511.

[40] J.M.Cline, W.F.Palmer and G.Kramer, *Phys.Rev.* **D40** (1989) 793.

[41] M.Voloshin and M.Shifman, *Sov.J.Nucl.Phys.* **45** (1987) 292 and *Sov.J.Nucl.Phys.* **47** (1988) 511.

[42] E.Eichten and B.Hill, *Phys.Lett.* **B234** (1990) 511;
H.Politzer and M.Wise, *Phys.Lett.* **B206** (1988) 681 and *Phys.Lett.* **B208** (1988) 504;
N.Isgur and M.Wise, *Phys.Lett.* **B232** (1989) 113 and *Phys.Lett.* **B237** (1990) 527;
B.Grinstein, *Nucl.Phys.* **B339** (1990) 253;
H.Georgi, *Phys.Lett.* **B240** (1990) 447;
J.D.Bjorken, *SLAC preprint*,SLAC-PUP-5278, 1990;
A.Falk, H.Georgi, B.Grinstein and M.Wise, *Nucl.Phys.* **B343** (1990) 1;
A.Falk and B.Grinstein, *Phys.Lett.* **B247** (1990) 406;
T.Mannel, *Talk at the LP-HEP Conference*, Geneva, 1991;
M.Neubert, *Talk at the LP-HEP Conference*, Geneva, 1991;

[43] M.Neubert, *Phys. Lett.* **B264** (1991) 455.

[44] H.Albrecht *et al.*, (ARGUS Collaboration), *Phys.Lett.* **B275** (1992) 195.

[45] E. de Rafael, J.Taron, *Phys.Lett.* **B282** (1992) 215.

[46] G.Altarelli and S.Petrarka, *Phys.Lett.* **B263** (1991) 325.

[47] P.Roudeau , *Plenary talk at LP-HEP Conference*, Geneva, 1991.

[48] H.Albrecht *et al.* (ARGUS Collaboration), *Phys.Lett* **B249** (1990) 359.

[49] H.Albrecht et al.(ARGUS Collaboration), *Phys.Lett.* **B234** (1990) 409;
H.Albrecht et al. (ARGUS Collaboration), *Phys.Lett.* **B255** (1991) 297.

[50] R.Fulton *et al.,* (CLEO Collaboration), *Phys.Rev.Lett.* **64** (1990) 16.

[51] G.Altarelli, N.Cabibbo, G.Corbo, L.Maiani and G.Martinelli, *Nucl.Phys.* **B208** (1982) 365.

[52] See for example M.B.Voloshin, M.A.Shifman, *JETP* **91** (1986) 1180,
M.B.Voloshin *et al.*, *Yad.Fiz.* **46** (1987) 181.

[53] A.J.Buras, M.Jamin and P.H.Weisz, *Nucl.Phys.* **B347** (1990) 491.

[54] M.I.Vysotsky, *Yad.Fiz.* **31** (1980) 1535.

[55] H.Albrecht *et al.*, (ARGUS Collaboration), *Phys.Lett.* **B192** (1987) 245.

[56] M.Artuso *et al.,* (CLEO Collaboration), *Phys.Rev.Lett.* **63** (1989) 2233.

[57] H.Albrecht *et al.* (ARGUS Collaboration), *Z.Phys.* **C55** (1992) 357.

[58] D.Acosta *et al.*,(CLEO Collaboration), *Contributed paper to XXVI International Conference on High Energy Physics, Dallas*, 1992.

[59] M.B.Voloshin, M.A.Shifman,*Sov.Phys. JEPT* **64** (1986) 698.

[60] S.Henderson *et al.*,(CLEO Collaboration), *Preprint CLNS* 91/1101, 1991.

[61] H.Yamamoto, *Talk at the SSCL*, July, 1992.

[62] A.Ali, T.Mannel, *Phys.Lett.* **B264** (1991) 447 and references therein.

[63] G.Cella *et al.*, *Phys.Lett.* **B248** (1990) 181.

[64] M.Schmidtler and K.R.Schubert, *Z.Phys.* **C53** (1992) 347 , and private communications.

[65] L.Wolfenstein, *Phys.Rev.Lett.* **51** (1983) 1945.

[66] For a recent review see, e.g., L.Maiani, *Rome preprint* n.798,(1991).

THE SEARCH FOR THE TOP QUARK
AT THE TEVATRON

Paul L. Tipton
University of Rochester
Rochester, NY 14627 USA

ABSTRACT

This paper reviews the prospects for discovering the t-quark at Fermilab's Tevatron $\bar{p}p$ Collider. With a wealth of experimental evidence in agreement that the top quark is heavy, above 91 GeV/c², it appears that Fermilab is the only facility able to produce t-quarks until the LHC and SSC era begins. The search techniques expected to be used by D0 and CDF until that time are summarized. Testing the Standard Model with a t-quark mass measurement, as well as other Electroweak parameters is also discussed.

INTRODUCTION

Since the discovery of the b-quark, a major goal in High Energy Physics has been to observe its weak isospin partner, the t-quark. Without the t-quark, flavor-changing neutral current (FCNC) decays of the b-quark would not be suppressed. FCNC decays are not observed in B meson decay, [1] thus ruling out most (if not all) models which do not contain a t-quark. In addition, LEP measurements of Electroweak parameters, when combined with measurements of the W mass performed at CDF and UA2, can be combined with radiative corrections to the vector boson masses. These data require a t-quark - b-quark mass difference which is not infinite, [2] thus implying the existence of a t-quark. Unless one appeals to new physics beyond the minimal Standard Model, the top quark must exist.

Experiments at PEP and Petra, Tristan, SLC and LEP have searched for the $t\bar{t}$ pairs produced in e^+e^- collisions. The highest upper limit from these experiments comes from LEP and is 45.8 GeV/c², [3] This result is independent of the decay mode of the t-quark. In $\bar{p}p$ collisions, $t\bar{t}$ pairs are produced through gluon fusion or quark annihilation. The cross section for $t\bar{t}$ production at the Tevatron energy of

$\sqrt{(s)} = 1.8$ TeV ranges from about 80 pb^{-1} (if the top mass is 100 GeV/c²) to about 1 pb^{-1} if the top mass is 220 GeV/c².[4]

In the minimal Standard Model the t-quark in a T meson will decay by the process $t \rightarrow Wb$. Since a W decays to a μ or an e 1/9 of the time, a $t\bar{t}$ event will have at least one muon or electron about 40% of the time. Without at least one semileptonic decay, the $t\bar{t}$ signal is lost in the QCD multijet background. The top events in which both t and $\bar{t}$ decay to e or μ are particularly well separated from background, with a signal-to-background ratio (S/B) in excess of one. When only one of the t-quarks decays semileptonically there is a sizable background from QCD produced W's recoiling against significant jet activity (W+jets). This background can be suppressed using some sort of a b-quark tag to select out the $t\bar{t}$ events, which have 2 b-quarks, from the W+jets events, which in the overwhelming majority of events will have no b-quarks. In the 1988-89 Tevatron Collider run, CDF searched for a Standard Model decay of the t-quark by looking for dilepton events ($ee, e\mu$, and $\mu\mu$). In addition, CDF looked for evidence of b-quark decays into μ in the inclusive high P_T lepton sample. There is one event in the signal region. If one assumes it is signal, a lower limit of 91 GeV/c²is set on the t-quark mass at the 95 % confidence level.

THE 1992-94 TEVATRON RUN

The CDF collaboration has significantly improved their detector since the 1989-1990 run. The upgrades to the muon system and the addition of a silicon vertex detector (SVX) are particularly relevant for the t-quark search. The muon system has been upgraded in two ways. First the central muon system was upgraded by the addition of a steel wall of over 2 hadronic absorption lengths and additional layers of muon chambers outside the previous muon system. This increases the total number of absorption lengths to over 8 in the region $| \eta | < 0.4$. Secondly, the central muon system was extended to provide muon identification in the region $0.63 < | \eta | < 0.9$. The SVX consists of two barrels in the region $z < 26 \; cm$. There are four layers of single-sided silicon wafers per barrel, providing precision $r\phi$ tracking information. The layers are arranged from $r = 3$ to $8 \; cm$. This covers a region out to $| \eta | < 1.9$ with all for layers. The impact parameter resolution of the device is better than 40 μm for tracks with $P_T > 1$ GeV/c. Using preliminary alignment constants, we obtain an impact parameter resolution of $\sim 13 \; \mu m$ for tracks with $P_T > 10 GeV/c$.

The D0 detector is commissioned and taking data. It is a non-magnetic detector designed to maximize hermeticity. The calorimeters are constructed primarily of depleted uranium with liquid argon as the ionizing medium. The electromagnetic calorimeter has over 20 radiation lengths and the hadronic calorimeter has over 6 interaction lengths. The central tracking system contains TRD's which are used for π/e discrimination. Outside the cryostats there is an additional ≈ 7 interaction lengths of instrumented magnetized iron used for muon detection and momentum determination.

The commissioning and turn-on of the D0 detector has gone remarkably well. Unfortunately for the D0 collaboration, the main ring beam pipe passes through the D0 detector. Currently, losses from the main ring make it impossible for D0 to take data during $\approx 20\%$ of the available Tevatron running. If this problem continues, one expects D0's integrated luminosity to be lower than that of CDF's by this fraction.

THE SEARCH FOR TOP USING DILEPTONS

As previously stated, top events in which both t and $\bar{t}$ decay to e or μ are rather clean with S/B is ≈ 1 or better. Both D0 and CDF will attempt to observe an excess over background of high P_t ($P_t > 20 GeV/c$) dileptons events. This sample should be clean enough to use as a simple counting experiment either for a discovery or a t-quark mass limit. If the t-quark mass is very heavy, above ≈ 160 GeV/c²,then dilepton backgrounds could overwhelm the signal. Fortunately by that mass the top events have jet activity from the b-quarks and from initial and final state radiation. One can require two leptons and additionally require two energetic jets and still be reasonably efficient on double semileptonic top decays. The background should be substantially reduced by this requirement preserving a good Signal-to-background ratio.

No D0 estimates of the yield of $t\bar{t}$ into dileptons for various t-quark masses are available at this time. The greater lepton coverage of the D0 detector should give a higher acceptance than the CDF detector. At higher t-quark masses the leptons are more centrally produced and the D0 advantage is minimized. A very preliminary D0 analysis on the first 0.9 pb^{-1} estimates that for a t-quark mass of 80 GeV/c² the observed cross section for $t\bar{t}$ into ee ($e\mu$) is similar to, but slightly larger than, the expected cross sections from the 1989 CDF analysis. [5]

The current Tevatron run has a 4 month shutdown planned for the summer of 1993. The luminosity goal is to have 25 pb^{-1} delivered to each experiment before the shutdown and 100 pb^{-1} delivered over the whole run, which should end sometime in 1994. CDF has set a goal of writing in excess of 80% of delivered luminosity to tape. So far in the run CDF is averaging slightly below this.

CDF has not specified what the expected rates for dileptons from $t\bar{t}$ events will be for the current run. We assume the rate will be the same as stated in the top search from the 1989 CDF data. This assumes the slight increase in efficiency with t-quark mass will be offset by tighter selection cuts to reduce background rates. Under these assumptions, the CDF expects 29,13,6 observed dilepton events from a top mass of 120, 140, 160 GeV/c² in 80 pb^{-1} *on tape.*

THE SEARCH FOR TOP IN THE LEPTON + JETS SAMPLE

We turn now to the search for $t\bar{t}$ events in which only one of the t-quarks decays semileptonically to an e or a μ. This decay mode has two advantages, the first being higher rate. The second is that one of the t-quarks decays hadronically with no neutrino, and is therefore (in principle) fully reconstructable. A signal in both single and dilepton samples would be reassuring evidence for a t-quark discovery. The signature for one t-quark decaying hadronically and the other decaying hadronically is a high P_T lepton, significant missing E_T, and significant jet activity. There is sizable background in an event sample passing these criteria, (the so called leptons + jets sample) from QCD produced W's recoiling against significant jet activity (W+jets). As stated previously, the W+jets background can be suppressed using a b-quark tag. The $t\bar{t}$ events will have 2 b-quarks and the W+jets events will have, in the overwhelming majority, no b-quarks.

Two techniques for b-tagging will be used by CDF. The first, tagging the b-quark decay via its semileptonic decay to a *soft* lepton, will presumably also be used

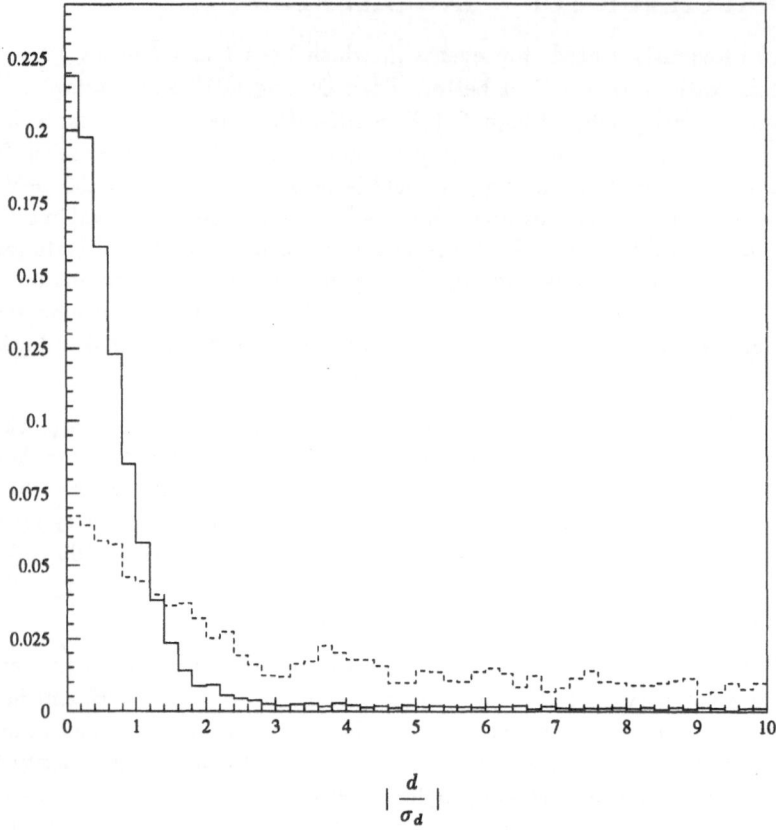

$$| \frac{d}{\sigma_d} |$$

Figure 1. Impact Parameter Significance S_d for t-quark mass = 120 GeV/c^2 Monte Carlo. The Dashed line is for tracks from b-flavored hadrons. The Solid line is for all other tracks. Distributions are normalized to area of 1.0.

by D0. CDF will use muons and electrons in the range 2 to 15 GeV/c. (The upper limit is imposed to keep the soft lepton search and the dilepton search from counting the same events.) Recall that CDF used a soft muon search in the 1989 data to help set the 91 GeV/c²top mass limit. CDF's extended muon system will improve the soft muon efficiency and the addition of a pre-radiator will help pion rejection for soft electron identification. Although the efficiency is a function of the t-quark mass, for masses between 100 and 150 GeV/c², the CDF soft lepton search should tag $\approx 25\%$ of $t\bar{t}$ events. Expected yields are shown in figure 3. D0's soft lepton search will most likely be limited to muons and will start at slightly higher muon P_T since the D0 hadron absorber is thicker.

The SVX detector gives CDF a unique way of tagging b-quarks, and thereby searching for top, through the measurement of impact parameters of daughter charged particles from B meson and baryon decays. Since the mean charged multiplicity of B meson decays is ≈ 5, the signature searched for is many (typically three or more) charged tracks which have significantly displaced impact parameters relative to the primary event vertex. These tracks will be correlated in space such that they are consistent with coming from a B hadron of ≥ 5 GeV/c². A natural variable to use in this type of an analysis is the impact parameter significance (S_d), which is the absolute value of the impact parameter d divided by σ_d, the total error in d. The

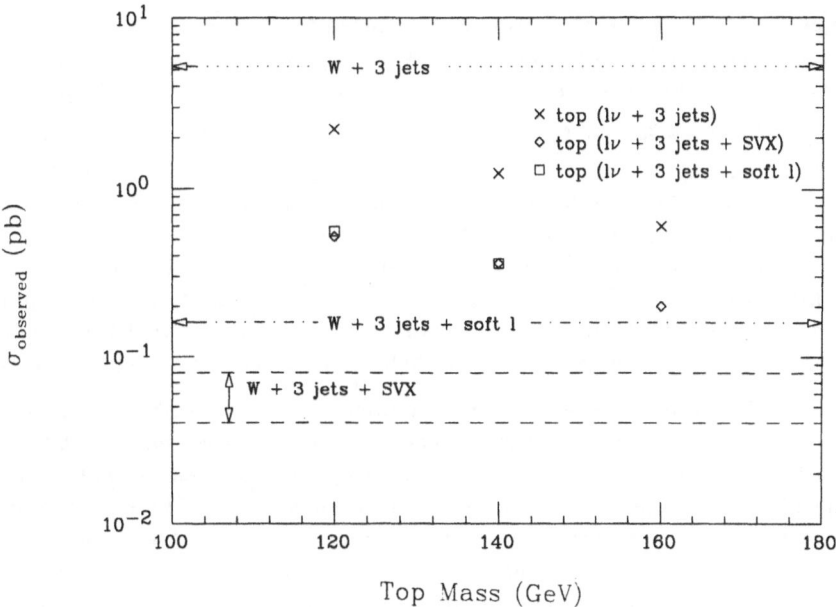

Figure 2. Expected observed cross sections for $t\bar{t}$ and W+jets using the high P_T lepton + 3 jets signature. The crosses are for top, before a b tag is applied. This should be compared to the dotted line for W+jets background. The diamonds are for top, after requiring an SVX b tag. Dashed lines bound the rate of background from W+jets+SVX. The squares are for top, after requiring a soft lepton b tag. The dot-dashed line is background from W+jets+soft lepton.

S_d distribution for tracks which are daughters of b-flavored hadrons falls slowly with increasing S_d. In contrast, S_d is sharply peaked towards zero for tracks which are not from b. The distributions from top MC with $m_{top} = 120$ GeV/c^2 are shown in fig. 1.

CDF has developed a few tagging techniques using the SVX for use in the top search. Significantly displaced tracks are selected by requiring a minimum S_d for each track. Minimum requirements on P_T are also made to reject very low momentum tracks which typically have long tails on the impact parameter distribution, and which preferentially come from the underlying event. Displaced tracks in an event are then grouped into candidate b tags by vertexing the tracks or by looking for correlations in $\eta\phi$ space. In the tagging method using vertexing, termed the jet vertexing method, displaced tracks are assigned to jets if the track is within 25° of the jet axis. The displaced tracks in a given jet are then vertex constrained. Tracks which contribute more than 20 to the total χ^2 of the vertex constraint are discarded and the vertex is refit without them. Jets with at least 2 tracks which form a vertex 5σ from the primary event vertex are tagged as b-jet candidates. Two other b-tagging algorithms try to exploit the $\eta\phi$ correlations without the use of vertex constants. The first simply requires that the displaced tracks be within a cone of 0.4 in $\eta\phi$ space. Cones with at least three displaced tracks are tagged as b-jet candidates. The second method, called the $d\phi$ clustering algorithm, relies on the fact that tracks from a secondary vertex form a sinusoid in $d\phi$ space, where d is the signed impact parameter, and ϕ is the usual azimuthal angle in the transverse plane. For forward boosted decays, we can approximate the relevant portion of the sinusoid with a straight line. Secondary vertices are found by linking displaced tracks with straight lines in $d\phi$ space. Events with at least three linked tracks are tagged as containing evidence for b-quark decay.

The event efficiencies for the various b tagging methods are quite similar and rise with increasing top mass. For $m_{top} = 120$ GeV/c^2, efficiencies of about 23% are obtained. For $m_{top} = 160$ GeV/c^2, the efficiency rises to $\sim 33\%$. The expected background rates from false tags in W+jets events are also similar at $\sim 0.7\%$ per event. This does not include b tags due to real c and b quarks in W+jets events. One of the first tasks for CDF will be to measure the rate at which gluon jets split into heavy quark pairs. The rate is predicted to be small enough so that this irreducible background does not dominate the top signal, even for very heavy t-quark masses.[6]

In a data sample of 80 pb^{-1}, the number of expected events using the SVX tag is ~ 42 for $m_{top} = 120$ GeV/c^2, falling to ~ 16 events for $m_{top} = 160$ GeV/c^2. The background from W+jets is ~ 6 events, which includes an estimate (based on reference 6) of tags due to real c and b in the W+jets events. The expected number of top events using the soft lepton tag is comparable to that given by the SVX tag. However, the expected background for this mode is ~ 13 events. The expected observed cross sections for top and background are shown in figure 2, where the event selection criteria requires at least 3 jets with $E_T > 10$ GeV, two of which must also have $E_T > 15$ GeV.

To conclude, assuming CDF and D0 collect ≈ 80 pb^{-1} in the 1992-1994 Tevatron run, a Standard Model top of less than ≈ 160 GeV/c^2 should produce at least 5 dilepton events on a reasonably small background, and ≈ 16 SVX b-tags, also on a small background.

BEYOND THE CURRENT TEVATRON RUN: PROSPECTS FOR TESTING THE STANDARD MODEL

If the t-quark is not discovered this Tevatron run, obviously the search will continue in future runs, and into the Main Injector era if necessary. One might ask if it is possible that a Standard Model t-quark could remain undiscovered until the LHC/SSC era. Current Electroweak parameters place an upper limit on the t-quark mass of roughly 225 GeV/c^2. The main injector will allow a luminosity upgrade such that 500 pb^{-1} can be delivered each calendar year. In a $1 fb^{-1}$ dataset, CDF expects 5 $e\mu$ events observed from a t-quark of 245 GeV/c^2. It seems likely then, that if the t-quark exists and decays in accord with the Standard Model, it will be discovered at Fermilab.

If the t-quark is discovered, focus will shift to a measure of its mass. Ultimately, a precision measure of the t-quark mass is desired to test the Standard Model. A convenient way of displaying the interdependence of the Boson masses, the t-quark mass, and the Higgs mass, is to look in the W mass-t-quark mass plane. Figure 3 shows such a plane, with the measured value of the W mass shown. Also shown are curves of allowed points which represent the minimal Standard Model for various assumptions of the Higgs mass, from $M_{Higgs} = 25$GeV/c^2 (top dotted curve) to $M_{Higgs} = 1000$ GeV/c^2 (lower dashed curve).

CDF estimates that its ultimate precision on the W mass will be about 50 MeV/c^2. To measure the t-quark mass, one looks at the leptons + jets events, where one t-quark has decayed semileptonically, the other hadronically. The W mass can

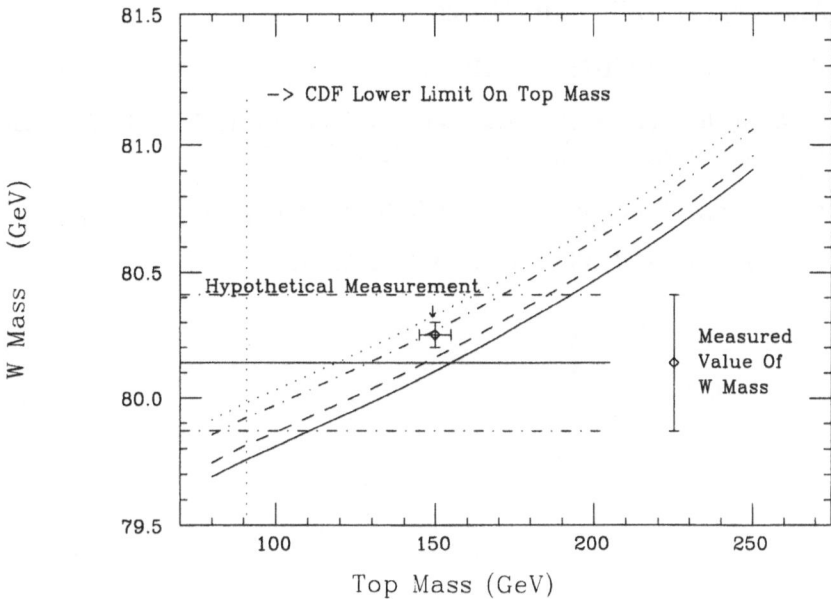

Figure 3. Standard Model Constraints on the W mass *vs.* *t*-quark mass plane. The Higgs masses used are 25, 100, 500, 1000 GeV/c², respectively from the upper to the lower curves. Also shown is the current measurement of the W mass and a hypothetical point with precision estimated for a main injector era dataset.

be imposed as a constraint. The resulting three-jet invariant mass distributions are broad, $\approx$ 20 GeV/c², but with high statistics the mean should be well determined. An estimate of the systematic uncertainties from the energy scale and backgrounds indicate that a 5 GeV/c² uncertainty on the *t*-quark mass is possible with 1 fb^{-1} for a *t*-quark mass up to 150 GeV/c². A hypothetical measurement of the W and *t*-quark masses with this precision is also shown in Figure 3. Such a measurement would likely give some useful information about the Higgs sector before the LHC and SSC turn on.

CONCLUSIONS

The Tevatron run currently in progress holds great promise for discovering or limiting the existence of a Standard Model *t*-quark in the mass range below $\approx$ 160 GeV/c². CDF and D0 should be able to cover the top mass range allowed by current measurements of Electroweak parameters in the main injector era. A "precision" mass determination will be possible with the use of *b*-quark tags in the single semileptonic event sample, where one *t*-quark can be fully reconstructed. Ultimate resolution of 5 GeV/c² is possible on the *t*-quark mass, up to a mass of 150 GeV/c². This should allow a glimpse into the Higgs sector before the LHC and SSC era.

REFERENCES

1. A. Bean *et al.* (CLEO Collaboration), Phys. Rev. **D35**, 3533 (1987).

2. Altarelli *et al.*, CERN-TH 6525-92.

3. Decamp *et al.*, PL B236, 511 (1990)

4. G. Altarelli *et al.*, Nucl. Phys. **B308**, 724 (1988); R. K. Ellis, *Fermilab-Pub-91/30T*; P. Nason *et al.*, Nucl. Phys. **B303**, 607 (1988).

5. R. Partridge, Proceedings of the APS DPF Meeting, Fermilab, Nov., 1992.

6. M. Mangano *et al.*, Phys. Lett. **B285** 160, 1992.

DETECTION OF DARK MATTER AND SOLAR NEUTRINOS

Michel Spiro

DAPNIA, Service de Physique des Particules
CE SACLAY
F-91191 Gif-sur-Yvette Cedex, France

INTRODUCTION

The estimate of the value of Ω, the ratio of the mean energy density in the universe to the critical energy density, is one of the main issue in modern cosmology. We can measure the components of Ω in various ways:
- from luminous matter (stars), Ω_{lum}
- from the dynamical behaviour of stars in spiral galaxies (galactic halos), Ω_{halo}
- from primordial nucleosynthesis (baryons), Ω_{bar}. We know from observations that the contributions from dust or gas to Ω are negligible. The estimates for the values of Ω_{lum}, Ω_{halo} and Ω_{bar} are shown in fig.1 and compared to the magic value $\Omega = 1$ which is the preferred value for aesthetical and theoretical reasons (to avoid fine tuning in initial conditions, and to agree with inflation theories). From all these values, one can draw two main conclusions:

1) All these estimates are below 1. However, the value $\Omega = 1$ is not at all excluded. The allowed range for the total contribution to Ω is $0.001 < \Omega_{tot} < 2$ where the lower limit comes from visible matter and the upper limit comes from the minimum estimate of the age of the universe combined with the measured expansion rate. To reach $\Omega = 1$, it seems unavoidable to invoke intergalactic non baryonic dark matter, such as:

WIMPs (Weakly Interacting Massive Particles: heavy neutrinos ν_h, or lightest supersymmetric particle LSP...)

or Light Neutrinos such as 30 to 100 eV ν_e or ν_μ or ν_τ (a light LSP could also be in principle possible).

2) The comparison of the allowed range for Ω_{lum}, Ω_{halo} and Ω_{bar} suggests that the halos of spiral galaxies, like our own galaxy, could be partly or totally made of MACHOs (Massive Astrophysical Compact Halo Objects) which is almost the only possibility left for baryonic dark matter: these MACHOs could be either aborted stars (brown dwarves, planet like objects), or star remnants (white dwarves, neutron stars, black holes) [1].

LEP + p$\bar{\text{p}}$ COLLIDER CONSTRAINTS ON WIMPS CANDIDATES WITH $\Omega_{\text{WIMP}} = 1$

Heavy Dirac or Majorana Neutrino ν_h

The LEP results on the Z^0 width, combined with the upper limits on the ν_e, ν_μ and ν_τ masses (10 eV, 250 keV and 35 MeV) excludes the possibility of any Dirac or Majorana neutrino in the mass range 35 MeV to 40 GeV. In particular, this excludes (fig.2) the 3 to 7 GeV range, which would naturally give $\Omega_{\nu h} = 1$ from the relic density which can be reliably estimated for heavy neutrinos in standard cosmology.

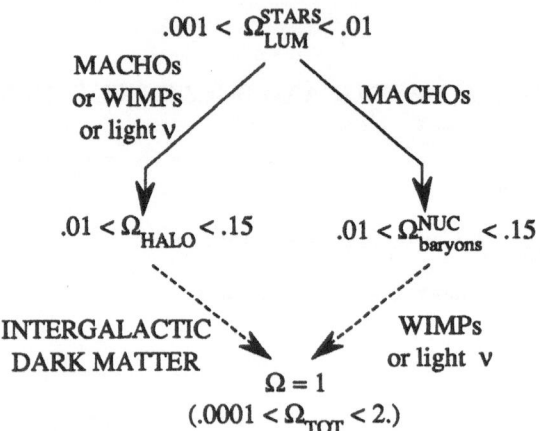

Figure 1. Present status of Ω determinations

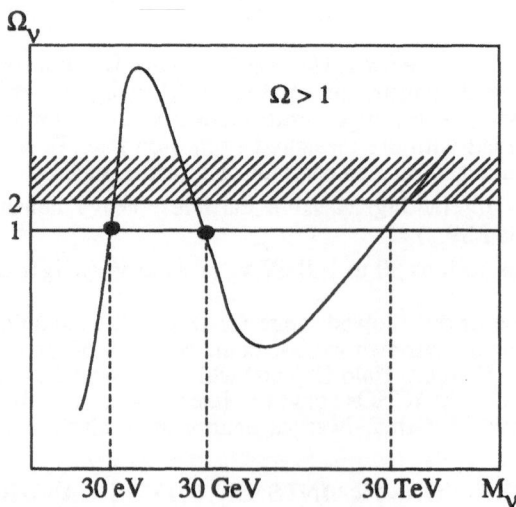

Figure 2. Ω_ν as a function of the Dirac ν mass

Lightest Supersymmetric Particle (LSP)

1) Boson

An illustrative candidate of that type is the sneutrino $\tilde{v}$. The contribution of a $\tilde{v}$ to the width of the Z^0 is half that of a standard neutrino. Again, from the LEP results one can exclude a $\tilde{v}$ with mass < 35 GeV at the 3σ level.

2) Fermion

The LSP could naturally be in that case the neutralino χ, a linear combination of the photino, the zino and the two higgsinos which are necessary in the minimal supersymmetric theory.

- Again, the residual cosmological density of neutralinos having survived annihilation can be reliably estimated. The decoupling time is determined by their low-energy annihilation cross section which depends mostly on the neutralino mass and the lightest scalar-fermion mass (fig. 3). From CDF ($p\bar{p}$) experiment), we know already that the sfermion masses are > 100 GeV. Demanding $\Omega_\chi = 1$, this implies that 15 GeV < M_χ < 1 TeV.

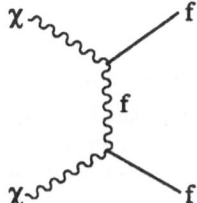

Figure 3. Annihilation diagram for neutralinos

- What do we learn from LEP? The Z^0 is not coupled to photino-photino, nor to Zino-Zino but does couple to two higgsinos. This implies that the higgsino masses are higher than 40 GeV although there are some loopholes in the derivation. Furthermore, the wino mass is greater than 40 GeV (again from LEP), which favours also a Zino mass greater than 40 GeV in any natural model. In summary LEP cuts already part of the parameter phase space for neutralinos below 30 GeV. This is bad news for direct detection experiments. As the neutralino masses get higher and higher, the rate of WIMPs scattering decreases.

- Furthermore, neutralinos can interact on quarks at low energy via two types of diagram, fig. 4 : the first one with squark exchange implies spin coupling and consequently no coherence factor, the second one via Z^0 exchange induces some coherence factor. Of course, this second diagram would be preferred by experimentalists due to the increase of the rate of elastic scattering. However, the chances that this process takes place are further reduced by LEP for < 30 GeV neutralinos.

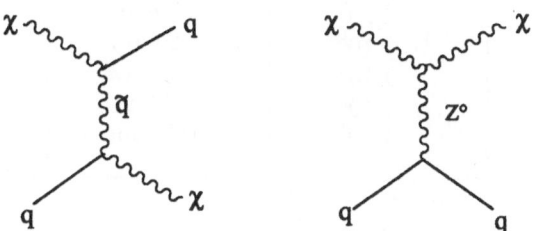

Figure 4. Elastic scattering diagrams for neutralinos

IS OUR GALACTIC HALO MADE OF MACHO's or WIMPs?

To answer this question, the idea [2] is to survey for several months, at 10% accuracy or better, the luminosity of a large number of stars in the Large Magellanic Cloud (LMC), see fig. 5, and search for stars that undergo a characteristic brightening due to a MACHO passing near the line of sight and producing a gravitational microlensing event: when a spatially-small massive object (a deflector D which is the MACHO) happens to lie near enough to the line of sight between a light source (a star S from the LMC) and an observer O, then the observer collects more light from the star than he would in the absence of this deflector.

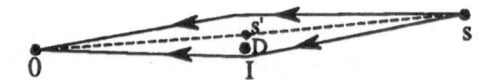

Figure 5. Microlensing of a LMC star (S) by a MACHO (D)

Of course, the observer, deflector and source stars are all moving (with typical galactocentric velocities of 200 km/s), so that the brightening of the star is time-dependent. The important characteristics of this light curve are its symmetry, unicity (the brightening occurs only once) and its achromaticity (the brightening is identical whatever the filter used by the observer).

The probability that a star undergoes a brightening due to microlensing with an amplification larger than 1.34 is simply the probability that the line OS crosses the zone of gravitational influence of area πR_0^2 associated to a deflector where $R_0^2 = 4GMd/c^2$ (M mass of the deflector) is the Einstein ring. For various values of deflector mass (MACHO masses), we give in table 1 the mean values of the excitation time (above 1.34 amplification) and the mean number of microlensing events expected for a survey of 10^6 stars in the LMC monitored during 10^7 seconds (4 months) [3].

Two groups , one in the US (Livermore, Berkeley), one in France (Saclay, Orsay, Paris, Marseille) are setting up experiments based on these principles. The experiments will be located in Chile and maybe also in Australia. This survey will be of interest for both particle physics and astrophysics. If MACHOs are indeed found, this will give a strong indication that non-baryonic dark matter is not clustered around galaxies, meaning that present searches for WIMPs are likely to be ineffective. On the other hand, if such compact objects are not found, then there would be an added incentive for all kind of WIMPs searches.

Table 1. Microlensing events characteristics

Deflector Mass ($M_\odot$)	Mean R_0 (km)	Mean μlensing time	NB of μlensing events
> 10	> 3 10^9	> 1 year	0.5
1	1 10^9	3 mths	1.0
10^{-2}	1 10^8	9 days	5
10^{-4}	1 10^7	1 days	50
10^{-6}	1 10^6	2 hrs	500
10^{-8}	1 10^5	12 mins	5000

DIRECT DETECTION

The hypothesis that dark matter particles are gravitationally trapped in the galaxy leads to the conclusion that, like stars, they should have a local Maxwell velocity distribution with a mean spread of 250 km/s. Then, the mean kinetic energy E_r received by a nucleus of mass M_n (in units of GeV) in an elastic collision with a dark matter particle of mass M_x is:

$$E_r = 2 \text{ keV } M_n M_x^2/(M_x+M_n)^2$$

The energy distribution is roughly exponential [4]. The expected event rate for elastic scattering on a given nucleus, assuming that 0.4 GeV/cm^3 is the local density of the halo (needed to account for the flat rotation curve of stars) depends only on the mass and interaction cross section.

Semiconductor Germanium and Silicon detectors have already set limits on cross sections and masses of dark matter particle [5]. Fig. 6 shows the exclusion plot for the mass and interaction cross sections of WIMPs obtained by these experiments.

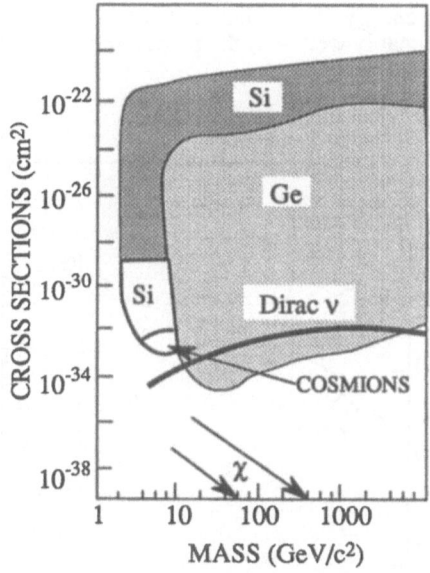

Figure 6. Exclusion zones for dark matter particles.

Dirac neutrinos with masses from 11 GeV to 3 TeV are excluded. This closes almost completely the remaining allowed window by cosmology (fig. 2).

Cosmions are particles which were invented to solve both the dark matter problem and the solar neutrino problem [6]. If they exist, their masses (4 to 10 GeV/c^2) and cross sections on hydrogen (few picobarns) are well predicted. A special silicon experiment [7] excludes nearly all of the mass range possible for cosmions with coherent nuclear interactions (fig. 7) [8].

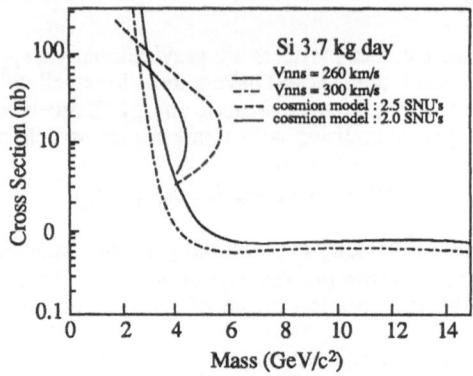

Figure 7. Exclusion plots for the silicon experiment [7]. Also shown are the expected curves where should lie the Cosmions with coherent scattering for a resulting neutrino flux of 2 and 2.5 SNU's [8].

These conclusions depend crucially on the knowledge of the response of the detector to nuclear recoils in the keV range; fig.8 shows the relative ionisation (silicon nucleus/electron) as a function of the kinetic energy. These data have been obtained by using pulsed neutron beams [9]. They agree reasonably well with the predictions of a statistical model by Linhard et al. (LSS theory [10]).

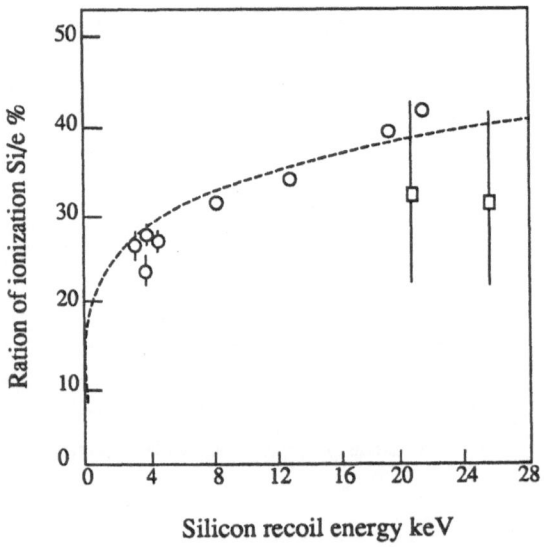

Figure 8. Ratio between the observed energy (equivalent electron energy) and the calculated recoil energy as a function of the Silicon recoil energy.

Finally neutralinos (fig.6) are presently out of reach (rates 2 to 3 order of magnitudes below the radioactivity background) from these conventional techniques. The recent success in measuring both the ionisation and heat from a silicon bolometer looks promising as a mean to reject background (fig.9).

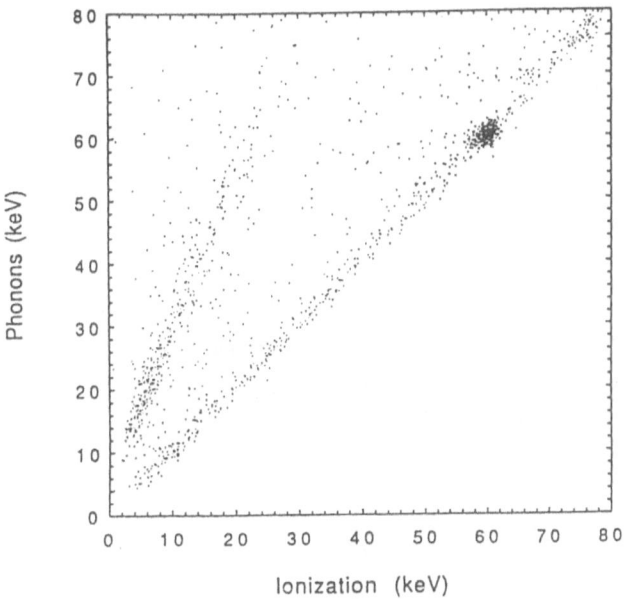

Figure 9. Observed events with Am/Be source, showing separation of photon and neutron events. (B.Sadoulet, private communication)

SEARCH FOR CHARGED DARK MATTER (CHAMPS)

A. de Rujula, S. Glashow and U. Sarid [11] have claimed that contrary to a common prejudice, dark matter halos could be made of charged particles C^+ and C^- (CHAMPS), provided that the mass of these particles is in the range 10 to 1000 TeV. Most of these particles should appear in form of superheavyhydrogen (C^+e^-) whose energy levels differ from ordinary hydrogen only by 1/2000 (relative difference in reduced mass) and in form of small neutral atoms (neutrachamps C^-p).

Search for C^+e^-

The predicted terrestrial abundance for the past 10^7 years is $2\ 10^{-12}$ (10 TeV/M). Note that below 10^4 TeV these particles should be constantly mixed by the currents in the ocean [12]. An experiment [13] designed to search for superheavy isotopes of hydrogen started in Paris two years ago. This search is based on the centrifugation mostly of sea water (efficient above 10 TeV), followed by atomic spectroscopy. It is then sensitive to masses of 10^4 TeV to 10^8 TeV. The laser spectroscopy is based on the 1S to 3S excitation. The results (fig. 10) were reported early this year [14]. The experiment provides now a limit at 90% confidence level for the relative abundance in sea-water of superheavy hydrogen (10 to 10^4 TeV) compared to ordinary hydrogen of few 10^{-15}. These results supplement those obtained by P. Smith et al.[15] from natural water by electrolysis and density measurement ($< 10^{-15}$) which are relevant however only if the evaporation-rain cycle would be the same for superheavy water as for light water.

A reported balloon experiment [16] using track etch detectors rule out the possibility that our halo is made of charged massive particles in the mass range 350 to $8.6\ 10^4$ TeV. All together these results rule out almost entirely the possibility that the halo of our galaxy could be made of CHAMPs.

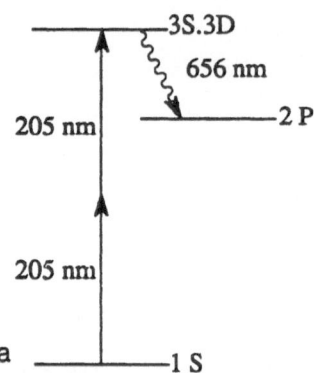

Figure 10a. Two photon excitation scheme.

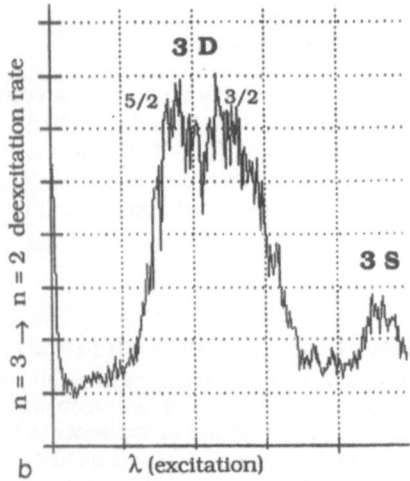

Figure 10b. Excitation curve in the deuterium region.

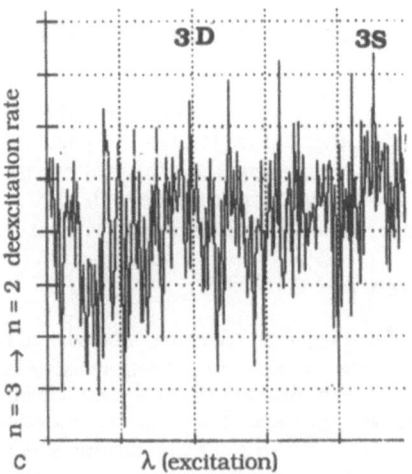

Figure 10c. Excitation curve in the region where the presence of superheavy hydrogen is expected.

Neutrachamps

The same previous experiment can also rule out masses of neutral C^-p composites between 100 and 4.10^4 TeV provided that they charge exchange with C or O nuclei with a cross section having a value in the interval 30mb to 30b. However recent calculations favour much higher cross sections (few thousand barns) [17].

THE CASE FOR LIGHT NEUTRINOS AND DETECTORS FOR THE COSMIC NEUTRINO BACKGROUND

Although they are not favoured by theories dealing with small scale structure formation (galaxy formation), 30 to 100 eV neutrinos are quite appealing to explain the nature of our halo [18].

The ν_e is now excluded due to the severe upper limit on the mass (10 eV)[19]. However the ν_μ and ν_τ are perfectly viable candidates. Even the limits on their masses coming from SN1987A [20] are not stringent enough.

Various theoretical proposals [21] were made in the past for the detection of the cosmic neutrinos. One involved effects on the endpoint of the Curie plot in tritium decay, another involved coherent effects of mechanical pressure of massless or massive particle, and other ones spin precession effects on polarized electrons and induced currents in superconductors. All these effects are extremely small and out of reach by present techniques.

Another method was recently presented [21]. It assumes a large magnetic moment for the ν_e as suggested by the observed (?) anticorrelation between the sunspot cycle and the solar neutrino flux. It is based on the bremsstrahlung in the coherent scattering of the neutrinos off free electrons in the detector material (1kton of highly conducting metal at 10 mK).

Since all this looks a little bit like science fiction, we can add another fascinating possibility. If our galactic halo is made of light neutrinos there should be a sharp resonant absorption line in the spectrum of ultrahigh energy intergalactic neutrinos reaching the earth. The detection of such a narrow line would be a proof of the neutrino halo. Its position would provide for the neutrino mass: $2 M_\nu E_\nu = (M_{Z0})^2$.

Obviously the direct (laboratory or next supernova) or indirect (oscillations) measurements of the ν_μ and ν_τ masses are of crucial importance in the context of the dark matter problem. A likely scenario could be that the ν_τ is much heavier (30 eV) than the ν_μ and that the ν_μ is much heavier than the ν_e. In that sense the deficit of solar neutrinos could be a hint for neutrino oscillations and then for neutrino masses. Even if the favored range of masses to explain the solar neutrino deficit is 10^{-4} to 10^{-2} eV, this could be the range of mass for the ν_μ with a much lower mass for the ν_e and a much higher one for the ν_τ (30 eV). Neutrino masses could then both solve the dark matter and solar neutrino problems.

THE SOLAR NEUTRINO PROBLEM AND NEUTRINO MASSES

The most firm and solid prediction we have on the solar neutrino flux is based on energy conservation and steady state of the sun. We know that these two well admitted assumptions imply that the total power radiated by the surface of the sun (the luminosity L_*) should be equal to the thermonuclear power generated by the fusion of hydrogen into helium.

For four protons to combine into an ^{4}He nucleus, two electrons must be involved in the initial state for electric charge conservation, and then two ν_e must be emitted in the final state. The overall reaction is then:

$$4 p + 2e^- \rightarrow {}^4He + 2\nu_e + 27 \text{ MeV}$$

where 27 MeV is the difference of the masses between the particles involved in the initial state and those involved in the final state (the energy of the neutrinos and the kinetic energy of the nuclei can be neglected in this approximate relation).

It is then easy to derive the total flux of neutrinos expected to reach the earth:

$$N_\nu \text{ (cm}^{-2} \text{ s}^{-1}) = 2L_* / (27 \text{ MeV } 4\pi \text{ d}^2) = 65 \; 10^9 \text{ cm}^{-2}\text{s}^{-1}$$

where d is the distance from the earth to the sun.

Gallium target detectors are so far the most appropriate to measure the total number of neutrinos. This is because:

1) of the very low threshold (233 keV) of the capture reaction $\nu_e + {}^{71}\text{Ga} \rightarrow {}^{71}\text{Ge} + e^-$

which makes Gallium target detector sensitive to the bulk of the solar neutrino energy spectrum

2) of the high natural abundance of the stable ${}^{71}\text{Ga}$ isotope (40%)

3) of the relatively easy identification of even a few radioactive ${}^{71}\text{Ge}$ atoms in a large quantity of Gallium (30 tons).

However the prediction on the total number of solar neutrinos is not enough to compute the capture rate of solar neutrinos on a given target nucleus. This is because the capture rate depends very much on the energy of neutrinos as this can be seen on fig. 11.

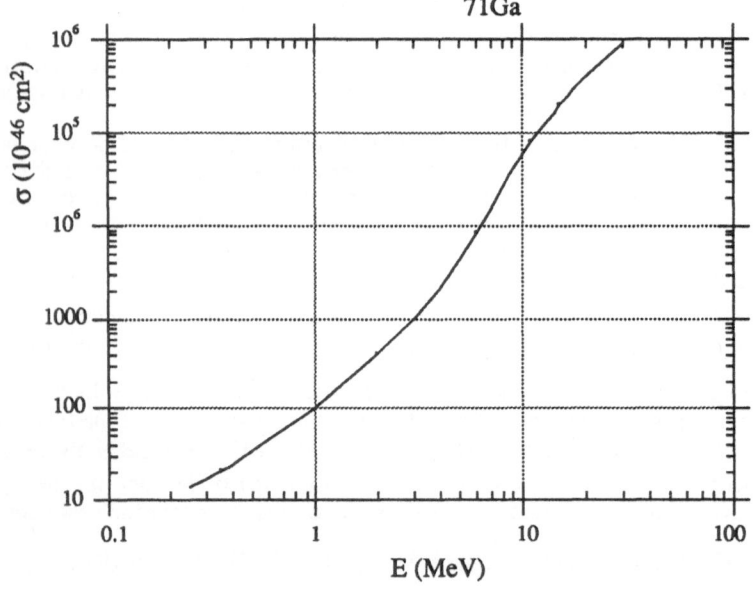

Figure 11. ν capture cross section on Gallium as a function of energy.

To compute the energy spectrum one needs to go through solar modelling and through the exact chain of reactions which combine hydrogen into helium. There are mostly three cycles of reactions: ppI, ppII and ppIII. They are shown in table 2.

In ppI the two neutrinos are coming from the pp reaction (2 ν_{pp})

In ppII one neutrino is a ν_{pp}, the other comes from the decay (through electron capture) of ${}^7\text{Be}$ (ν_{7Be})

In ppIII one neutrino is a ν_{pp}, the other comes from the decay of ${}^8\text{B}$ (ν_{8B}).

The energy spectrum of the ν_{pp}, ν_{7Be} and ν_{8B} are well known from nuclear physics. They are shown in fig. 12.

Table 2. ppI, ppII and ppIII cycles

Reactions	Neutrinos	Cycle number
$p + p \rightarrow {}^2H + e^+ + \nu_e$ ${}^2H + p \rightarrow {}^3He + \gamma$	ν_{pp}	
${}^3He + {}^3He \rightarrow {}^4He + p + p$		I
${}^3He + {}^4He \rightarrow {}^7Be + \gamma$		II and III
${}^7Be + e^- \rightarrow {}^7Li + \nu_e$ ${}^7Li + p \rightarrow {}^4He\ {}^4He$	ν_{Be}	II
${}^7Be + p \rightarrow {}^8B + \gamma$ ${}^8B \rightarrow {}^8Be^* + e^+ + \nu_e$ ${}^8Be^* \rightarrow {}^4He + {}^4He$	ν_B	III

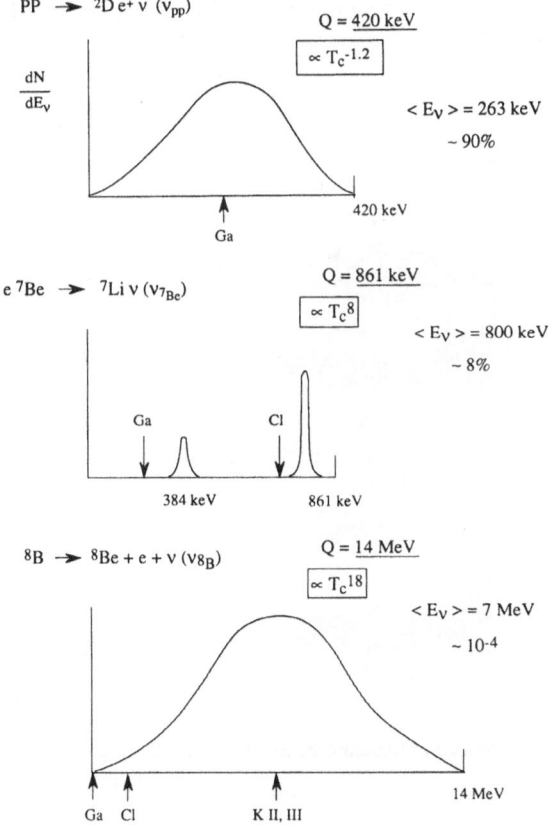

Figure 12. ν_{pp}, ν_{7Be}, ν_{8B} energy spectra .

177

The ν_{pp} spectrum extends from 0 to 450 keV. Only the Gallium experiments are sensitive to those neutrinos through the reaction $\nu_e + {}^{71}Ga \rightarrow {}^{71}Ge + e^-$ which has a threshold of 233 keV.

The ν_{7Be} are monoenergetic with a line at 860 keV. Both the Gallium and Chlorine ($\nu + {}^{37}Cl \rightarrow {}^{37}Ar + e^-$ threshold 820 keV) are sensitive to to those neutrinos.

Finally the ν_{8B} neutrino spectrum extends from 0 to 15 MeV. All the presently running experiments (Gallium, Chlorine and Kamiokande) are sensitive to them. The Kamiokande experiment is based on the detection of the recoil electron in the elastic scattering of a ν_e with an experimental threshold of about 7 MeV on the energy of the recoil electron.

From the Solar Standard Model [22] the ν_{pp}, ν_{7Be} and ν_{8B} intensities are computed to be 90%, 8% and 10^{-4} of the total flux. Although the relative intensity of the ν_{8B} neutrino is very small, they contribute significantly to the capture rate, even in the Gallium experiment, due to their high energy. Notice however that the ν_{7Be} and ν_{8B} fluxes are highly sensitive to the ingredients of the SSM. If , for instance, one changes the input parameters, with, as a result, a change in the central temperature Tc prediction, it has been shown that the ν_{8B} flux witll vary as Tc^{18}, the ν_{7Be} as Tc^8 and the ν_{pp} flux only as $Tc^{-1.2}$.

The predictions of the SSMs are shown in fig.13 , for the Gallium experiments, in terms of SNUs (Solar Neutrino Units). One SNU corresponds to a capture rate of 10^{-36} per second and per target nucleus (in this case for Ga). We see that the Bahcall et al. SSM which is generally considered as giving high SNU values predicts fluxes only slightly higher than the Turck-Chieze et al. SSM expectation which is generally considered as giving low SNU values. So one might say that the predictions of the SSM for Gallium experiments are rather firm. Notice also, that although the ν_{pp} are expected to represent 90% of the total flux of solar neutrinos, their contribution to the capture rate amounts only to 70 SNU out of 132. This is due to their low energy.

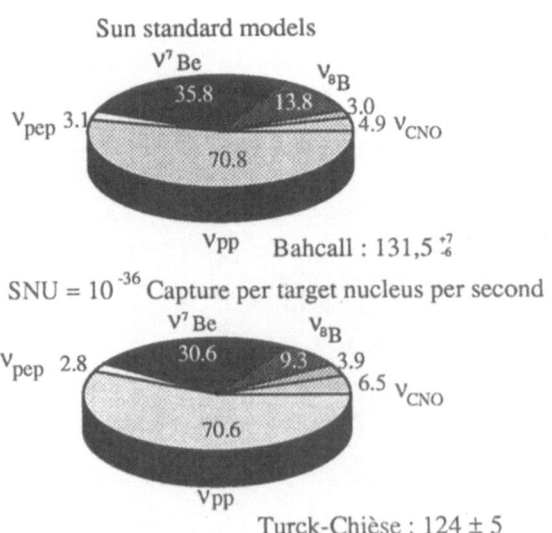

Sun standard models

$\nu^7 Be$ ν_{8B}

35.8   13.8   3.0

ν_{pep} 3.1   4.9 ν_{CNO}

70.8

ν_{pp} Bahcall : 131,5 $^{+7}_{-6}$

SNU = 10^{-36} Capture per target nucleus per second

$\nu^7 Be$ ν_{8B}

ν_{pep} 2.8   30.6   9.3   3.9

                        6.5 ν_{CNO}

70.6

ν_{pp}

Turck-Chièse : 124 ± 5

Figure 13. Predictions of Standard Solar Models for Gallium experiments.

Are these predictions right?

ν_{8B} flux

The Kamiokande experiment uses a water Cerenkov detector. The basic process is neutrino scattering on electrons which then give detectable Cerenkov light. They measure two quantities, the energy of the recoil electron and its direction. A clear peak can be seen in fig. 14 in the direction of the sun and the excess in that direction is then taken as coming from solar neutrinos.

However the flux of ν_{8B} they measure [23] is only .46 ± .08 of the Bahcall et al. SSM so .5 10^{-5} of the total flux.

ν_{7Be}

Since 1967 Davis and co-workers have performed a pioneering experiment by extracting ^{37}Ar from a tank of 615 t of tetrachloroethylene C_2Cl_4. The ^{37}Ar decays by electron capture. The resulting hole in the K shell can give X rays and Auger electrons with a total energy of 2.5 keV. The counter of 0.5 cm^3 volume is designed to measure this electron. The half life of the decay is 35 days.

A typical run consisted of a 50 days exposure of the big tank followed by an extraction of the Argon atoms which are then introduced in the small counter. The counting lasts for 260 days. For the period 1970-1984 the data were analyzed to give 339 counts of ^{37}Ar. This gives an uncorrected ^{37}Ar counting rate of 5 per run. The data are analyzed by a maximum likelihood method assuming a flat background (as a function of time) plus a ^{37}Ar decaying component.

The result [24] is 3.6±.4 times lower (fig.15) than expected in the Bahcall et al. SSM.This implies, taking into account the fact that the experiment is sensitive to both the ν_{7Be} and ν_{8B} components and taking into account the Kamiokande result (reduction of a factor 2 on the ν_{8B} component) that the ν_{7Be} flux is lower by a factor > 4 than the prediction of Bahcall et al. SSM.

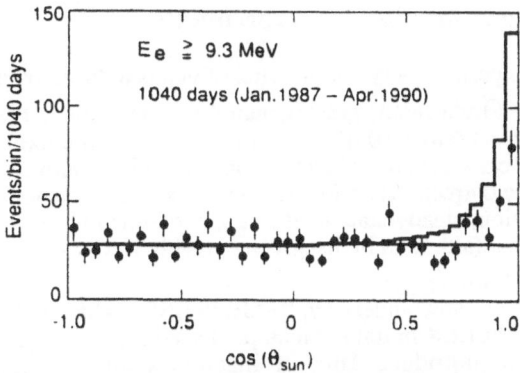

Figure 14. Counts of the Kamiokande detector plotted against the angle of the electron to the sun's direction.

These deficits are the basis of the solar neutrino problem. The reductions are very hard to reconcile with any modification of the SSM since we expect that any reduction on the ν_{7Be} component should be accompanied by a stronger reduction for the ν_{8B} component. Neutrino masses and mixing could reconcile these reductions with the SSM through ν_e, ν_μ oscillations. However, before invoking new physics in the neutrino sector, the results of the gallium experiments were eagerly awaited. The expectations are much less sensitive to SSM and we can derive absolute lower limits for the capture rate based only on energy conservation and steady state of the sun.

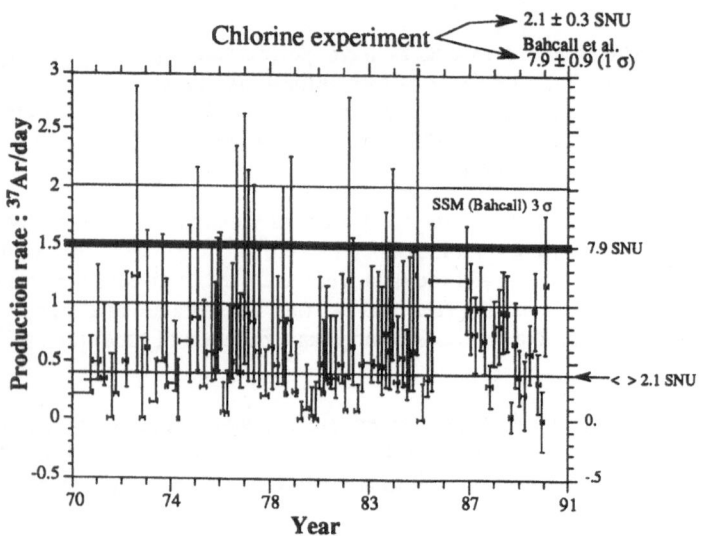

Figure 15. Results of the Chlorine experiments for all runs.

Consistent predictions for Gallium experiments

Since we know experimentally that the flux of ν_{8B} is reduced by a factor 2 and that the flux of ν_{7Be} is reduced by a factor greater than 4 we can deduce that the number of ν_{pp} should be increased to 1.06 to 1.10 of the SSM to conserve the total number of neutrinos which insuring energy conservation. The predictions for the Galium experiments can then be derived by applying the corrections on fig. 13. One obtains expectations which are consistent with energy conservation, steady state of the sun, experimental results of the chlorine and Kamiokande experiment (although in disagreement with SSM predictions) and which range from 80 to 105 SNUs [28].

Two experiments are now underway, SAGE in USSR which published the first results in January 1991 and GALLEX in Italy, which published their first results in June 1992. The recipes are the same: introduce 1mg of inactive stable germanium in the 30t of Gallium,expose the Gallium to solar neutrinos in a low background environment, extract by a chemical method the sola ν-produced ^{71}Ge atoms together with the inactive Germanium, transform into a counting gas (GeH$_4$), fill a proportional counter and count the decays of ^{71}Ge (11 d half life). The main difference is that the SAGE experiment uses metallic liquid Gallium target while the GALLEX experiment uses an acidic aqueous Gallium Chloride solution. This induces important differences in the chemistry.

SAGE

The Soviet-American Gallium Experiment is located in the Baksan Valley in the Caucasus mountains (Russia) under about 4700 meter water equivalent. The expected rate for 30 t target and 132 SNU is 1.2 [71]Ge atom created per day. Taking into account all the efficiencies, one expect only 3 counts per run (a run is 4 weeks exposure) due to [71]Ge K electron capture ([71]Ge + K e- $\rightarrow$ [71]Ga + ν + X-rays + Auger electrons). Fig.16 shows the cumulative number of counts in the K-peak region as a function of time for the five runs they published. A [71]Ge signal should give a curvature due to the finite lifetime of [71]Ge (11d) over the linar increase due to the flat background as a function of time.Most of the runs have preferred values of 0 SNU. Altogether they published [25] an upper limit of 55SNU (68% C.L.) and 79 SNU (90% C.L.). More recently they announced the results they obtained in the last runs when they increased the total mass of Gallium from 30t to 60t [26]. This is shown on fig.17. A signal seems now to emerge.

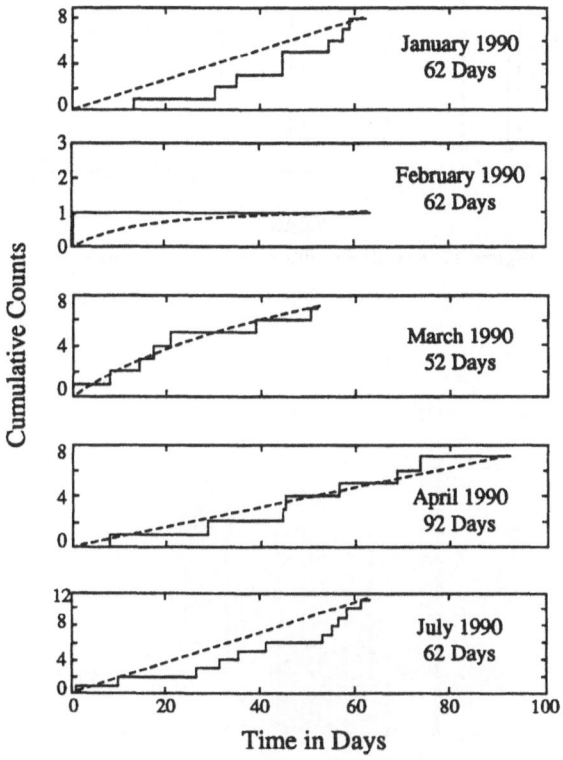

Figure 16. Cumulative number of counts in the K-peak region as a function of time in the SAGE experiment.

GALLEX

This is an European experiment located in the Gran Sasso Underground Laboratory in Italy. Fig. 18 shows a schematic view of the detector. The 30 tons of Gallium are in the form of a solution of $GaCl_3$ acidified in HCl. The Ge atoms form the volatile compound $GeCl_4$. At the end of 3 weeks exposures, these molecules are swept out by bubbling a large flow of inert gas (N_2) through the solution. The experiment is sensitive to both K-shell and L-shell electron captures in the decay of [71]Ge atoms. Seven counts are then expected after each run, in the K and L regions. The data used in the analysis consist of 14 runs taken from May

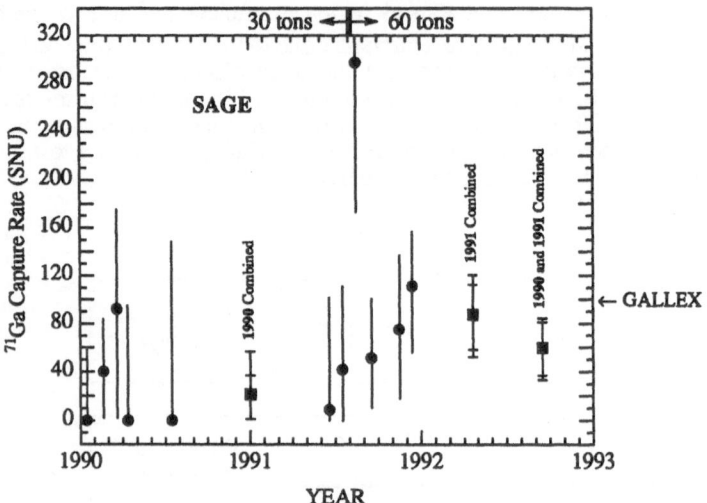

Figure 17. Results for all runs of the SAGE experiment.

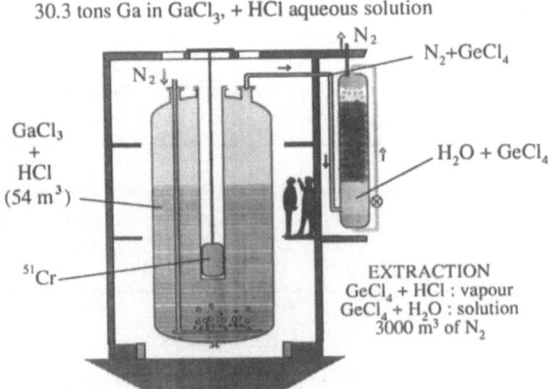

30.3 tons Ga in GaCl$_3$, + HCl aqueous solution

Figure 18. Schematic view of the GALLEX detector.

1991 to May 1992. They are now published [27]. Fig. 19 shows the results for all runs together with the combined result of 83 ± 20 SNUs. There is compelling evidence for a signal: the peaks in energy at 1.2 keV and 10 keV for L and K electron capture are seen, the 11.3 half life of ^{71}Ge is well identified over a flat background. Fig. 19 shows the results for all runs together with the combined result of 83 ± 20 SNUs.

The SSM is unable to account for the deficit of solar neutrinos as observed by the Chlorine and Kamiokande experiments. However, on the basis of these experiments it is impossible to decide whether these discrepancies come from new physics in the neutrino sector or wrong ingredients in the Solar Standard Models. The Gallium experiments are in a much better position to do so. First, the predictions of the SSM are more stable to changes in the ingredients (120 to 140 SNU) and second it is impossible to have predictions below 80 SNU from basic simple principles. Consistent predictions for Gallium experiments which agree with these basic principles and with the deficits of solar neutrinos observed by the Chlorine and Kamiokande experiments are in the 80-105 SNU range .

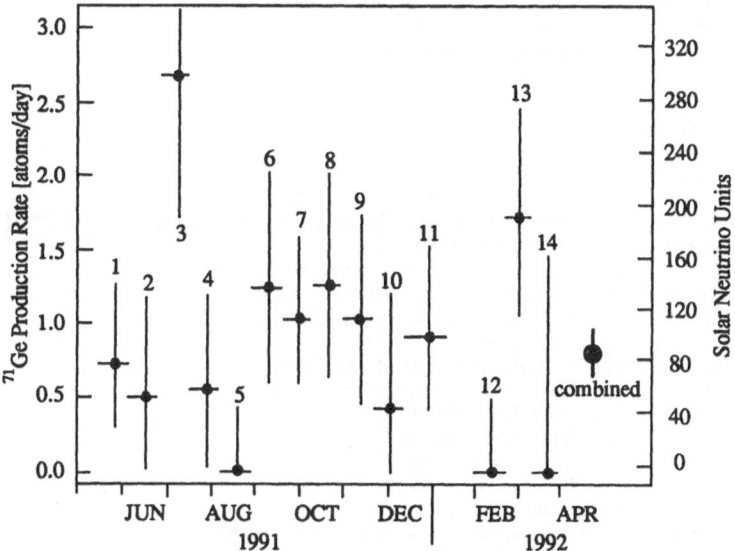

Figure 19. Results of the GALLEX experiment for all runs.

Taken at face value, the GALLEX and SAGE results disagree. It might be that, with the last results of the SAGE experiment, they could be reconciled. In any case these results are still preliminary. The experiments await calibration with artificial neutrino sources. This should be done in 1993. It is clear that if the final result, with better statistical accuracy and with a calibration to ensure that there are no unknown sources of inefficiency, is in the 80 - 90 SNU vicinity, the solar neutrino problem will not give a compelling evidence for neutrino masses. More experiments and more solar models will be needed. On the other hand if the Gallium experiments converge towards a number which is significantly below 80 SNUs, neutrino masses and neutrino oscillations would be the best way to explain the solar neutrino deficits. No firm conclusions on neutrino masses can yet be drawn from the present status of solar neutrino experiments and solar modelling. This may not be the case hopefully in few years from now.

CONCLUSION

We are still desperately investigating the dark matter and solar neutrino problems..

ACKNOWLEDGEMENTS

I thank J. Rich for many helpful discussions, N. Lelievre for editing the manuscript.

REFERENCES

1. B. Carr, The Quest For The Cosmological Constant, Moriond 1990, Editions Frontières.
2. B. Paczynski, Ap. J. **304** (1986) 1.
3. A. Milsztajn, Proceedings of the Rencontres de Moriond 1990, edited by O. Fackler and J. Tran Thanh Van, Publisher Editions Frontieres, p. 481.
   M. Moniez, Orsay preprint, LAL 90-20, 1990 (Moriond Astrophysics, Les Arcs 1990).
4. J. Rich, in Dark Matter (Moriond 1988), edited by J. Audouze (Editions Frontières), p. 43.
   J. R. Primack, B. Sadoulet and D. Seckel, Ann. Rev. Nucl. Part. Sci., **38** (1988) 751.
   P.F. Smith and J.D. Lewin, Phys. Reports, **187** (1990), 203.
5. S. P. Ahlen et al., Phys. Lett. **B195** (1987) 603.
   D. O. Caldwell et al., Phys. Rev. Lett. **61** (1988) 520.
   J. Rich, R. Rocchia and M. Spiro, Phys. Lett. **B194** (1987) 173.
6. J. Faulkner and R.L. Gilliland, Ap. J. **299** (1985) 994.
   D. N. Spergel and W.H. Press, Ap. J. **294** (1985) 663.
   W. H. Press and D. N. Spergel, Ap. J. **296** (1985) 679.
   R. L. Gilliland et al., Ap. J. **306** (1986) 7038.
7. D. O. Caldwell et al., Phys. Rev. Lett. **65** (1990) 1305.
8. J. Kaplan, F. Martin de Volnay, C. Tao and S. Turck-Chièze, to appear in Ap. J. **378** (1991) 315
9. G. Gerbier et al., Phys. Rev. **D42** (1990) 3211.
10. J. Lindhard et al., Mat. Fys. Medd. Dan. Vid. Selsk. 33 n° 10 (1963).
11. A. De Rujula, S. Glashow and U. Sarid, Nucl. Phys. **B333** (1990) 173.
12. W. S. Broecker and T. H. Peng, Tracers in the sea, Eldigio Press, 1982.
13. B. Pichard et al., Phys. Lett. **B193** (1987) 383.
14. M. Spiro, Proceedings of the Rencontres de Moriond 1990, edited by O. Fackler and J. Tran Thanh Van, Publisher Editions Frontieres, p. 489.
    P. Verkerk et al., Phys. Rev. Lett., **68** (1992) 1116
15. P. Smith et al., Nucl. Phys. **B206** (1982) 333.
    T. K Hemmick et al., Phys. Rev. **D41** (1990) 2074.
16. S. W. Barwick et al., Phys. Rev. Lett. **64** (1990) 2859.
17. J. L. Basdevant, private communication.
18. R. Cowsik and J. Mc Clelland, Phys. Rev. Lett. **29** (1972) 669.
    J. L. Basdevant, Preprint IAP- July 1984.
19. J. Wilkerson, to be published in the Proc. of Neutrino '90 (Geneva 10-15 June 1990).
20. J. A. Grifols and E. Masso, Phys. Lett. **242** (1990) 77.
    G. Raffelt and D. Seckel, Phys. Rev. Lett. **60** (1988) 1793.
    K. J. F. Gaemers et al. Phys. Rev. **D40** (1989) 309.
21. A. Loeb and G.D. Starkman, preprint IASSNS-AST90/10 submitted to Phys. Rev. Lett. References therein.
22. J. N. Bahcall and R. K. Ulrich, Rev. Mod. Phys. **60** (1988) 297
    J. N. Bahcall, 1989, Neutrino Astrophysics (Cambridge Univ. Press, Cambridge, England)
    S. Turck-Chieze et al., Astrophys. J., **335** (1988) 415.
23. K. Hirata et al., Phys. Rev. Lett., **66** (1991) 9
24. R. Davis et al., 1990 in Proc. Int. Conf. "Neutrino 90", Geneva
25. Abazov et al., Phys. Rev. Lett., **24** (1991) 3332
26. V. Gavrin, Astroparticle Conf., Blois, 1992
27. P. Anselmann et al., Phys. Lett. **B285** (1992) 376
28. M. Spiro and D. Vignaud, Phys. Lett., **B242** (1990) 279

RECENT DEVELOPMENTS IN TRACKING DETECTORS

D.H. Saxon

Department of Physics and Astronomy
University of Glasgow
Glasgow G12 8QQ
Scotland

DIRECTION AND MOMENTUM MEASUREMENT

Tracking is the determination of particle four-vectors (p_x, p_y, p_z, E). In vertex detection we are also concerned with the distances particles travel between their production and decay. We shall turn to this aspect later. Meanwhile, let us remind ourselves of the basics of the determination of the four vector, which we re-write as (p, θ, ϕ, m) where $p_z = p \cos\theta$, $p_x = p\sin\theta\cos\phi$, $p_y = p\sin\theta\sin\phi$, $E^2 = p^2 + m^2$. The measurement of the angles (θ, ϕ) is relatively easy. To measure p we need to measure the angle of deflection in a magnetic field. Measurement of mass, *ie* the establishment of the particle identity, is the hardest of all. Thus time of flight measure $1/\beta$ (velocity $v = \beta c$), ionisation loss measure $1/\beta^2$ and $\ell n \; \gamma$ ($\gamma = (1-\beta^2)^{-1/2}$), Cerenkov counters measure β in the region of a threshold. We shall discuss below just the question of identifying electrons, either by their interactions in calorimeters, or by the emission of transition radiation X-rays, which depend on $\gamma = E/m$.

Motion in a Magnetic Field

The motion of a particle of charge, q, in electric and magnetic fields is given by

$$\frac{d\mathbf{p}}{dt} = q \; (\mathbf{E} + \mathbf{v} \times \mathbf{B})$$

(1)

Consider $\mathbf{E} = 0$. Then $\mathbf{p}. \; d\mathbf{p}/dt = 0$, *ie* $d(p^2)/dt = 0$ and $|\mathbf{p}|$ is a constant of the motion. Also $\mathbf{B}.d\mathbf{p}/dt = 0$. Therefore in a uniform magnetic field the component of $\mathbf{p}$ parallel to $\mathbf{B}$ ($p_L = \mathbf{p}.\mathbf{B}/B$) is a constant and only $\mathbf{p}_T = \mathbf{p} - p_L\mathbf{B}/B$ varies.

Figure 1(a) shows an element of track of length ds projected on to the plane perpendicular to $\mathbf{B}$. ($\mathbf{B}$ into the paper). Then $\mathbf{v} = d\mathbf{s}/dt$, so $d\mathbf{p}/dt = q \; d\mathbf{s}/dt \times \mathbf{B}$.

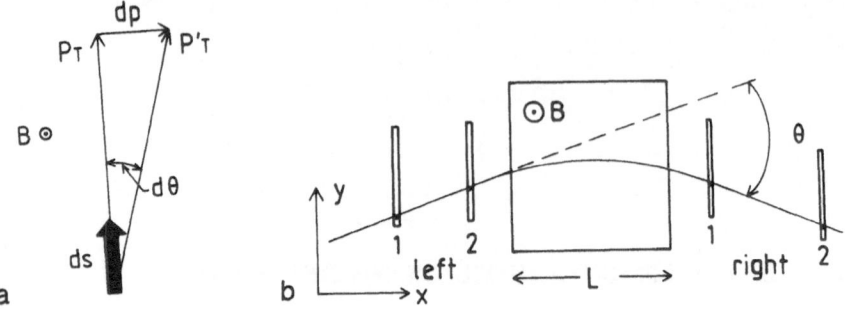

Figure 1. (a) Charged particle moving in a magnetic field (b) magnetic spectrometer.

One may cross out dt on both sides. Then the new value of $\mathbf{p_T}$, $\mathbf{p'_T} = \mathbf{p_T} + \mathbf{dp}$ is equal in magnitude to $\mathbf{p_T}$, but is rotated through $d\theta = dp/p_T = qds_TB/p_T$ (ds_T = component of ds perpendicular to $\mathbf{B}$). This represents rotation around a circle of radius $R = ds_T/d\theta$ and hence $p_T = qBR$. This equation is correct in MKS units. If we measure p in GeV/c, q in units of the proton charge and B and R in Tesla and metres it becomes $p_T = qBR$, where c = velocity of light = 0.29979 Gm/s. So finally

$$p_T = -0.29979qBR \tag{2}$$

(The sign convention is explained below). The angle turned through in traversing a distance L (measured along the particle trajectory)

$$\theta = -0.29979 \, qBL/p_T \tag{3}$$

Momentum Measurement by Deflection by Magnetic Field

Figure 1(b) sketches a magnetic spectrometer with straight line trajectories measured before and after deflection by a field of magnitude B over a distance L. Then the particle trajectories in the regions before and after the magnet are $y = m_1 x + c_1$ and $y = m_2 x + c_2$. Let there be a set of N measurements in the region before the magnet at $\{x_i\}$, each providing a position measurement with an error $\{y_i, \sigma_i\}$. Then to find m and c in this region we set up

$$\chi^2 = \sum_i (d_i(x_i, y_i, v)/\sigma_i)^2 \tag{4}$$

where $d_i = y_i - mx_i - c$ is the distance of the measured point from the fitted line and v is the parameter vector $v_1 = m$, $v_2 = c$. Then at the best fitted values of v,

$$g_\alpha = \partial\chi^2/\partial v_\alpha = 0 \text{ for } \alpha = 1,2. \text{ Now } g_\alpha = \sum_i 2d_i(\partial d_i/\partial v_\alpha)/\sigma_i^2.$$

The two equations ($\alpha = 1,2$) can be written in matrix form as

$$
\begin{bmatrix} \sum_i 2\, x_i^2/\sigma_i^2 & \sum_i 2\, x_i/\sigma_i^2 \\[2mm] \sum_i 2\, x_i/\sigma_i^2 & \sum_i 2/\sigma_i^2 \end{bmatrix} \begin{bmatrix} m \\[2mm] c \end{bmatrix} = \begin{bmatrix} \sum_i 2\, x_i y_i/\sigma_i^2 \\[2mm] \sum_i 2\, y_i/\sigma_i^2 \end{bmatrix}
$$

$$(5)$$

Symbolically $Gv = y$. Whence $v = G^{-1} y$ is the required fit. Note that G contains information about the measuring planes that provide hits. The actual measured values occur only in y. One may expand the χ^2 around the solution to get $\chi^2 = \chi_0^2 + \frac{1}{2}\,\Delta V^T G\,\Delta V$, (since the first derivatives are zero here) and see that the error matrix on m and c, $\Sigma = [G/2]^{-1}$, corresponding to an increase in χ^2 by 1. In the case where all the errors σ_i are equal to σ one finds

$$
m = \frac{<xy> - <x><y>}{<x^2> - <x>^2}
$$

and its error

$$
\sigma_m = \frac{\sigma}{\sqrt{N}\,\sqrt{<x^2> - <x>^2}}
$$

$$(6)$$

as expected $\sigma_m \sim N^{-1/2}$, and for the smallest σ_m one should group the measurements into two clusters at opposite ends of the longest possible baseline, to maximise $<x^2> - <x>^2$ (provided of course the association of hits to tracks is not ambiguous). Then for small angles of deflection in the magnet one can write $\theta = m_2 - m_1$, and for its measurement error $\sigma_\theta^{m2} = \sigma_{m1}^2 + \sigma_{m2}^2$, and from (3)

$$
\frac{\sigma_p^m}{p} = \frac{\sigma_\theta\, p_T}{0.3 B L}
$$

If in addition there is multiple scattering in the material within the magnet it contributes an additional error $\sigma_\theta^s = (0.015\,|\theta|\,/\,p\beta)\,\sqrt{x/x_0}$ where there is a depth x of material of radiation length x_0. This error must be added in quadrature to σ_θ^m, $(\sigma_\theta^2 = \sigma_\theta^{m2} + \sigma_\theta^{s2})$. As an example, consider the measurement of muon momenta using magnetised iron using L metres with $B = 1.7$T. (To go above this value saturation of iron requires substantially more ampere-turns). Then $x_0 = 0.0176$m. Suppose we use two measuring planes each side, 30 cm apart, with 100 μm precision. Then $(\sigma_p/p)^2 = (0.65.10^{-3} p/L)^2 + (0.21)^2/L$.

Figure 2(a) sketches the momentum dependence for $L = 1$ and $l = 4$m. The general form of this curve holds for all momentum measurements using magnetic fields. For almost all interesting momenta the second, scattering, term dominates. For $L = 1$m, we are limited to 21% error. To achieve a 10% error, 4m of iron are needed.

Tracking Inside a Magnetic Field

Consider a magnetic field in the $+z$ direction and set $x = r\cos\phi$, $y = r\sin\phi$. The circular trajector of a track as projected on to the x–y plane is commonly described by three parameters, d_0 - the distance from the origin ($x = y = 0$) to the point of closest approach,

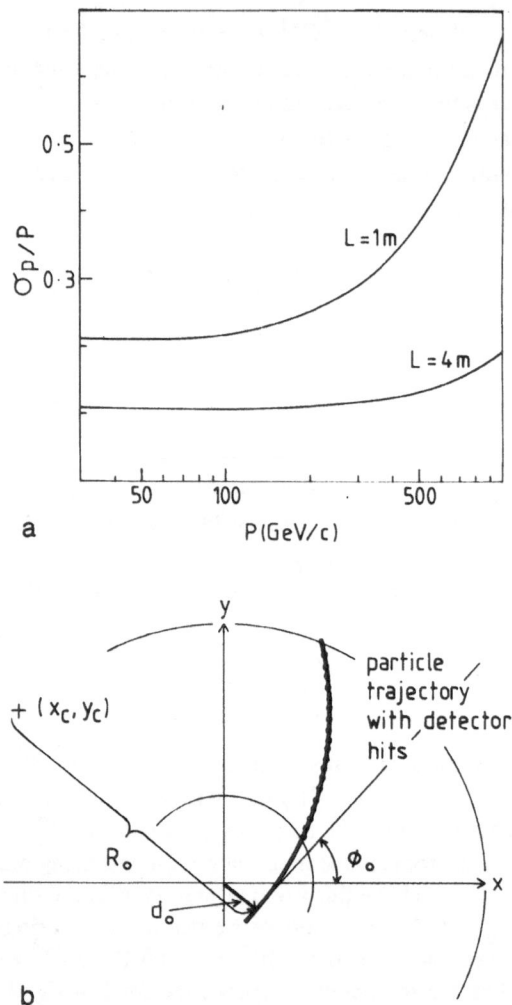

Figure 2. (a) Spectrometer fractional momentum resolution plotted as a function of momentum for normal incidence in 1 m and 4 m of magnetised iron (see text for specification). (b) Coordinate system for description of a track within a magnetic field.

ϕ_0 - the direction of motion of the particle at this point, and R_0 the track radius. (see fig. 2(b)). We need a sign convention for d_0 and R_0 and pick that d_0 is positive if d_0, $(\mathbf{p} \times \hat{z}) > 0$, (this convention is valid as $R_0 \to \infty$), and $R_0 > 0$ if the trajectory bends to increasing ϕ. This is consistent with the sign in equations (2) and (3)). Then the centre of the circle is at $(x_c, y_c) = (-(R_0 - d_0)\sin\phi_0, (R_0 - d_0)\cos\phi_0)$ and the equation of the circle, $(x - x_c)^2 + (y - y_c)^2 = R_0^2$

can be rewritten as

$$(r^2 - d_0^2)/(2(R_0 - d_0)) - r\sin(\phi - \phi_0) = d_0. \tag{7}$$

The distance of a point, such as a measurement, from the circle,

$$d = R_0 - \sqrt{(x - x_c)^2 + (y - y_c)^2}$$

(The sign convention is that $x = y = 0$ has $d = d_0$). This expression is troublesome to evaluate as $R_0 \to \infty$ and leads to numerical instabilities in computations. We therefore rewrite it in the form

$$d = \kappa (r^2 - d_0^2)/2 + r\sin(\phi - \phi_0) + d_0 \tag{8}$$

where $\kappa = -1/(R_0 - d_0)$. We have made two approximations here, $|d| << |R_0|$ and $|2dd_0| << r^2$. Both are very well satisfied for all measurements to be associated with tracks, whether or not d_0 is small. Note that it is exact for points on the circle, which satisfy equation (7). One finds from equation (8) immediately that these have $d = 0$.

We chose as our circle fit parameter vector $\mathbf{v} = (\kappa, d_0, \phi_0)$ and set up a χ^2 using equation (4). The second derivative matrix

$$G_{\alpha\beta} = \frac{\partial^2 \chi^2}{\partial v_a \partial v_\beta} = \sum_i \frac{2}{\sigma_i^2} \left[d_i \frac{\partial^2 d_i}{\partial v_\alpha \partial v_\beta} + \frac{\partial d_i}{\partial v_\alpha} \frac{\partial d_i}{\partial v_\beta} \right]$$

All terms are easily calculated. One minimises χ^2 by an iterative method. Let $\mathbf{v}^n$ be the nth iteration value of $\mathbf{v}$. Then

$$\chi^2(\mathbf{v}) = \chi^2(\mathbf{v}^n) + \mathbf{g} \cdot (\mathbf{v} - \mathbf{v}^n) + \tfrac{1}{2} (v_\alpha - v_\kappa^n) G_{\alpha\beta} (v_\beta - v_\beta^n) + \dots$$

and $\mathbf{g}(\mathbf{v}) = \mathbf{g}(\mathbf{v}^n) + G(\mathbf{v} - \mathbf{v}^n)$.

At the minimum of χ^2, $\mathbf{g} = 0$. This should occur at

$$\mathbf{v}^{n+1} = \mathbf{v}^n - G^{-1}\mathbf{g}$$

and we predict here that $\chi^2 (v^{n+1}) = \chi^2 (v^n) - \frac{1}{2} G^{-1} g$. One continues until the expected improvement in χ^2 falls below a cutoff value. This method[1] is, to my knowledge, the only method of circle fitting that is numerically stable as $R_0 \to \infty$.

Accuracy of Momentum Measurement by Circle Fit

The error matrix on (κ, d_0, ϕ_0), $\Sigma = 2G^{-1}$. Consider the case r_i, $R_0 \gg |d_0|$, all d_0 small, then, setting $S_n = \sum_i r_i^n \, \sigma_i^{-2}$,

$$
\Sigma^{-1} = \begin{pmatrix} S_4/4 & S_3/2 & S_2/2 \\ S_3/2 & S_2 & S_1 \\ S_2/2 & S_1 & S_0 \end{pmatrix}
$$

Hence

$$
\Sigma_{\kappa\kappa} = \frac{4 \left(S_2 S_0 - S_1^2 \right)}{\left(S_4 S_2 S_0 - S_4 S_1^2 - S_3^2 S_0 + 2 S_1 S_2 S_3 - S_2^3 \right)}
$$

The expression looks ungainly. But we can take advantage of an invariance. The error on κ should be independent of a translation of the system. (Prove it!). Therefore we move the origin of coordinates such that $S_1 = 0$. For a symmetrical system this makes $S_3 = 0$ also. In the case where all the σ_i are equal, setting $\rho = r - <r>$ we find

$$
\Sigma_{\kappa\kappa} = 4\sigma^2 / \left(N (<\rho^4> - <\rho^2>^2) \right).
$$

In the case of a set of N measurements spaced uniformly over a distance L this gives

$$
\Sigma_{\kappa\kappa} = 720 \, \sigma^2 \, (N-1)^3 / (L^4 (N-2) \, N(N+1)(N+2)),
$$

giving Gluckstein's formula for momentum resolution[2]

$$
\frac{\sigma_p}{p} = \frac{p_T \, \sigma}{0.3qBL^2} \, \sqrt{\frac{720}{N+4}} .
$$

If the points are gathered in three clusters, 25% at each end, the rest in the middle (cf the L3 muon system) one obtains

$$
\sigma_p/p = p_T \, \sigma/(0.3qBL^2)\sqrt{256/N} .
$$

If one wishes to design a solenoidal measurement system for a particular performance, one can now make design choices. Thus for $\sigma_p/p = 0.1$ at 50 GeV/c one could choose the set (B, L, N, σ) = (2.0T, 0.31m, 4, 12μm) or (0.6T, 1.22m, 30, 120m). Note that the first set is more demanding not only on point precision but also on alignment. If the errors due to detector misalignment are to be below 50% of the random errors the alignment precisions are 7μm and 45μm in the two cases.

Lastly we note that for tracks with $d_0 = 0$, the propagation in the z-direction is very simple:

$$\phi = \phi_0 + (z - z_o) \; 0.3 \; (-q) \; B/(2p_z) \tag{9}$$

The tracks are straight lines in the (ϕz) plane and $d\phi/dz$ measures p_z directly. This formula is exact for all p_z (ignoring scattering). In addition for high momentum tracks one may approximate $r = (z - z_o) \tan\theta_0$, where $\tan\theta_0 = dr/dz$ at $r = 0$.

Progressive Track Fitting

Complex modern detectors often offer sets of precise position measurements in separate groups with substantial scattering throughout the system. A scatter to the right, say, at a certain point would introduce correlated offsets at all subsequent measurements. For an isolated scattering plane the introduction of the scattering angle as a fit variable provides an adequate method. Equation (4) is modified by the addition of a term θ^2/σ_θ^2 to the χ^2 for scattering at a single layer at radius a and one displaces the trajectory by angle θ for each measurement at $r_i > a$. θ is then a variable to be found by searching for the minimum of χ^2.

With many scattering planes, or with continuous scattering throughout the system, this method becomes unwieldly. A formalism is required which does not involve the inversion of huge matrices, (dimension increasing by one for each scattering plane,) and which makes the best use of the available information in the form of measurements and quantities of scattering material. The technique of *Kalman filtering* is increasingly used. The track fit is *progressive*. Trajectories are to be reconstructed based on information from measurements at a sequence of measuring surfaces traversed by particle trajectories. The surfaces have serial numbers 1,2, ...k....They do not have to be parallel, or planar, though for ease of discussion here we consider a set of parallel measuring planes at $z_1, z_2,....z_k,...$ Suppose that at a certain moment a fit exists based on information from measurements at z_1, z_2, ...z_{k-1}. We can extrapolate this through the field (allowing for energy loss) and make a prediction of the track parameters at $z = z_k$. The prediction will have errors based on the accuracy of the existing track fit and on the multiple scattering between z_{k-1} and z_k. The effect of the multiple scattering is to increase the extrapolated error matrix. This is inverted to get a new second derivative matrix.

We then increase the information by adding to the second derivative matrix the new measurements at $z = z_k$ to obtain the best fit based on z_1 to z_k. (See figure 3). The method can be described as a *Predictor-Corrector* method. The fit proceeds along the track in such a way that at the finish the best track information is available at all points, and one can see immediately the improvement in χ^2 by removing each measurement individually. This provides easy rejection of one bad point.

Kalman filtering is properly the subject of a separate text. The interested reader is referred to the work of Regler and Frühwith[3].

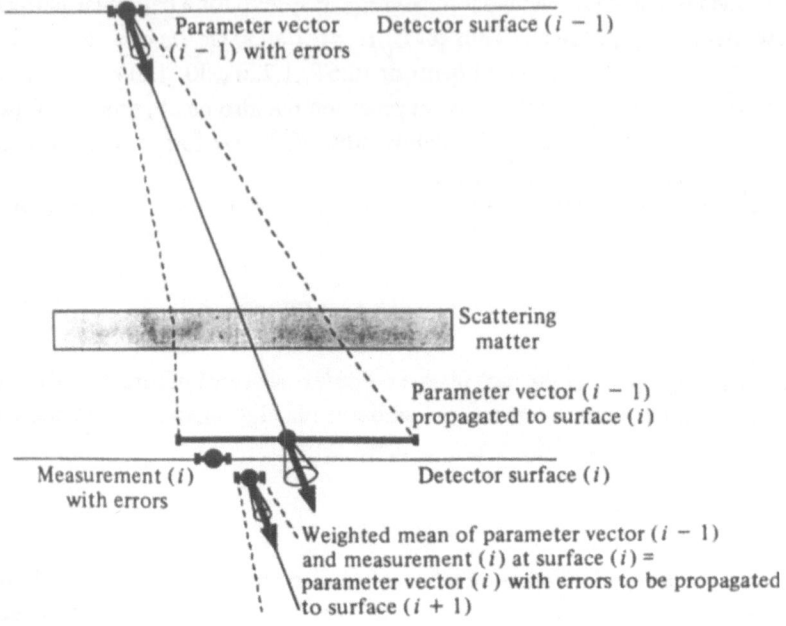

Figure 3. The Kalman Filter : prediction, error propagation and weighted mean (from ref. 3)

FAST TRACKING DETECTORS FOR HIGH REPETITION RATE COLLIDERS

Vector Tracking Detectors

The beam-crossing intervals at HERA are 96ns. At SSC/LHC 15 or 30ns are planned. These times are much shorter than at LEP (up to 22μs) or SLC (8ms). Long drift paths and long readout times are excluded. The detector also needs a number of qualities - strong pattern recognition; self-calibration by data; ready rejection of out of time events; good position resolution and two-track resolution; good three dimensional precision, without excessive ambiguities; and the detector should contribute to the trigger. We consider now "vector" tracking detectors in solenoidal magnetic fields as a strategy to achieve these goals. For a more extensive review, see reference 4.

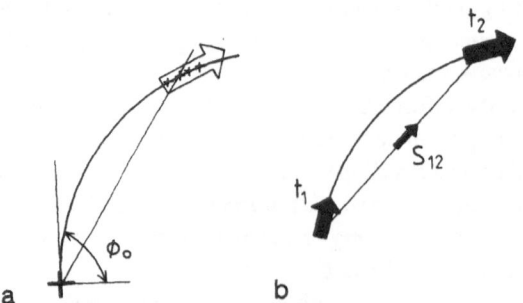

Figure 4. (a) Measurement in a single superlayer measures momentum and direction if the vertex is assumed, (b) linkage of vectors without assuming a production vertex.

The basic idea of the vector chamber, or stereo-superlayer chamber is to combine the three-dimensional precision given by stereo wires, (equivalent to $\sigma_z = \sigma_{r\phi}/\sin\alpha$, where α is the stereo angle,) with robust track finding in which one first reconstructs a short straight-line vector in each superlayer and then links these vectors into circular tracks. Figure 4 illustrates vector linkage in two ways. Each superlayer provides a momentum and direction measurement if the production vertex is assumed known, giving (κ, ϕ_0). One makes a scatter plot of (κ, ϕ_0) for vectors in all axial layers and identifies clusters with tracks. An alternative method which does not make any assumption about the production point is illustrated in figure 4(b). Two superlayers measure track direction unit vectors t_1 and t_2. Define $\eta_{12} = t_1 \times S_{12} + t_2 \times S_{12}$ where S_{12} is a unit vector joining the centres of the track vectors t_1 and t_2. Then if the two vectors are produced by the same track they are tangent to a common circle. In this case they make equal angles with the chord joining them and $\eta_{12} = 0$. Using only sum and product calculations, (suitable for fast real-time computation) one can test whether η_{12} is consistent with zero. The error on η_{12} can easily be estimated using equation 6. One finds $\sigma_\eta = \sigma/L\sqrt{24/N}$ where L is the baseline for measurement of each individual vector of N points. Track linkage is unambiguous (at the n S.D. level) if $\sigma/d < L/nS \sqrt{N/}\,6$ where d is the double-hit resolution and S is the distance between the two superlayers concerned. If σ is sufficiently small, one may make links across a gap of several superlayers.

This chamber strategy has been followed by Mark 2, SLD, CDF and ZEUS following the original initiative of Mark 3. Mark 2 and SLD both use "low" magnetic fields (B = 0.45, 0.6T) with 72 and 80 layers grouped into 12 and 10 superlayers with axial and 3° to 4° stereo. Within each superlayer are constructed a set of jet drift cells. Figure 5(a) illustrates the Mark 2 drift cell, with 6 sense wires offset with left right stagger. Electrons liberated by passing particles drift in at 50μm/ns and track candidates found. For points in a cell on a candidate track one calculates $(T_1 + T_3 - 2T_2 - (T_4 + T_6 - 2T_5))/8$ using the drift times on the six wires. As figure 5(b) shows, each good track falls in one of two peaks, depending on which side of the sense-wire plane it lies. The width of the peak indicates the resolution, on average 175 μm.

The SLD chamber follows similar principles but with two refinements. The electric drift field is controlled very precisely by a full box of wires, closing the cell ends and forming a grid on either side of the sense-wire planes. This allows them to use a non-saturated gas (ie one in which the drift velocity depends strongly on the electric field) and a much slower drift velocity of 9 μm/ns. This improves the position resolution, and the improved field shaping gives better two track resolution of 1 mm. In addition to the precise z-resolution given by stereo, every sense wire provides a three-dimensional space point using charge division along the sense wire.

Drift Chambers for use in High Magnetic Fields

Chamber Layout. The higher particle energies at the TeVatron and HERA force the use of higher magnetic fields (1.5T, 1.8T) for measurement of particle momenta in CDF and ZEUS.

Consider a naive model of electron drift through a gas at constant velocity. The total force on it

$$F = e(E + v \times B) - ev/\mu = 0$$

where the last term represents the viscous drag of the material. Examining figure 6(a) and taking the component of F at right angles to v one sees that the direction of motion is rotated away from E through angle θ where $\sin\theta = vB/E$. (**B** at right angles to **E**). Thus the demand of high B (for precision) and high v (for repetition rate) forces θ large or E large (with penalties in chamber construction). For example for a gas with v=50 μm/ns. $B = 2T$, $E = 1.4$ kV/cm one finds $\theta = 45°$. In practice, the electron velocity is altered continuously by collisions and acceleration in the field. The naive formula underestimates the rotation angle (the "Lorentz" angle) by 10 to 20%. Figure 6(b) gives results for two gases.

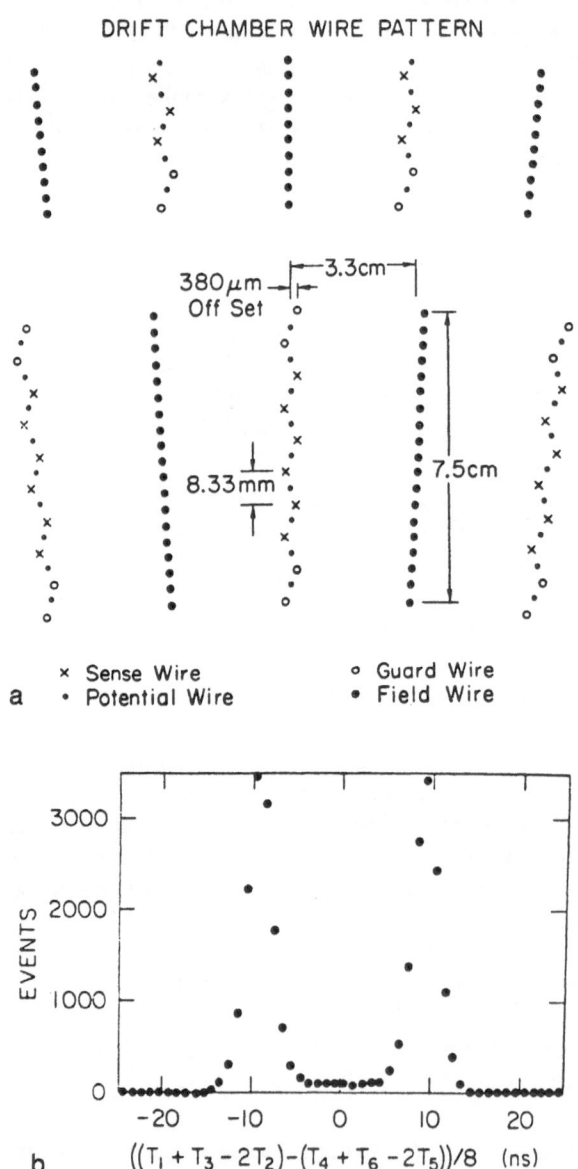

DRIFT CHAMBER WIRE PATTERN

Figure 5. (a) Mark 2 drift cell layout (b) drift time sums for tracks passing through cells.

Both CDF and ZEUS have chosen to use a Lorentz angle of 45°, and to achieve an azimuthal electron drift by rotating the line of sense wires by 45°. Figure 7 shows the ZEUS layout, with five axial and four stereo superlayers, each cell having eight sense wires, located, for reasons of electrostatic stability, in a straight line without stagger at the mid point of a line joining two corresponding high voltage wires[5]. Figure 8 illustrates the rejection of left-right ambiguities, false tracks rotated by $\tan^{-1} 2 = 63°$. In each superlayer and fail to link to other layers,) and of out of time tracks. An error of 96 ns in the time origin produces a jog of 9.6 mm (at a drift velocity of 50 μm/ns) each time a track crosses a line of sense or high voltage wires. This same geometry allows wire by wire calibration of time origins, and of drift velocity.

Signal Processing. The maximum drift distance of 2.5 cm corresponds to a drift time of 500 ns, several times the inter-crossing interval, but much less than the microseconds needed for a trigger decision. It is thus necessary to build a "pipelined" data flow and trigger

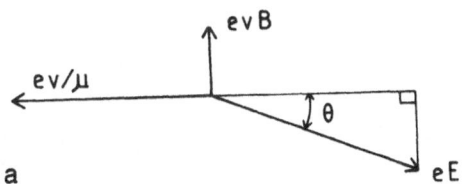

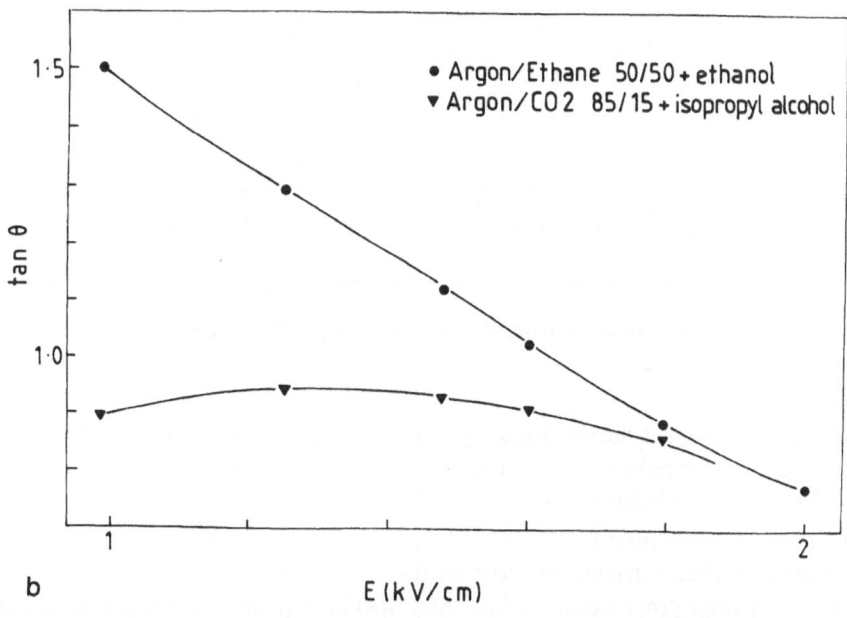

Figure 6. (a) Forces on an electron moving through a gas, (b) measurements of Lorentz angle for B = 1.83T - Ar/C2H6 50/50 bubbled through ethanol, and Ar/CO$_2$ 85/15 bubbled through isopropyl alcohol.

system. On-chamber preamps drive 45 m of coaxial cable which, via a postamplifier, feeds an 8-bit 104 MHz FADC. This has two outputs - a wire hit-flag that is in level-1 trigger logic, and a digital pipeline with 512 steps of 9.6 ns. After 5μs a level-1 trigger decision is received and the data are either copied into a primary buffer (at a rate below 1 kHz) or discarded.

The data-flow in the pipeline is tremendous: 4608 channels giving 8 bits every 9.6 ns= 0.5Tbaud. At the primary buffer the flow is 30 Mbaud. To reduce it further before onward

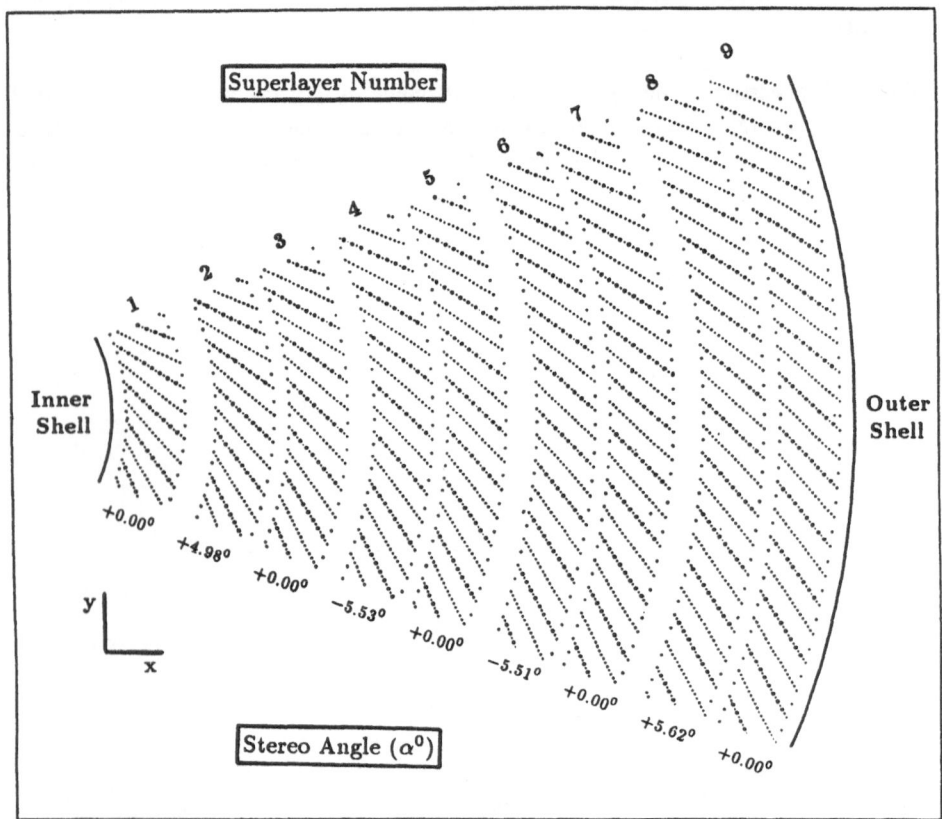

Figure 7. Layout of an octant of the ZEUS Central Tracking Detector.

transmission one analyses the raw FADC profiles to extract hit information (time and pulse height). Each chamber pulse on the FADC has a rise time peaking in about 18 ns but falls in about 100 ns with the $1/t$ shape characteristic of the charge induced on the sense wire as the positive ions drift away from it. We use a digital filter[6] to deconvolute this standard shape and discover the electron arrival time that caused it.

The digital filter processes the FADC bias $S(n)$ in sequence to give a response of the form

$$R(n) = AS(n) + BS(n-1) + CS(n-2) - DR(n-1) - ER(n-2).$$

(The sum and product maths is well-suited to implementation on a digital signal processor.) With appropriate constants this looks for an 18 ns risetime with a low Q-value ($Q = 0.5$)

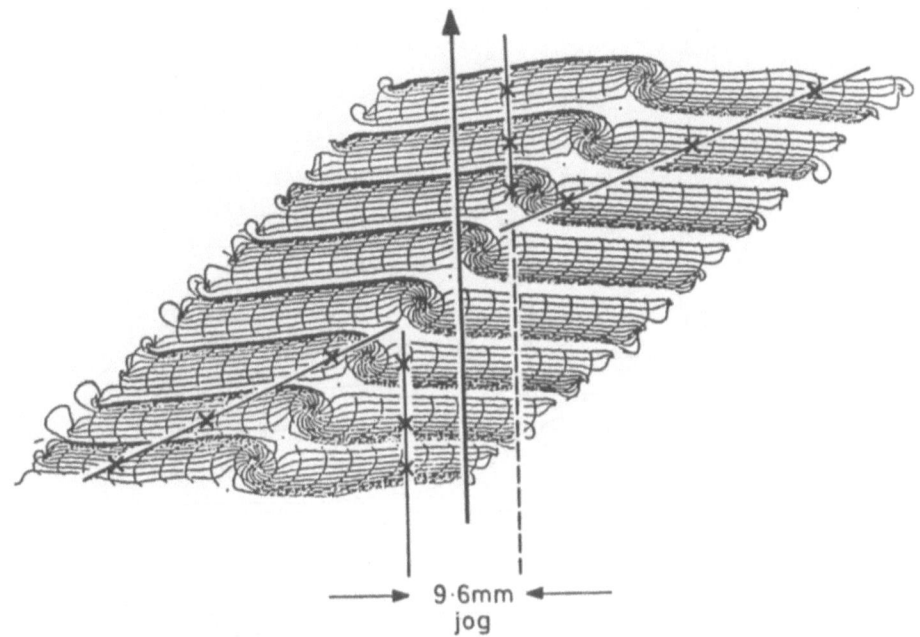

Figure 8. A Zeus drift cell, showing electron drift directions, equal drift time contours, (50 ns intervals) and the hits reconstructed for a throughgoing track one beam-crossing out of time.

designed to accept pulse-to-pulse variation and to prevent ringing. Figure 9 shows pure and filtered signals on test beam data, for a track passing through eight layers. The filtered pulses are then analysed for drift time (using a constant-fraction algorithm) and dE/dx (based on pulse height.) The filtering maximises the use of the information available. Position resolutions are improved by about 10%, and two track resolution from 2.1 mm to 1.6 mm. The electron dE/dx resolution for a track traversing 72 layers is 6% or better.

On the three innermost axial superlayers the position of each pulse along the wire is measured using the time difference between the signals seen at the two ends of the chamber, measured using a constant fraction discriminator at each end driving the start and stop of a time-to-amplitude converter[7]. A time difference resolution of 200ps = 3 cm is obtained. Note that the time differences of two samples of the same avalanche can be measured very accurately. The pulse-to-pulse variation which dominates the position resolution does not enter here. Since the maximum drift time (13 ns over 2m) is less than the two-hit resolution (1.6 mm = 32 ns drift time) every hit can have its z-coordinate measured in time to supply information to the level-1 trigger. The Central Tracking Detector can therefore be used as part of the event trigger to localise the interaction point along the beam-line in order to suppress the dominant beam-gas background.

EVENT RECONSTRUCTION AT pp SUPERCOLLIDERS

Event Characteristics and Detector Requirements

At SSC (40 TeV cm energy, luminosity $\mathcal{L}=10^{33}$ cm^{-2} s^{-1}), or LHC (16 TeV, $1.7.10^{34}$ cm^{-2} s^{-1}) soft hadronic collisions are produced in profusion. (10^8 s^{-1} at $\mathcal{L}=10^{33}$ cm^{-2} s^{-1}).

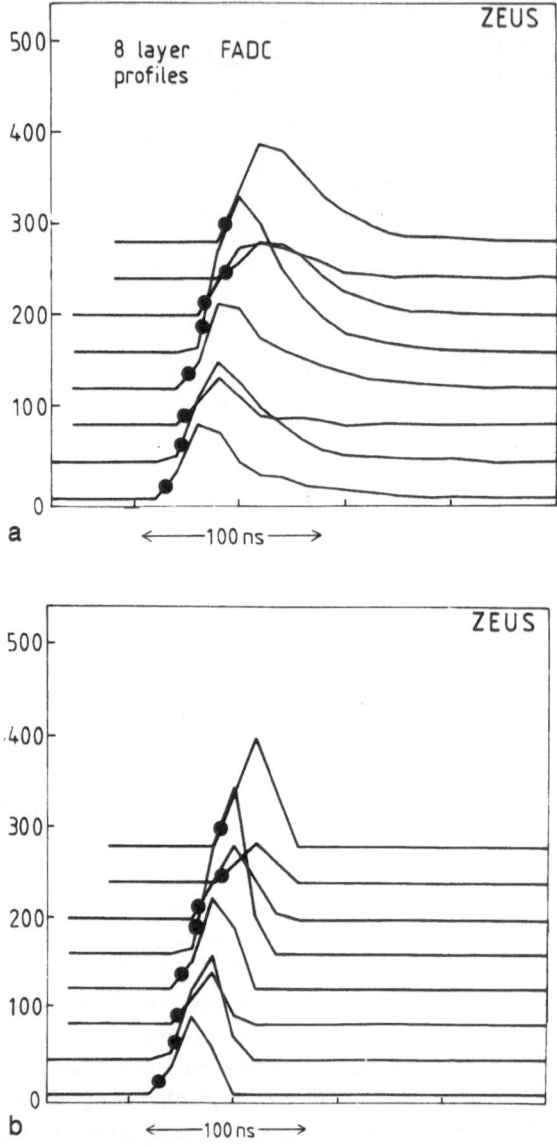

Figure 9. FADC profiles, (a) pure (b) filtered, for a track traversing eight sense layers, indicating the times assigned by a constant fraction algorithm.

Each of these produces a spectrum of particles which are limited in p_T, $<p_T> = 0.6$ GeV/c) (see figure 10) and flat in rapidity, with the density of 1.0 particles per unit of rapidity per unit of azimuth. (The appendix demonstrates the flat rapidity distribution characteristic of parton collisions.) Thus a detector spanning ± 2.5 units of rapidity sees on average 31 charged particles. For magnetic tracking with $B = 2$T, $r = 1.8$m an average of four of these form loops which orbit several times, congesting the detector, before escaping from the detector end caps. A TeV jet-event by contrast contains say 200 tracks. One needs to resolve angular differences of 0.5 to 1 mrad. Thus for operation at $\mathcal{L} = 1.10^{33}$ cm^{-2} s^{-1} and 15 ns crossing a detector needs to accept say 220 tracks. At $\mathcal{L} = 4.10^{34}$ one expects $200 + 40 \times 31 \times 1.5 = 2060$ tracks. At the lower luminosity one might wish to resolve two or three production vertices along the beam, and to tag b-decays by vertex characteristics. At the higher luminosity individual events are not resolved. One requires only to see isolated

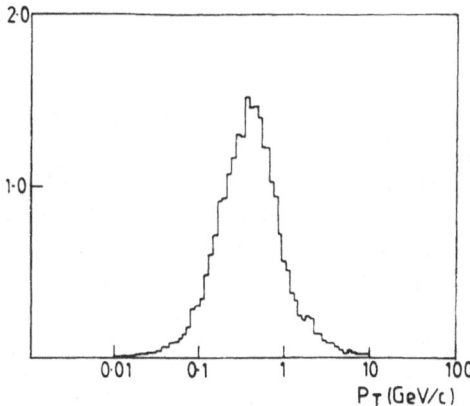

Figure 10. Transverse momentum distribution of particles produced in soft hadronic collisions at the LHC.

leptons and jets. A measure of detector congestions is the "occupancy", defined as the fraction of detector cells occupied, equal to (as long as the occupancy is low) the fraction of hits lost. One calculates the occupancy using the formula

$$\text{occupancy} = \frac{\text{(hit frequency) x (busy time of detector)}}{\text{No of parallel cells}}$$

Thus if one designs for constant occupancy of 1% at $\mathcal{L} = 1.7.10^{34}$ (= 1000 tracks for busy time = crossing time), then 100,000 cells per layer are needed. One easily arrives at detectors with several million elements. We note two features. At a radius of 30 cm, say, 100,000 azimuthal divisions give a position resolution (= strip width/$\sqrt{12}$) of 19 μm, so that one is naturally led to record hits on strips, rather than drift times - this helps detector speed and simplicity. Secondly, since the particles radiate from the origin the detector element size can grow as r^2.

The physics aims of an inner tracker at maximum luminosity are to reconstruct gauge bosons via leptonic decays. This demands the reconstruction in three dimensions of high-p_T isolated leptons; that e and μ are identified with background rejection at the 10^{-4} level; that $e^{\pm}$ signs are measured, say $\sigma(p)/p = 0.3$ at $p_T = 500$ GeV/c; and that the detector contributes to the level-2 electron trigger. Muon identification is accomplished using and behind the calorimeter. We discuss electron identification in the next section. At lower luminosity one focuses on (higher cross-section) processes characterised by quark and lepton flavours, such as $H^o \to b\bar{b}$; $gg \to t\bar{t} \to W^+ W^- b\bar{b}$, $t \to H^+ b$, $H^+ \to \tau^+ \nu$. In this case flavour tagging using vertex quantities becomes important. SDC in addition wish to reconstruct particles within jets, to understand the QCD input and background to gauge processes, and add level-1 information to the tracking trigger. Note that at high luminosities beam-gas interaction backgrounds are negligible and the beam lifetime is dominated by beam-beam collisions. Therefore essentially all the radiation dose limiting the detector lifetime comes from the physics reactions we produce in the detector.

The issues discussed in this and the following sections are explored a little further in references 8 and 9.

Finding High-p_T Electron Events

A calorimeter trigger, requiring a high energy deposit in an electromagnetic calorimeter cell, (corresponding to $p_T > 20$ GeV say,) is sufficient to reduce the rate to the 10 kHz or so needed for an LHC/SSC trigger, even at luminosities above 10^{34}. This trigger will select events with high-p_T e or γ, but also predominantly QCD jets with large π^0 content. The immediate function of the inner tracking detectors is to reduce this trigger rate at level-2 by looking for a high-p_T track pointing out this calorimeter cell, and to identify this as an electron. We discuss here two possible methods; first, identifying an electron by E(calorimeter) $= p$(tracking) and secondly by using a transition-radiation signature for electrons.

Precision Tracking. In the first method one uses a superlayer of Si strips 200 μm wide and 10 cm long at radii of 70 to 90 cm (four φ-layers in a field of 2T) to reconstruct high p_T tracks, as shown in figure 3(a). Setting a threshold at say $p_T = 30$ GeV/c one achieves efficiency above 90% for isolated electrons of $p_T > 35$ GeV/c, even in the presence of a 10^{34} luminosity background, with a rejection factor for jets of 50 to 100. One then refines the electron signature by demanding E(cal) $= p$ (track).

The $E = p$ test runs into possible difficulties because of material in the tracking detectors. If a second superlayer is used at mid-radius to optimise the sagitta measurement (cf fig. 3(b)) it will act as a source of bremsstrahlung for electrons and the ideal Gaussian energy response of the calorimeter (using a cluster of 3 x 3 cells around the cell with the highest energy,) is degraded by a long tail going down to very low shower energies. However this can be rescued because one knows where the radiated photon went (emitted in the track direction at the traversed matter) and can add its energy into the shower cluster. Further, one can refit the electron track allowing for the change in curvature here. Poppleton has shown that for a mid-tracking layer of 2% of a radiation length if one demands 98% efficiency for identifying an electron of 10 GeV one can tighten the E/p cut from $E/p < 1.7$ to $E/p < 1.2$ with a substantial reduction in QCD jet background, even at a luminosity of 2.10^{34}. Figure 11 illustrates the improvement. Note that even at this low amount of material, it is the material which dominates the resolution, and hence the background rejection: improved tracking or electromagnetic calorimetry does not help.

Transition Radiation Detection. An independent technique utilises the radiation emitted by particles traversing material boundaries. This "transition radiation" (emission depends on $\gamma = E/m$ of the particle.) Low-Z materials, (CH_2 foils or foam are commonly used) give the best yield of X-rays without self absorption. The RD6 collaboration[10] have studied detectors using multilayers of 4mm diameter single wire drift tubes ("straw-tube" detectors) embedded in CH_2 foam or foils. To convert and detect T.R.photons a Xe: CO_2: CF_4 (70:20:10) gas is used. Figure 12(a) shows the probability that the energy deposit in a straw exceeds 5 keV as a function of $\gamma = E/m$. The energy scales for e and π are displaced by a factor $m_\pi/m_e = 264$.

The straw detectors are very simple in construction and give uniform gains to within ±2%. The choice of materials and gases gives a negligible drop in gain even at a charge deposit for 5C/cm of wire. Let us calculate the detector lifetime. A pion gives a typical energy deposit of 2 keV = 70 electrons. At a gas gain of 10^4 this is 0.11 pC. 1 cm of wire at a distance of 50 cm at 90° subtends $d\eta d\phi = (1/50) \times (0.4/.50) = 1.6.10^{-4}$ and hence $1.6.10^{-4}$ particles/interaction pass through. One year $(10^7 s)$ at $\mathcal{L} = 4.10^{34}$ gives $10^7 \times 4.10^{34} \times 10^{-25} = 4.10^{16}$ interactions, (total inelastic cross section = 10^{-25} cm^2) and hence a dose of $0.11.10^{-12} \times 1.6.10^{-4} \times 4.10^{16}$ C/cm = 0.7 C/cm. To accumulate 5 C/cm would take 7 years. Tests up to this point have shown negligible damage, whereas other combinations of gas mixtures, construction materials and sealants and gas system lubricants can produce spectacular declines in performance at far lower doses.

The straws are read out using two thresholds. One (~ 200 to 500 eV) is efficient for minimum ionising particles. The other (5 keV) preferentially detects electrons. The electron signature is a track right through the detector with hits in, say, 40 layers of which some 22%

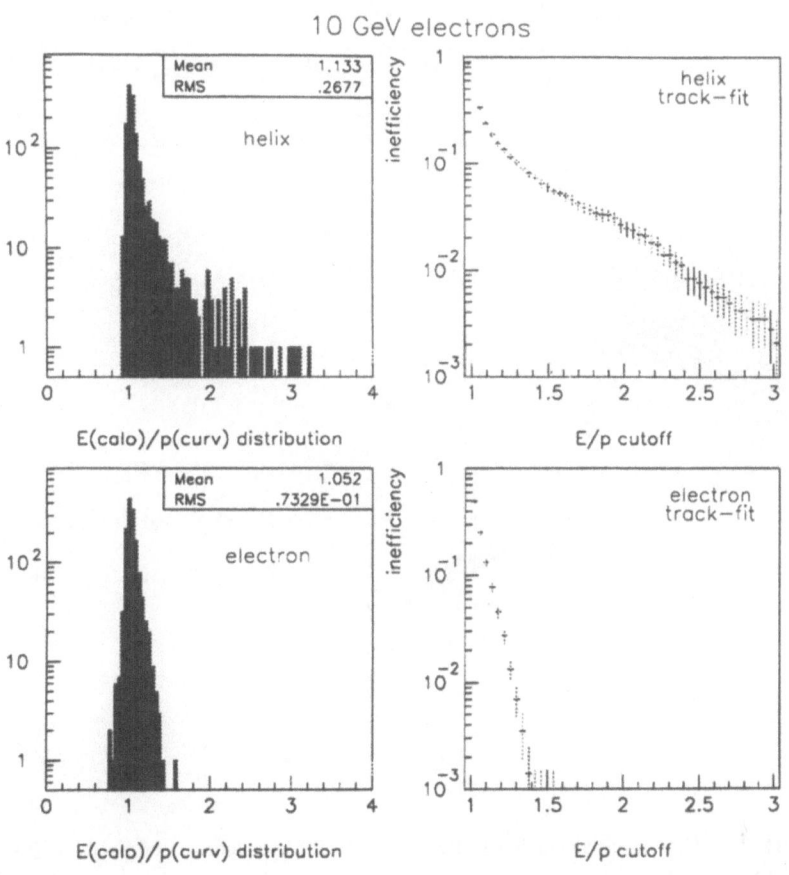

Figure 11. *E/p* distributions for electrons of p_T = 10 GeV/c using simple helix fit and fit allowing for bremsstrahlung photons, together with fraction of electrons lost as a function of E/p cut off.

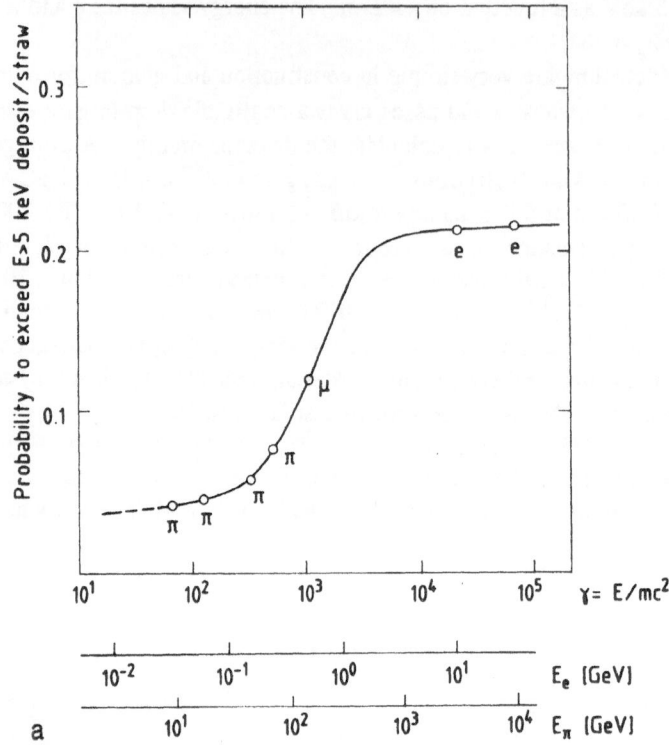

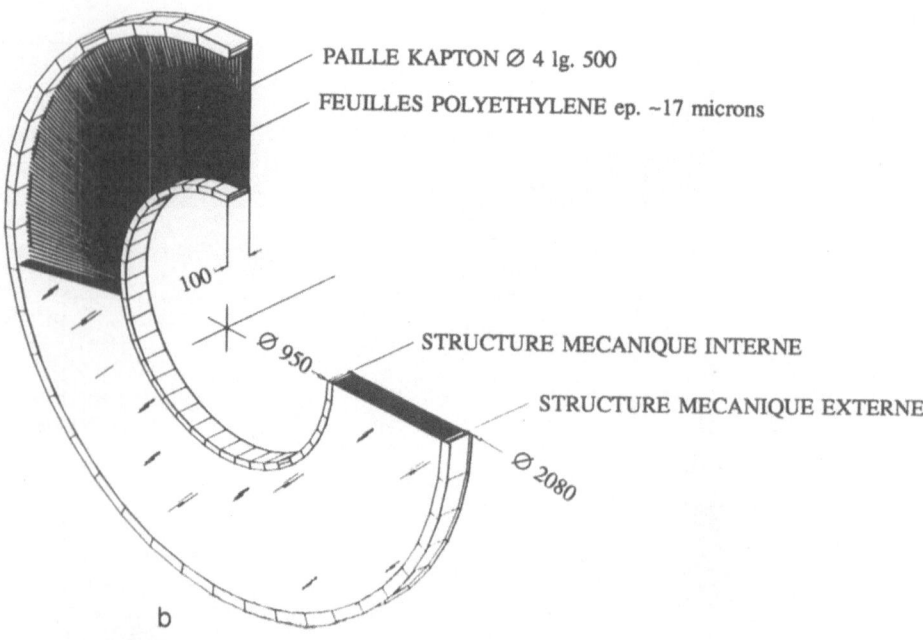

Figure 12. (a) TRD X-ray absorption in Xe - CO_2 (50:50) as a function of $\gamma = E/m$ (b) layout of forward TRD wheel. The real detector has a sequence of up to ten such wheels.

are at the high threshold[1]. One detector layout that has been proposed for tracking at medium rapidity ($1 < \eta < 2.5$) consists of a sequence of detector wheels, each of which has CH_2 foils and radial straws located at fixed z. The density of straws is arranged such that a particle traverses 40 straws in passing from $r = 50$ to $r = 90$ cm at all rapidities. Referring to equation (9) we see that tracks from pp collisons, close to the nominal axis, appear as straight lines in the (ϕz) plane. (Electrons from photon conversions in the material form sine waves in this plane.) The slope of a track, $d\phi/dz = 0.3$ $(-q)$ $B/2p_z$ and the change in ϕ from track start to finish $= 0.3$ $(-q)$ $B\Delta r/2p_T$ where Δr is the radial extent of the tracker. Such isolated electron tracks are readily found by a level-2 trigger processor even in the presence of 80 minimum bias events. The probability with 80 pile-up events to make a fake track of ($p_T > 10$ GeV) which points at a calorimeter cluster ($\Delta\eta\Delta\phi = 0.2 \times 0.06$) is 0.07% for this detector. In addition the TRD electron signature provides a pion rejection factor of 20 at 20 GeV at the same luminosity, and much better at lower luminosities.

Microstrip Gas Counters: (MSGCs). We have described so far gaseous detectors using anode separations of 5 cm (drift chambers) and 4 mm (TRD straws.) To run at high luminosities without using high TR thresholds to reduce the occupancy one seeks a further reduction in detector granularity. Is this possible using a gaseous detector with avalanche multipliation? Not with wires held in place by tension - at close spacings electrostatic instabilities develop. But we can do it if we use not free-space wires, but metal strips etched on to a glass, plastic or silicon substrate[11]. The basic construction is illustrated in figure 13(a). Particles pass perpendicularly through this detector, depositing ionisation in the gas. This drifts under an electric field to anode strips at say 200 μm pitch on a substrate. Cathode strips in between provide additional electric field lines to form a gas avalanche of gain up to 1.10^4 (using an anode cathode voltage of 700 - 800 volts). The positive ions produced in the avalanche do not drift back into the ionisation volume, but almost all are captured by the anode strips.

Excellent performance results. For 6 keV X-rays from ^{55}Fe a pulse height of 12% FWHM has been seen. Position resolutions of 40 μm have been seen in Xe-DME and 47μm in Ar-DME using pulse height on centre of gravity methods. Using thin substrates (~ detector pitch) and strips the orthogonal coordinate can be read out using induced pulses, with similar resolution. Induced signals have also been used to give pixel readout. The gain is constant up to rates exceeding 10^5 mm^{-2} s^{-1} (15 cm at $\mathcal{L} = 4.10^{34}$).

Of course there are some problems, as with any new technology. Most are associated with the charging up and possible migration of ions in the substrate. Depending on the substrate resistivity (10^9 to 10^{15} ohm-cm) rapid short-term loss of gain is seen as positive ions hit the insulating substrate and distort the electric field causing the avalanche gain. A solution is to supply a back electrode charged to a potential such that no field lines cross the dielectric surface between the anode and the cathode strips. Then no charge will be deposited. One can adjust the surface conductivity to 10^{11} ohm/square either at the surface by vacuum evaporation or gas implantation (causing high electric fields and risk of breakdown) or by choice of substrate material (more DC current = more noise.) Long-term high rate stable operation is achieved with certain glass substrates.

This is a very active field of development. Rapid progress is expected. Compared to silicon detectors the process is cheaper, and larger areas can be made. But one should recall not only the overheads on a gaseous system but that in a 67 MHz rad-hard multistrip detector well over half the cost is the readout, which is largely independent of the detector technology.

---

[1] Electron track finding looks very simple if one uses only hits of 5 keV or more. 95% of all pion hits vanish and the detector looks very empty

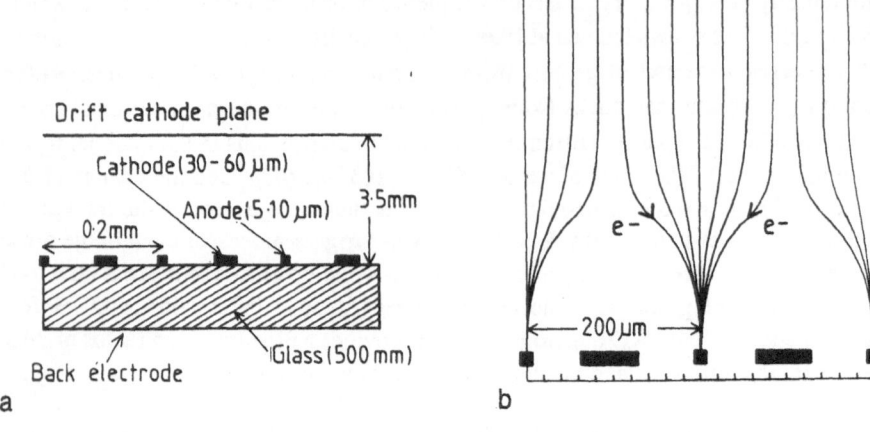

Figure 13. (a) Construction of a microstrip gas counter (b) electron drift lines.

Scintillating Fibre Detectors. Scintillating fibres form an alternative technology for supercollider tracking. We discuss two developments: first, the use of fibres of 500-1000 μm diameter, and secondly the use of 30-60 μm diameter fibres. Consider first a fibre of diameter 800 μm in which a minimum ionising particle releases 800 photons. The fibre of refractive index n_1 is clad in a glass of index n_2, giving a critical angle $\sin\theta = n_2/n_1$, and the fraction of light which is trapped and propagates down the fibre towards the photomultiplier is $0.5 (1 - n_2/n_1)$, typically 3% for $n_2 = 1.49$, $n_1 = 1.58$. Perhaps 50% is lost in attenuation to a photomultiplier, which is therefore struck by 12 photons. So called 'Visible Light Photon-Counters'[12] offer a quantum efficiency of 80%, a risetime of about 10 ns and a gain of 10^4, with a pixel structure over the surface of this solid state device which runs at a temperature of 7K.

A different approach is adopted by Kuroda *et al*[13] using conventional photomultipliers, but with the anode divided into say 228 independent pixels. The intrinsic speed of this device is at the 200 ps level and this allows high speed multiplexing. The detector fibres of 0.5 mm diameter form a 5 mm thick layer and are grouped on the photocathode so that neighbouring pixels correspond to a 250 μm shift across the surface for normally incident tracks. Large angle (*ie* low-momentum) tracks deposit their ionisation across several pixels and give lower pulse heights. Thresholding the signals therefore provides a rejection factor against low-momentum tracks. With this high speed, the detector is proposed as part of a level-1 tracking trigger for supercollider experiments.

Leutz *et al*[14] have investigated the use of much smaller diameter fibres. This is justified if the photoelectron yield is higher and if cross-talk between fibres is suppressed. The solution to both problems is to use a scintillator dopant with large Stokes shift. PMP is used as a scintillator. This has no overlap in wavelength between its emission and absorption bands and large concentrations can be used. Compare POPOP + p-terphenyl in which the small Stokes shift forces it to be dilute. In addition the two-stage wavelength shift involves a radiative transfer with 300 μm absorption length. Thus the intermediate light would travel across ten 30 μm fibres before reabsorption and the localisation would be lost. The wavelength shifting in PMP is by vibration and is local, so narrow fibres can be used. Tests with 30 μm fibres show 2.4 hits/mm of track. A resolution of 35 μm plus non-transmission tails is found in track-fitting. They propose a detector layout with three axial shells. Each shell is 1 cm thick and has 60 z-layers, 60 each plus and minus stereo layers and 60 z-layers again within this thickness. A track scores 25 hits in this shell, giving a three-dimensional vector.

The information flow is tremendous. For a detector with three shells at 16, 30 and 45 cm radius readout both ends there are 8.10^9 pixels to be stored in a pipeline every 15 ns (a rate of 67 Pbaud). They devised an ingenuous pipeline memory in which electrons are emitted from a photocathode and drifted with energies about 4 eV in an $E = 0$ region along the magnetic field lines of the detector. (The magnetic lines preserve the photocathode image by preventing transverse movement.) In 0.5 μs they drift 60 cm and are reflected by an electric potential gradient, arranged so that the more energetic electrons travel further before reflection so that all the photoelectrons from a given crossing arrive back close to the photocathode simultaneously. If the event is not wanted they are lost. Otherwise a large voltage pulse accelerates them back through the system where they can be read out using for example a CCD array.

To date scintillating fibres have been used successfully in non-magnetic detectors. Experience tells us that control of cross-talk (in readout) and noise are vital to avoid excessive ambiguities.

VERTEX PHYSICS AT HIGH-ENERGY COLLIDERS

A good vertex detector should be designed with several features: good point precision, accurate alignment, minimal scattering, minimal extrapolation, good two track resolution and excellent track finding. For use at LHC/SSC it needs high rate ability, high radiation tolerance and needs to manage its own heat output. A detector which measures the z-coordinates as well as $r\phi$ has a distinct advantage in many of these needs[15].

The importance of good track finding cannot be overestimated[16]. The signature for heavy flavours or tau leptons is that tracks do not appear to come from the main production vertex. Photon conversions, K^0-decays, secondary interactions in material, multiple scattering, detector noise, and in particular track finding errors - false associations of hits with tracks - all conspire to mimic this behaviour. Any sample of heavy flavour candidate tracks is polluted by low-grade spurious or erroneously recorded tracks[16,17]. Point precision, lever arm, and especially redundancy to provide constrained track finding are the weapons against ambiguity. The detector layout suggested for precision by equation (6) is the worst for track-finding errors. A more continuous layout provides stronger χ^2-test constraints. An added confusion affecting tau and charm events is that these are more congested in the vertex detector than typical events, because of the low Q-values of the decays.

Decay Length and Impact Parameter

Figure 14 illustrates the definition of "impact parameter", δ, the distance by which an extrapolated track misses the true (or nominal) production vertex. δ is defined to be positive if it corresponds to a positive decay time for the parent particle. Negative values can arise through measuring or track-finding errors, and if the primary vertex is displaced from its assumed position, such as the centre of the beam-spot (see also ref.17).

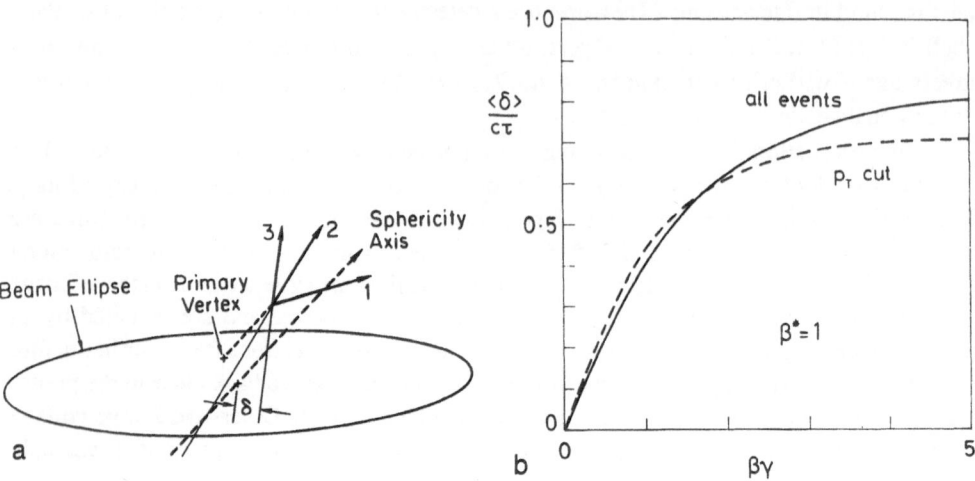

Figure 14. (a) Definition of impact parameter, δ (b) < δ > plotted against Lorentz boost for an emitted electron.

The decay length $\ell = \beta\gamma ct$ where β is the parent velocity and t its decay time (the γ factor arises from time dilation.) The angle between the parent and daughter tracks is, for large γ, proportional to 1/γ, so that $\delta \sim \beta ct$. Figure 14(b) plots the average over many events, $<\delta>/ct$ against the boost βγ, assuming β = 1 for the daughter particle in the rest frame of the parent. $<\delta>/ct$ tends to 1 for very large βγ. However the large values of δ arise from the few particles which are slow in the laboratory. These cannot be used because of multiple scattering. After a cut such as $p_T > 1$ GeV/c with respect to the jet axis for b-decay one finds $<\delta> \sim 0.7\ ct$ (~ 270 μm) for βγ > 3, (ie p > 15 GeV/c), with a very slow dependence on βγ.

One sees now the virtues of tagging b-events by the use of impact parameter. This quantity can be calculated track by track, (on average 5.5 tracks from b-decay,) and low momentum or poorly reconstructed tracks rejected without losing the event. We do not need all the b-decay tracks and so do not need to identify a particular decay mode. We do not need to reconstruct the b-decay point. We do not need to know the b-momentum. As a protection against track-reconstruction errors, it is common to demand at least two or three tracks with δ exceeding a cut. Note that the most common b-decay time is zero. In this case δ is small (zero within errors) for all tracks. Therefore a b-tag method based on decay geometry can never reach 100% efficiency.

If one reconstructs the decay vertex explicitly by intersecting tracks[1], the fractional error on the lifetime is the same as that on mean impact parameter, the longitudinal error increases as 1/θ (figure 14(a)), and hence is proportional to γ, which cancels the benefit from time dilation.

Detector Layout

Precision on impact parameter is therefore a significant figure of merit for a vertex detector. For a detector with two layers of measuring error, σ, at radius r_1 and r_2, with material of t radiation length (beam pipe plus detector) located at r_1, one can show from equation (6) that

$$\sigma_\delta^2 = \sigma^2 \frac{r_1^2 + r_2^2}{(r_1 - r_2)^2} + r_1^2 \left(\frac{0.015}{p\beta}\right)^2 t \sin^{-3}\theta$$

The best vertex detector therefore has σ, r_1, t small and $(r_2 - r_1)/r_1$ large, coupled with internal geometric stability, monitoring and good track finding and two track resolution.

The properties of the solid state vertex detectors presently at high energy colliders are given in table 1. One sees the advantages of a small value of r_1. At LEP the X-ray flux from the electron beams forces the detectors out to a larger radius than at SLC or the TeVatron. The detectors with two coordinates of Si readout have more robust track-finding. The quality of the physics is much influenced by this. The better the quality of the data the less the background will be and the higher the yield of b-quarks. Note that table 1 must be read with care. The OPAL performance is, for example, improved by the gaseous vertex detector outside. Impact parameter resolutions are given for high momentum tracks. In some cases these numbers are those expected to be achieved after various alignments etc.

Table 1. Properties of vertex detectors

	Aleph	Delphi	Opal	SLD	CDF
Inner radius (mm)	65	63	61	29	29
Outer radius (mm)	115	110	75	41	81
Coord meas.	rϕ,z	rϕ	rϕ	rϕ,z	rϕ
	Orthog.Strips			*Pixels*	
No of layers	2	3	2	4	4
Readout pitch (μm)	100	50	50	20,20	60,55
Signal/noise	15	17	22	20	11
Point resolution (μm)	12,13	8	5	5,5	10
Impact param. res (μm)	20,29	21	18	12,41	12

b-tagging

Figure 15(a) illustrates this by showing the impact parameter distributions in an LHC simulation of $gg \rightarrow t\bar{t} \rightarrow W^+W^- b\bar{b}$. By demanding two tracks above a certain impact parameter cut one can obtain a background suppression factor of 10, say, with a 50% yield of b-jets. This yields improve t-quark mass resolution using the Wb ($W \rightarrow$ jets) decay.

Bedeschi[19] has proposed a tagging method that makes better use of the event information. Close to the production point we may set R_0 to infinity and rewrite equation (7) as $r\sin(\phi - \phi_0) = d_0$. Then if a track comes from a decay vertex at (r_v, ϕ_v) we have

$$r_v \sin(\phi_v - \phi_0) = d_0.$$

This equation is satisfied by all tracks which come from this vertex. Therefore on a plot of d_0 against ϕ_0 the tracks lie on a straight line ($\phi_v - \phi_0$ is small). Figure 15(b) illustrates this. Two linear clusters are seen. The one on the right is confused by a secondary charm decay. This (d_0, ϕ_0) topology allows us to develop a tagging strategy as follows:

Find the primary event vertex in (r, ϕ). Move the origin of coordinates here. Cut all tracks consistent with it within three standard deviations. Then look for linear clusters in (d_0, ϕ_0) space.

CDF hope for a background rejection factor of 30 and a high b-tagging efficiency. With a pixel or crossed strip detector, the method can be generalised to three dimensions with resulting performance improvements.

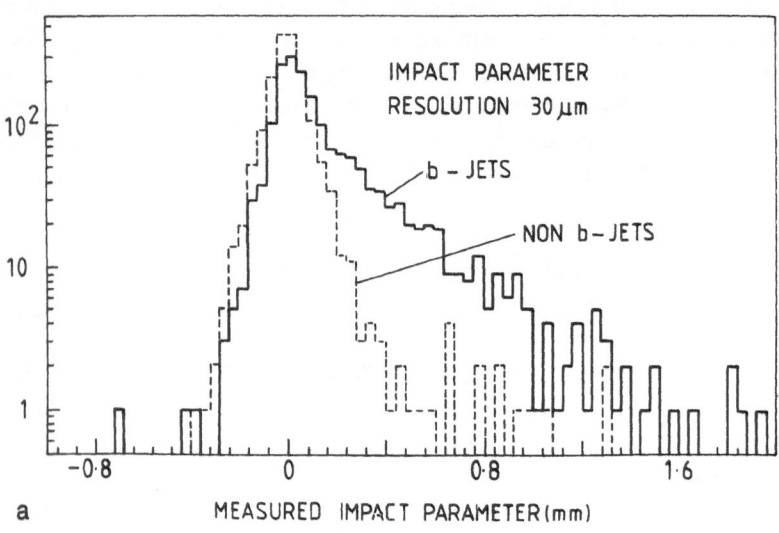

a

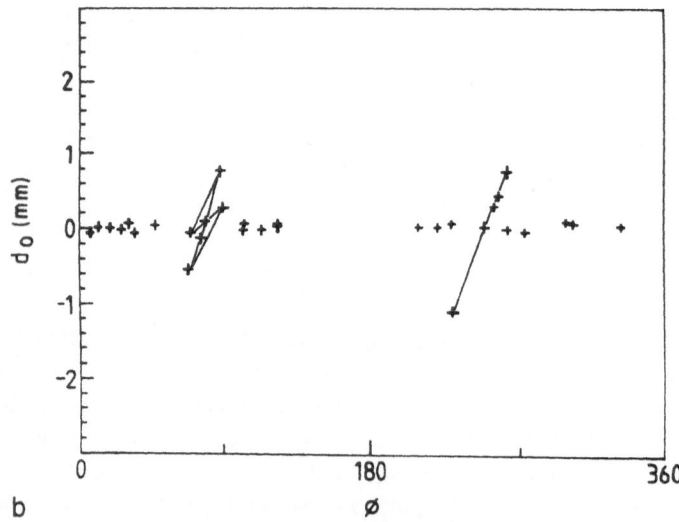

b

Figure 15. (a) LHC simulation (EAGLE expression of interest). tt events with $t \rightarrow Wb$, one $W \rightarrow e\nu$ or $\mu\nu$ with p_T (e,μ) > 40 GeV as tag : look at jets in the opposite hemisphere, p_T (jet) > 40 GeV and p_T (track) > 2 GeV (b) scatterplot of d_0 against ϕ_0 (after ref. 19).

Radiation Hardness

Figure 16(a) shows the results of calculations for LHC of radiation damage per year at LHC at $\mathcal{L} = 1.10^{34}$ due to charged particles, (following an inverse square law from the production point,) and neutrons back scattered from the calorimeter (isotropic to first order). The two curves are normalised on the assumption that one neutron does 8 to 10 times the damage caused by one minimum ionising particle[20]. Curve (1) shows the damage per detector element, assuming constant occupancy. We see that small radii are referred, but note that the beam line will be an additional source of neutrons. We mention here some recent results on the knowledge of radiation damage to detectors, but ignore the equally necessary issue of radiation hard electronics.

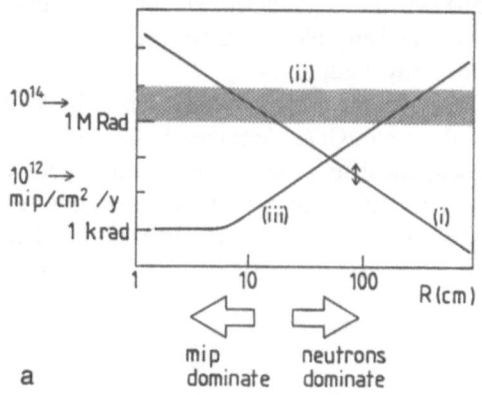

a

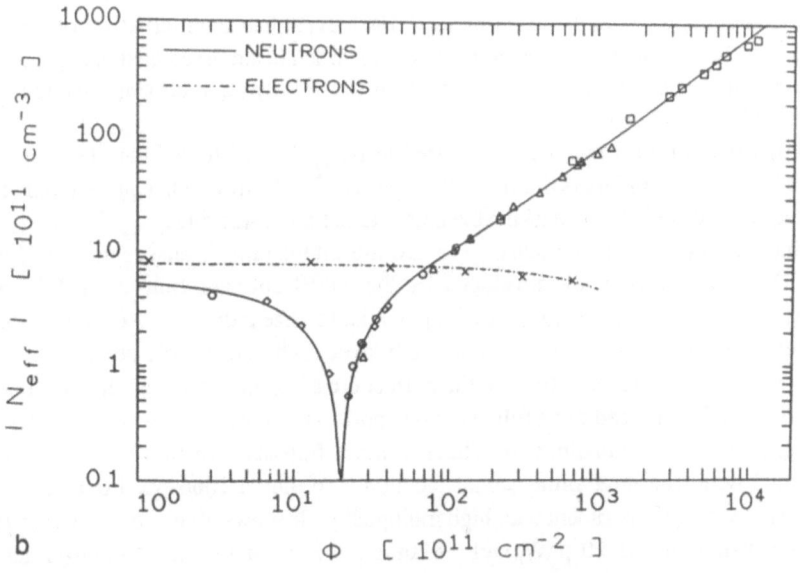

b

Figure 16. (a) Silicon detector degradation as a function of radius due to [(i) minimum ionisation particles (ii) neutrons (no calorimeter shielding by CH_2) (iii) relative damage per detector element assuming constant occupancy)] (b) absolute value of effective dopant concentration versus normalised fluence for neutrons and electrons (from ref. 20).

The main cause of radiation damage in silicon detectors is the displacement of atoms by non-ionising recoils. These cause lattice damage, and create energy levels in the band gap. These enhance leakage currents, and result in charge trapping. This bulk damage converts n-type Si to p-type, the inversion taking place at an integrated neutron flux of 2.10^{13} cm^{-2}. (see figure 16(b), taken from ref. 20). The interstrip resistance is reduced, both on n and p surfaces, but remains usable. So long as one is prepared to raise the depletion voltage, the detector continues to operate. Particularly helpful is the annealing of lattice defects with time. It seems that the leakage current and equivalent noise current after 10 LHC-years are less than twice those after one year, and the damage saturates with time. If necessary one can cool the detector during data taking to reduce leakage currents, but anneal at room temperature between runs.

Silicon charge collection is a little slow by SSC/LHC standards. Figure 9(a) could as well stand for the time development of a Si pulse as for the ZEUS central tracking detector. The RD20 collaboration have developed pipelined readout and (in their case analogue) signal filtering in a similar way, and are able to localise a signal to one crossing in a given strip. Signals 30 ns away are clearly resolved.

A medium which offers advantages over silicon in both speed (4 ns at base) and radiation hardness, is the use of GaAs detectors[21]. These are much more robust against radiation, showing negligible damage up to 10^{14} neutrons per cm^2. (The pulse height is reduced by a factor three at 7.10^{14} n cm^{-2}. Leakage current is not a factor in GaAs detectors. Rather the issue is loss of signal due to charge trapping at lattice defects.

Pixel Detectors

The SLD experiment have installed a four-layer vertex detector as specified in table 1. Some 480 CCDs have a total of $1.2.10^8$ pixels each 20×20 μm. The detectors are read out serially taking 150 ms but is dead timeless. Each event has about 40 kbyte of noise plus a similar number of counts from beam-background. Such beam-associated background would saturate a strip detector, but causes no problems to the fully three-dimensional track-finding supported by a pixel detector.

The r.m.s. impact parameter measured in (r, ϕ, z) is $\sqrt{2}$ times that measured in (r,ϕ) only, and the track finding is dramatically superior[15]. The device has just started (1992) to take serious data and offers considerable potential in flavour-tagging.

Such readout time, being sensitive to new hits all the time, is unacceptable to LHC/SSC use. One solution is being developed by the RD19 collaboration at CERN. Si pixel detectors, currently 75×500 μm but with potential for size reduction, are bump-bonded to a readout chip, so that an amplifier sits directly over each detector element. This amplifier pixels notify x and y readout lines of the correct crossing time at which the event occurred. When it is desired to read out a full event one polls only those x and y with time flags set to see which (x,y) combinations also have a pixel flag set. In this way fast readout is combined with true three-dimensional location. To date a 1006-pixel detector has been demonstrated in an experiment with high multiplicity. It shows 50-electron noise at 10 MHz. Power consumption is 20 μW/pixel. Manufacture in an industrial environment looks encouraging. More sophisticated devices are just around the corner.

CONCLUSIONS

There are promising tracking technologies available for the future in gas, scintillating fibre and solid state devices. Lifetimes exceeding 10^{14} n cm^{-2} are achievable, provided readout electronics can be hardened to the same degree. Detectors of 10^7 elements with low occupancies are easily envisage.

These complement the performance of calorimeters both in electron identification and muon measurement, and in flavour tagging, providing the trigger rate reduction required for Higgs sector spectrometry.

Techniques are advancing rapidly. A 1995 design will outperform a 1993 design of the same cost. One should plan carefully one's strategy for deciding on a particular technology in any given experiment.

APPENDIX - Flat rapidity distribution in parton-parton collisions

Consider two beams (e^+e^- or pp) of fixed energy colliding head-on and shedding collinear bosons ($\gamma;g$) with a spectrum $dN/dE = k/E$. (Note the $1/x$ form of the gluon structure function of the proton.) The two bosons collide head-on with energies E_1 and E_2, giving for the resulting system $E = E_1 + E_2$, $p_L = E_1 - E_2$. Then the mass of the system produced, $m^2 = E^2 - p^2 = 4E_1 E_2$, rapidity $y = \frac{1}{2}$ ℓn (E_1/E_2).

Now $d^2N/dE_1 dE_2 = k^2/E_1E_2$. We use the Jacobian $dm^2 dy = (m^2/E_1E_2)\, dE_1 dE_2$ to find $d^2N/dm^2 dy = k/m^2$, independent of y.

This flat y-distribution is characteristic of all processes, soft or hard, involving collisions of wee partons. We remind ourselves that for massless particles $E = p$, and $y = -\ell$n tan $\theta/2$.

$y = 0, 1, 2, 3, 4$, occur at $\theta = 90°, 40°, 15.4°, 5.7°, 2.1°, 0.8°$.

REFERENCES

1. D.H. Saxon, Three-dimensional track and vertex fitting in chambers with stereo wires, *Nucl. Instr.* A234:258 (1985)

2. R. Gluckstern, Uncertainties in track momentum and direction due to multiple scattering and measurement errors, *Nucl. Instr. Meth.* 24:381 (1963)

3. M. Regler and R. Frühwirth, Reconstruction of charged tracks *in:* "Techniques and Concepts in High Energy Physics V", T. Ferbel, ed., Plenum Press, New York (1990)

4. D.H. Saxon, Multicell drift chambers, *Nucl. Instr. Meth.* A265:20 (1988)

5. C.B. Brooks *et al.*, Development of the ZEUS central tracking detector, *Nucl. Instr. Meth.* A283:477 (1989)

6. R.W. Hamming. "Digital Filters", 2nd edition, Prentice Hall, Englewood Cliffs (1985)

7. N. Harnew *et al.*, Vertex triggering using time difference measurements in the ZEUS central tracking detector, *Nucl. Instr. Meth.* A279:290 (1989)

8. H.F-W Sadrozinski, A. Seiden, A.J.Weinstein, Tracking at the SSC/LHC, *Nucl. Instr. Meth.* A279:223 (1989)

9. D.H. Saxon, Physics reach and detector choice at future hadron colliders, *Nucl. Instr. Meth.* A315:1 (1992)

10. V.A. Polychronakos *et al.*, Integrated transition radiation and tracking detector for LHC, CERN/DRDC/91-47 (1991)

11. A. Oed *et al.*, *Nucl. Instr. Meth.* A263:351 (1988)
     F. Angelini *et al.*, *ibid* A283:755 (1989) *Nucl. Phys.* 23A:254 (1991)
     F. Hartjes *et al.*, CERN 89-10: 455 (1989)

C. Budtz-Jorgensen *et al.*, *Nucl. Instr. Meth.* A310:82 (1991)

R. Bouclier *et al.*, CERN-PPE/92-53 (1992)

12. Solenoidal Detector Collaboration. Technical Design Report SDC-92-201: p.4-82ff (1992)

13. K. Kuroda *et al.*, Readout of optical scintillation fibres by a position sensitive photomultiplier, *Nucl. Instr. Meth.* A260:114 (1987)

14. C. D'Ambrosio *et al.*, Present status and future programme of scintillating fibres for central tracking, *in:* "Proc. Large Hadron Collider Workshop," G. Jarlskog and D. Rein, eds., CERN 90-10, Vol.3, p.255 (1990)

15. C.J.S. Damerell, Vertex detectors, *in:* "Techniques and Concepts in High Energy Physics IV," T. Ferbel, ed., Plenum Press, New York (1987)

16. D.H. Saxon, Measurement of short-lived particles at PETRA, *Helvetica Physica Acta*, 60:646 (1987)

17. L. Lyons and D.H. Saxon, Measurement of short-lived particles in high energy physics, *Rep. Prog. Phys.*, 52:1015 (1989)

18. D.E. Klein *et al.*, (DELCO Collab.), *Phys. Rev.*, D37:41 (1988)

19. F. Bedeschi, Top quark physics with b tagging at LHC, *in:* "Proc. Large Hadron Collider Workshop," G. Jarlskog and D. Rein, eds., CERN 90-10, Vol.3, p.268 (1990)

20. E. Fretwurst *et al.*, Radiation hardness of silicon detectors for future colliders, DESY 92-060 (1992)

21. S P Beaumont *et al.*, Gallium arsenide microstrip detectors for charged particles, CERN-PPE/92-51 (1992)

EXPERIMENTAL CHALLENGES AT
FUTURE HADRON COLLIDERS

James Siegrist

Superconducting Super Collider Laboratory *
2550 Beckleymeade Avenue
Dallas, Texas 75237

ABSTRACT

An overview of the experimental program at future hadron colliding beam machines is presented. Special emphasis is given to the Tevatron upgrade program and Superconducting Super Collider physics with the detector designed by the Solenoidal Detector Collaboration (SDC). Expected physics reach of the new machines is outlined. Expected physics performance of the proposed SDC design is detailed.

INTRODUCTION

The Standard Model

The standard model has the leptons ($e\mu\tau\nu$) and quarks ($udcsbt$) as its "fundamental" particles. Electroweak and strong interactions among these fundamental particles are mediated by the gluons (strong) and the γ, Z^0, and $W^\pm$ (electroweak). The standard model also features an electroweak symmetry-breaking (Higgs) sector and several parameters such as the quark and lepton masses and interaction coupling strengths. Theoretical extensions of and alternatives to the standard model reduce to the standard model at low ($\ll M_W$) energy, since all experimental results are "fit" by the standard model.

However, many residual problems and questions about the standard model framework are unresolved. These include confirmation of the top quark existence, the detailed nature of electroweak symmetry breaking, the pattern of masses and mixing of the fundamental particles, the origin of CP violation, the reason for the existence of

---

* Operated by the Universities Research Association, Inc., for the U.S. Department of Energy under Contract No. DE-AC35-89ER40486.

generations of quarks and leptons, whether the forces are unified at a more fundamental level, the origin of the gauge symmetries, *etc.* Many of these questions will be directly or indirectly addressed by the physics programs proposed for the future hadron colliders.

The Standard Model at Hadron-Hadron Colliders

The total hadron-hadron cross section is large, ranging from around 65 mb at the Fermi National Accelerator Laboratory (FNAL) Tevatron at $\sqrt{s} = 1.8$ TeV to of order 100 mb expected for the proposed Superconducting Super Collider (SSC) experiments at $\sqrt{s} = 40$ TeV. The soft collision processes that produce many low-energy particles and give rise to most of the total cross section (so-called "ln(s)" physics) will not be discussed here. These lectures concentrate on the hard-scattering processes among proton constituents that lead to events containing few, energetic products. The ln(s) physics generates a background of soft particles from which the hard constituent collisions must be separated. Since collision rates for the proton constituents can be more or less reliably estimated, hard collisions in these machines provide a laboratory for looking for new phenomena and testing standard model predictions at high Q^2.

For colliding beam machines, the total event rate R is given by $R = \mathcal{L} \cdot \sigma_{TOT}$, where $\mathcal{L}$ is the machine luminosity and σ_{TOT} is the total hadron-hadron cross section. For the Tevatron, luminosities achieved so far are in the range $\sim 10^{30}$ cm^{-2}sec^{-1}. For the SSC, luminosities are expected to be in range $\sim 10^{33}$ cm^{-2}sec^{-1}. For a total cross section of 60 mb, this luminosity yields a total event rate of 60 kHz at FNAL and 60 MHz for the SSC. The event size for a typical collider experiment, such as the Collider Detector at FNAL (CDF), is of order 80 kbytes of data. Electronic data acquisition systems cannot yet be constructed with the capacity to record all this information. Even if the data could be recorded, computational resources and storage capacity to fully analyze the resulting event sample do not exist.

The inability to record all events at hadron-hadron colliders is not a fatal flaw in such experiments because the interesting hard-scattering processes occur at a much lower rate. Hadron-hadron experiments require a "filter" to reduce the number of events and information per event to be recorded. For the multipurpose high p_t collider experiments, filter decisions are made based on a search for final states containing the fundamental entities in the standard model. This filtering process begins in the design of the detection hardware, through the online "trigger" system, and on into the offline data analyses.

Standard Model Tests in Hadron-Hadron Colliders

The procedure for testing the standard model by means of the event sample selected on the basis that each event contains one or more of the fundamental standard model objects begins with checks on the understanding of the detector performance for simple standard model processes, such as high p_t jet production, high p_t lepton production from W, Z decay, direct photon production, *etc.* Once a reasonable understanding of the detector performance has been achieved, tests for simple standard model extensions (quark compositeness, $Z + X$ production, *etc.*) can be made based on limits on deviations from the predicted behavior. Once detailed understanding of standard model predictions has been achieved, limits on new phenomena can be set by studying the details of event configurations, searching for exclusive final states, and placing limits on standard model forbidden final states.

The ability of multipurpose detectors to find and measure such events is built in at a fundamental level. The ability to measure the "energy flow" from the quark or gluon fragmentation to hadrons is achieved via calorimetric techniques. The ability to identify final states containing electrons and muons allows searches for "odd" combinations of particles (*e.g.*, $e\mu$ events). Final states containing neutrinos are detected by the imbalance in the transverse energy flow. The result is a powerful apparatus that is able to fully visualize events and carry out many different types of standard model tests simultaneously. This results in an interlocking network of tests, all within one apparatus. The multipurpose detectors are therefore necessarily complex and expensive, and they require a large collaboration to fully exploit their physics potential.

Future Machines and Their Physics Goals

In the past, the bulk of what we know about particle physics has been found by direct production of the particles in high-p_t experiments at high-energy accelerators. This fact explains why high-p_t experiments have first priority in the world HEP program. Over the coming decade, the two highest-energy machines in the U.S. will be the FNAL Tevatron and the SSC; in Europe, the Large Hadron Collider (LHC) is planned. Physics goals for the Tevatron upgrade program and for the SSC are detailed below. The LHC program is similar to that of the SSC, as discussed in the next section in more detail.

The upgrades planned for the 1990s for the Tevatron[1] are centered around the construction of a new Main Injector that will allow integrated luminosities of order 500 pb^{-1} per year to be delivered to the experiments (CDF, D0) from proton-antiproton collisions at 2-TeV center-of-mass energy. The instantaneous luminosity will be in the range 0.5–1×10^{32} cm^{-2}sec^{-1}, with the time between bunch crossings in the 132–396 nsec range. The new main injector will allow $\bar{p}$ accumulation rates up to 20 mA/hour. In addition, 120-GeV extracted beam will be available while running the Tevatron in collider mode. The CDF experiment will be upgraded with new electronics and triggering hardware to handle the higher luminosity, expanded muon coverage, and secondary vertex information from the new silicon vertex detector. The D0 experiment will also upgrade the electronics for the higher luminosity and will upgrade the tracker to withstand the increased radiation levels. These experiments will provide direct experience in the high-rate, high-p_t environment relevant to the SSC/LHC. The main physics goals of the Tevatron upgrade program are to find the top quark and measure its mass, and to precisely ($\delta \sim 50$ MeV) measure the mass of the W boson. The combination of the top quark and W boson mass measurements severely constrains the standard model, as we shall see later. The Tevatron with CDF and D0 will provide the means to directly search for new physics at the highest possible mass scales until the construction of the SSC/LHC is completed at the end of the decade.

The SSC is scheduled for completion near the end of 1999. The machine[2] will deliver 10^4 pb^{-1} per year integrated luminosity from pp collisions at 40-TeV center-of-mass energy. The instantaneous luminosity will be 10^{33} cm^{-2}sec^{-1}, with 16 nsec between bunch crossings. Two complementary high-p_t detectors are planned at turn-on, along with several smaller experiments. The Solenoidal Detector Collaboration[3] (SDC) design emphasizes charged particle tracking, hermetic calorimetry, good lepton identification and measurement, robust vertex detection capability, and redundancy in the particle measurements. The Gamma, Electrons, and Muons (GEM) design[4] features identification and precise measurement of photons, electrons, and muons. The design has inherent capability to operate at luminosities higher than the SSC design. With completion this summer of the SSC magnet string test,[5] all major technical hurdles to construction of the machine have been passed.

The foremost physics goal of the SSC is to fully explore the electroweak symmetry-breaking mechanism that generates particle masses. Detailed studies of top quark production and decay will be possible at SSC. The search horizon for new phenomena such as new heavy gauge bosons, quark and lepton substructure, and supersymmetric particles will be vastly expanded by the SSC. Most of the current problem areas with the standard model will be directly or indirectly addressed by the SSC. The central issue of electroweak symmetry breaking stands out as the most important feature of the SSC experimental program.

Physics vs. Luminosity Picture

Figures 1–7 illustrate the physics capabilities of the SSC/LHC.[6] The number of events produced by various low-background processes is plotted as a function of $\sqrt{s}$ and luminosity. Detector efficiencies and acceptances are ignored here in order to qualitatively gauge the physics reach provided by these machines.

Figure 1 shows the luminosity required to produce 20 events from a high-mass Higgs decaying to two Z^0, and each Z^0 subsequently decaying to charged leptons (e or μ), for various Higgs masses. In this mode, the SSC at 40 TeV and 10^{33} luminosity reaches beyond the 800-GeV Higgs range, while the LHC at 10^{34} luminosity and 10 TeV reaches just short of 800 GeV.

Figure 2 shows a similar plot for production of the intermediate-mass Higgs decaying to ZZ^* with subsequent Z decay to e or μ for 40 events produced. The machines reach down to the 120-GeV range in this mode. No interesting limits are set on these processes at the Tevatron, as the expected production cross sections are too small.

Figure 3 shows as a function of $\sqrt{s}$ the mass for which 50 heavy quark pairs are produced per running year for various integrated luminosities. Even relatively massive heavy flavor pairs are copiously produced at the SSC and LHC. The current upper limit on the top quark mass[7] is $M_{\text{top}} > 91$ GeV from the CDF experiment at the Tevatron.

Figure 4 shows the luminosity required to produce 100 Z' bosons with standard model couplings. The current CDF limit[8] is $M'_Z > 412$ GeV at 95% C.L. The SSC/LHC will greatly expand the search range, to more than 8 TeV for SSC and ~ 4 TeV for LHC.

Figure 5 shows the luminosity required to establish limits on various scales of quark compositeness. The current CDF limit[9] is $\Lambda_c > 1.4$ TeV at 95% C.L. The SSC should extend the limit to ~ 20 TeV, and the LHC to ~ 10 TeV.

Figure 6 shows the luminosity required to produce 10^4 gluino pairs/year. The gluino mass limit[10] is ≥ 200 GeV from CDF. The SSC/LHC should extend the limit to ~ 1.5 TeV.

Figure 7 summarizes[11] a number of SSC physics processes as a function of mass. Note that from a physics point of view, the Tevatron high-p_t program feeds smoothly into the SSC program—many of the physics topics being pursued today at FNAL will continue to be pursued at SSC with much greater risk of discovery. Important additional topics encompassing electroweak symmetry breaking become accessible at the SSC for integrated luminosities above 10^{39} cm^{-2}. The current round of FNAL high-p_t experiments provides direct experience in the high-rate, high-p_t environment relevant to the SSC.

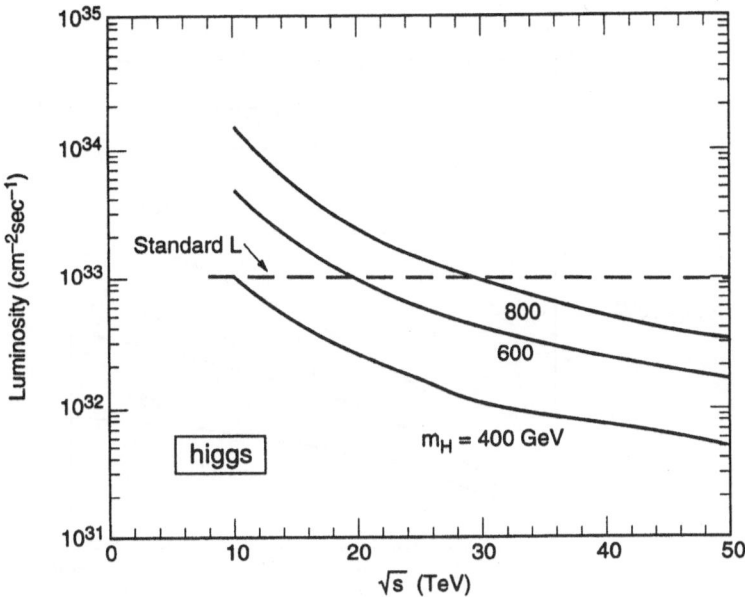

Figure 1. Luminosity required to produce in one year 20 Standard-Model Higgs bosons that decay to four charged leptons (e, μ) as a function of c.m. energy. (mass of t quark = 85 GeV.) Results are shown for Higgs masses of 400, 600 and 800 GeV.

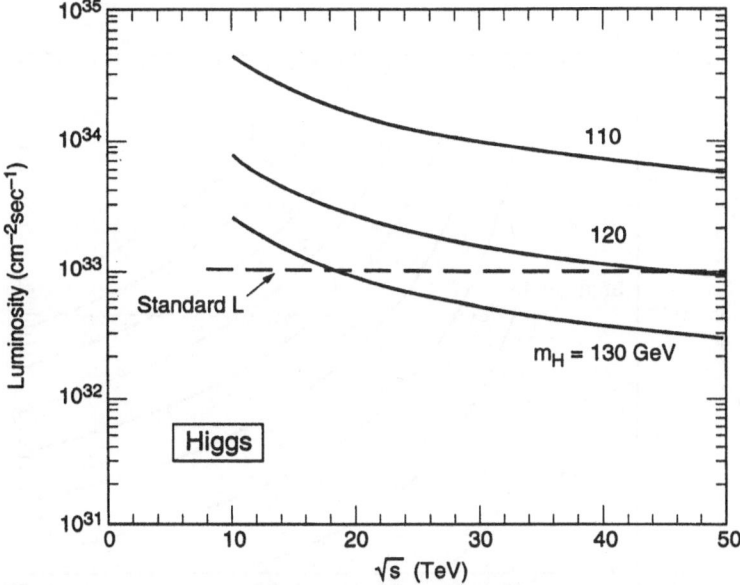

Figure 2. Luminosity required to produce in one year 40 intermediate mass Higgs bosons that decay into four charged leptons (e, μ) as a function of c.m. energy. (Mass of t quark = 85 GeV.) Results are shown for Higgs masses of 110, 120 and 130 GeV.

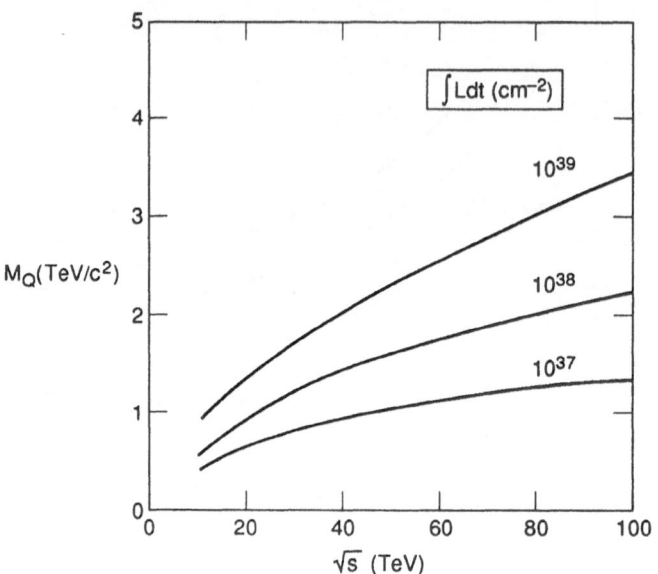

Figure 3. As a function of integrated luminosity, quark mass M_Q reached for 50 produced $Q\bar{Q}$ pairs.

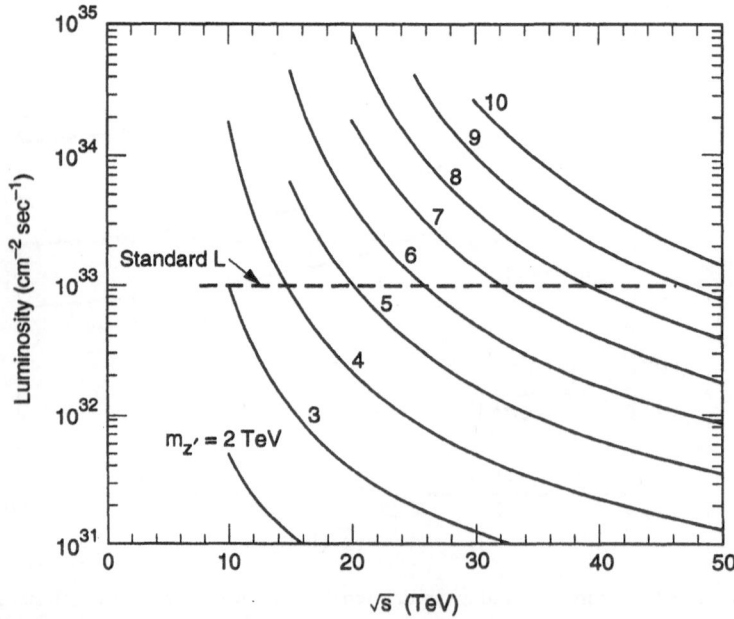

Figure 4. Luminosity required to produce in one year 100 Z' bosons, with Standard Model couplings, as a function of c.m. energy. Results are shown for Z' masses in the range from 2 to 10 TeV.

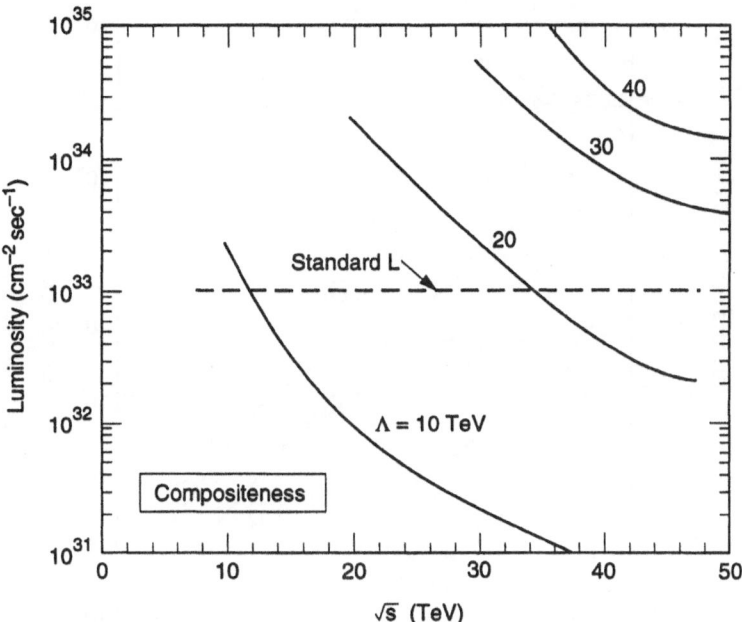

Figure 5. Luminosity required to establish in one year quark compositeness scales of 10, 20, 30, and 40 TeV as a function of c.m. energy.

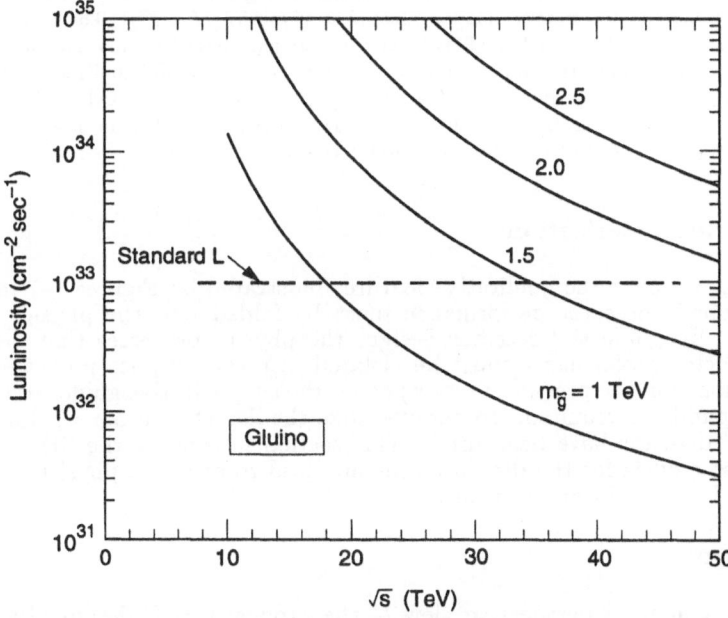

Figure 6. Luminosity required to produce 10,000 pairs of gluinos per year as a function of c.m. energy. Results are shown for gluino masses in the range of 1.0 to 2.5 TeV.

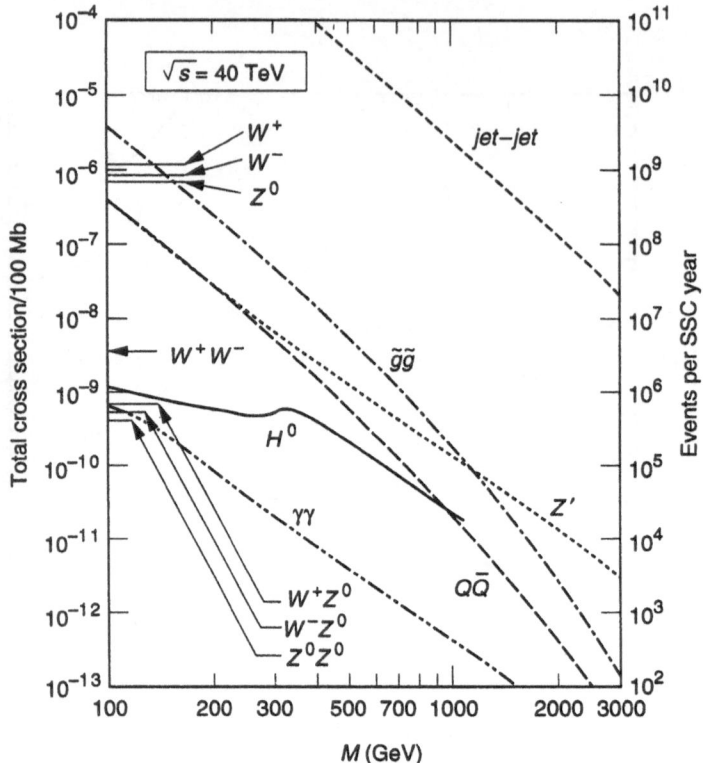

Figure 7. Examples of total cross sections at the SSC. The *jet-jet* and γ-γ cross sections are for wide-angle jet or γ pairs ($|\eta_{\text{jet}}|$ and $|\eta_\gamma| < 2.5$) with invariant mass greater than M. For heavy-quark pair (gluino pair) production, the cross section is evaluated at $M = M_Q$ ($M = M_{\tilde{g}}$). The Higgs cross section assumes $M_{\text{top}} = 150$ GeV. The left scale is total cross section divided by 100 mb, the approximate total pp cross section, and so the numbers are approximate production probabilities per collision. The scale on the right is the number of produced events per year under "standard conditions," defined as operation at $\mathcal{L} = 10^{33}$ cm^{-2}s^{-1} for 10^7 s. These rates must be further downrated by branching fractions and experimental cuts to obtain measurable rates.

Detector Parameterizations

In order to make the qualitative features illustrated by Figures 1–7 more quantitative, realistic detector performance must be folded into the physics capability assessment. To guide the detector design, the physics processes that demand the highest detector performance must be isolated. Certain physics processes (such as those mentioned in the previous section) can serve as qualitative guides to the general detector capabilities required. To parameterize the detector response, simple models of the SDC detector have been used. This section introduces the SDC design and the response models for the detector that are used to estimate physics performance capabilities in the following sections.

SDC Design[11]

Figure 8 shows a perspective view of the proposed SDC design. Particles emanating from the beam collision point near the center of the detector pass in turn through the beam pipe, a silicon-strip tracker, an "outer" tracker of straw-tube design, the solenoid coil, the highly segmented electromagnetic (Em) and hadronic (HAD) calorimeters, the muon detectors, and the iron toroids of the muon detection system. Figure 9 shows a detailed view of a quarter-section of the detector.

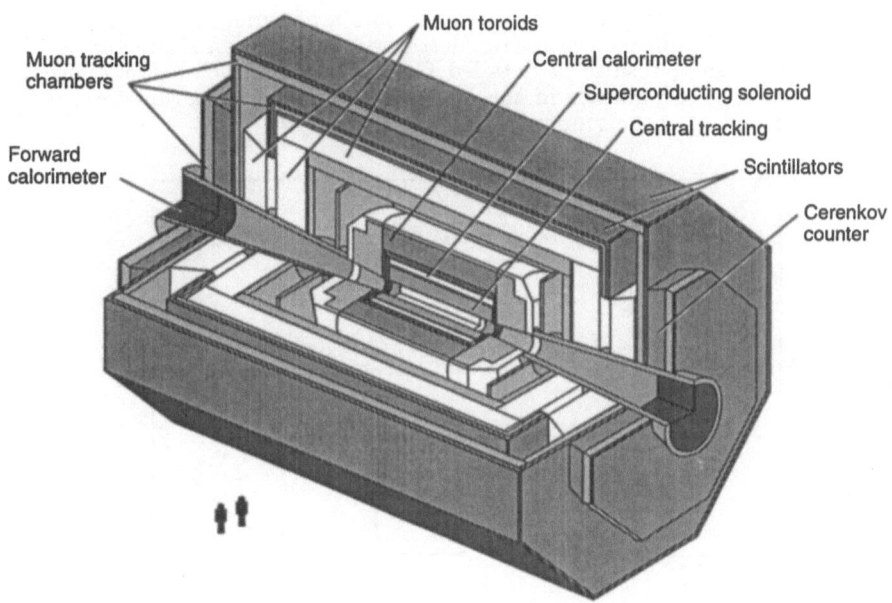

Figure 8. Isometric view of the preliminary baseline SDC detector configuration.

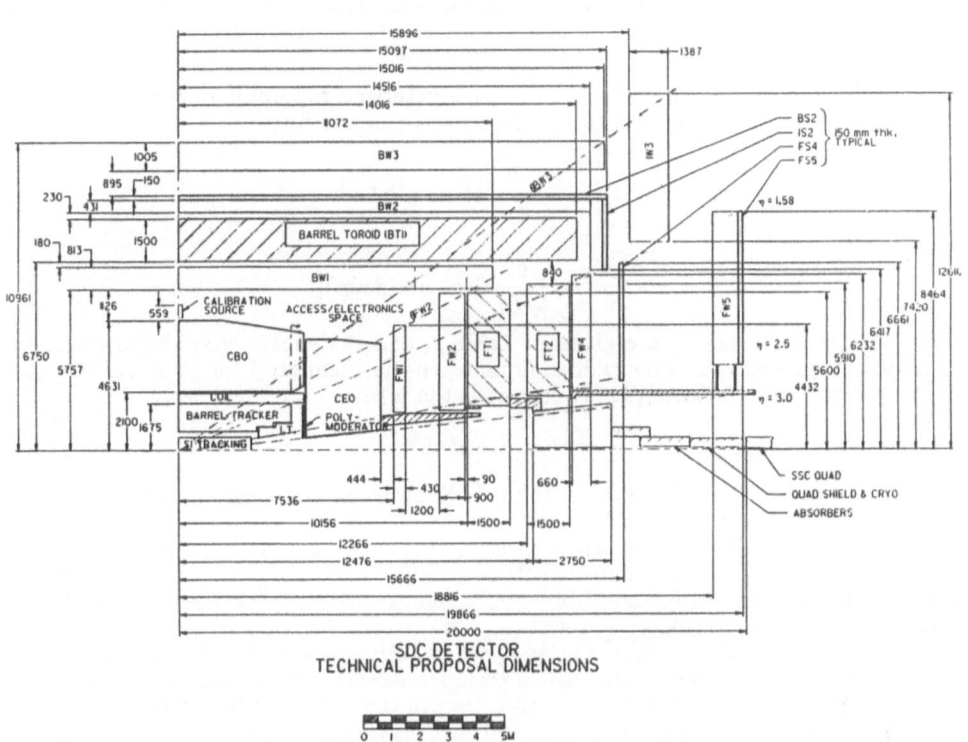

SDC DETECTOR
TECHNICAL PROPOSAL DIMENSIONS

NOTE: DIMENSIONS ARE IN MILLIMETERS

Figure 9. One quarter section view of the SDC design.

The model for the charged particle tracker uses a parameterized resolution that includes the effects of multiple scattering, detector resolution, and detector misalignment. When necessary, results of detailed GEANT simulations are employed, including generation of secondaries in the tracking volume. The parameterized track resolution is plotted against pseudorapidity η in Figure 10.

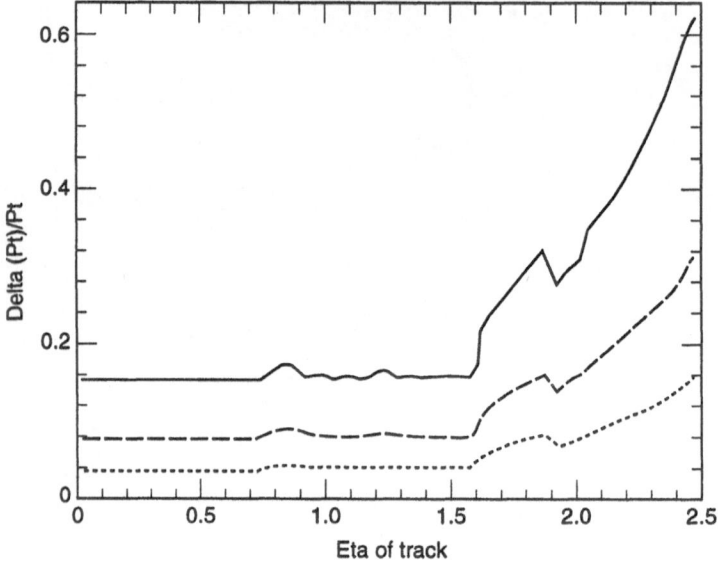

Figure 10. The resolution of the baseline tracking system as a function of η for several p_t values. The solid curve is for $p_t = 1000$ GeV, and the dashed (dotted) is for $p_t = 250$ (100) GeV.

Resolutions for the calorimeters are parameterized for single particles from EGS studies (Em) and CALOR89 studies (HAD) to be of the form

$$\frac{\sigma(E)}{E} = \frac{a}{\sqrt{E_t}} \oplus b \quad \text{(barrel)}$$

or

$$\frac{\sigma(E)}{E} = \frac{a}{\sqrt{E_l}} \oplus b \quad \text{(endcap)},$$

where the parameters a, b are given in Table 1, and $\oplus$ denotes that the terms are to be added in quadrature. The hadron nonlinearity is accounted for by a π/e response parameterized to fit CALOR89 studies with the form

$$\pi/e = \alpha - \beta/E^{0.15},$$

with the parameters α, β shown in Table 1. Shower shapes were parameterized from EGS (Em) and ZEUS data (HAD).

The tracking resolution parameterized for the muon system is shown in Figure 11. The parameterization includes the effects of multiple scattering in both the calorimeter and iron toroid, detector resolution, and alignment effects for the combined inner tracker and outer muon detection system.

A global efficiency of 85% for analyses requiring an isolated lepton or photon (including trigger efficiency and all selection criteria) was assumed. Results from current experiments (CDF) are used to estimate the expected background rejections against dominant backgrounds. For $p_t \gtrsim 20$ GeV, this rejection is of order 10^{-5} against the two-jet background.

Jets are reconstructed by using the transverse energy deposition in calorimeter cells. Seed towers with $E_t > 5$ GeV are used to define the initial jet axis. All cells above a threshold of 100 MeV inside a cone in η-ϕ space are collected into the jet. Studies with ISAJET, using the single-particle calorimeter response parameters mentioned previously, lead to a jet energy resolution of

$$\frac{\sigma(E)}{E} = \frac{0.61}{\sqrt{E}} \oplus 0.016.$$

Table 1. Calorimeter Parameters.

Parameter	Barrel	Endcap	Forward
Coverage	$|\eta| < 1.4$	$1.4 < |\eta| < 3.0$	$3.0 < |\eta| < 6.0$
Radius of front face (m)	2.10		
z position of front face (m)		4.47	12.00
Compartment depth			
EM (+ Coil)	1.1	0.9	
HAD1	4.1	5.1	13.0
HAD2	4.9	6.0	
EM resolution			
a	0.14	0.17	0.50
b	0.01	0.01	0.05
HAD resolution			
a	0.67	0.73	1.00
b	0.06	0.08	0.10
HAD nonlinearity			
α	1.13	1.16	1.16
β	0.31	0.38	0.38

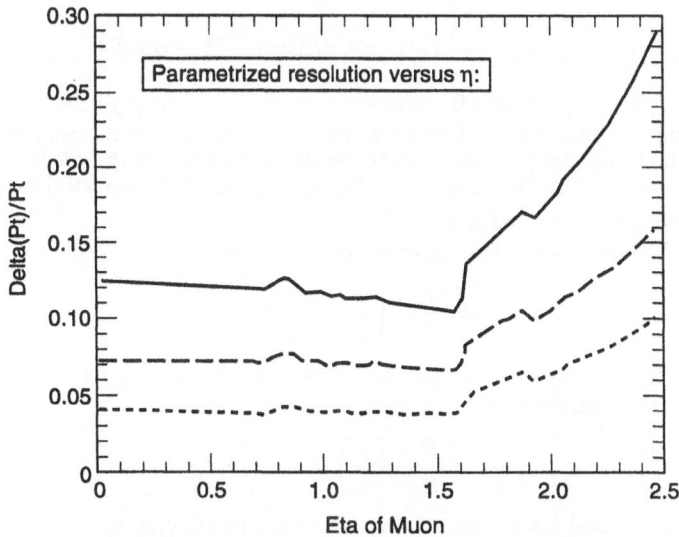

Figure 11. The resolution of the combined baseline tracking and muon system as a function of η for several p_t values. The solid curve is for $p_t = 1000$ GeV, and the dashed (dotted) curves are for $p_t = 250\ (100)$ GeV.

STANDARD MODEL PHYSICS AT HADRON COLLIDERS[12]

The bulk of the hard-scattering cross section arises from strong interactions among the proton constituents and is accurately predicted by QCD. To illustrate the nature of the calculations, Figure 12 shows a schematic diagram of the collisions among pp constituents. The partons are distributed within the proton with some distribution of fractional momentum characterized by the structure function f. The incoming parton interaction is described by the elementary scattering cross section $\hat{\sigma}$, and the outgoing parton lines materialize into the observed hadrons according to some fragmentation function D. In the final state, the two high-p_t partons fragment into hadrons having limited transverse momentum with respect to the directions of the scattered partons (jets). The total p_t of the dijet system is limited, and the two jets are produced back-to-back in azimuthal angle ϕ (see Figure 13).

In QCD, the inclusive jet yield is obtained by a folding integral over the parton distributions, each contribution being weighted by the relevant parton-parton differential cross section

$$E\frac{d^3\sigma}{d^3p} = \sum_{ij} \int dx_1 dx_2 f_i(x_1, Q^2) f_j(x_2, Q^2) \delta(\hat{s} + \hat{t} + \hat{u}) \frac{\hat{s}}{\pi} \sum_k \frac{d\sigma(ij \to k)}{d\hat{t}},$$

where i, j, k sum over parton species. The differential cross section can be written in terms of the matrix element M as

$$\frac{d\sigma}{d\hat{t}} = \pi\alpha_s^2(Q^2)\frac{|M|^2}{\hat{s}^2}.$$

The matrix element $|M|^2$ can be calculated in QCD, and the structure functions, along with their scaling violations, can be measured and parameterized in deep inelastic lepton-nucleon scattering. Figure 14 shows a comparison of a calculation with recent CDF data on jet production. Figure 15 shows the coordinate system for this problem. The momentum transverse to the beam direction is $p_t = p\sin\theta$. The variable $\hat{s}$, $\hat{t}$, $\hat{u}$ are given by:

$$\hat{s} = (p_1^\mu + p_2^\mu)^2 = x_1 x_2 s = \frac{1}{2}x_t s(x_1 \tan\theta/2 + x_2 \cot\theta/2)$$

$$\hat{t} = (p_1^\mu - p^\mu)^2 = -\frac{1}{2}x_t s x_1 \tan\theta/2 = -\frac{\hat{s}}{2}(1 - \cos\theta^*)$$

$$\hat{u} = (p_2^\mu - p^\mu)^2 = -\frac{1}{2}x_t s x_2 \cot\theta/2 = -\frac{\hat{s}}{2}(1 + \cos\theta^*),$$

where the initial parton $p_t = 0$, all masses are zero, $x_1 = 2p_1/\sqrt{s}$ and $x_2 = 2p_2/\sqrt{s}$ are the initial parton momentum fractions, and $x_t = 2p_t/\sqrt{s}$ is the scaled transverse momentum of the outgoing parton. Figure 15 shows the definition of θ^*, the parton-parton center-of-mass scattering angle; the lab system has been boosted by an amount $P_L = \frac{\sqrt{s}}{2}(x_1 - x_2)$ along the beam axis.

Other useful variables are the rapidity y defined by

$$y = \frac{1}{2}\ln\left(\frac{E + P_L}{E - P_L}\right),$$

with E and P_L the energy and longitudinal momentum of, $e.g.$, the parton-parton system. The pseudorapidity is defined by[13]

$$\eta = \frac{1}{2}\ln\left(\frac{P + P_L}{P - P_L}\right) = -\ln\tan\frac{\theta}{2}.$$

A change of variable and integration over azimuthal angle yields

$$E\frac{d^3\sigma}{d^3p} = \frac{1}{2\pi p_t}\frac{d^2\sigma}{dp_t d\eta}.$$

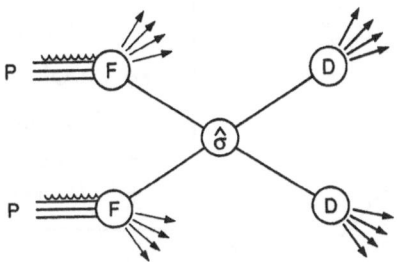

Figure 12. Schematic illustration of *pp* collisions. *F* represents the proton structure function, $\hat{\sigma}$ the elementary parton-parton scattering cross section, and *D* the fragmentation function on the outgoing parton line.

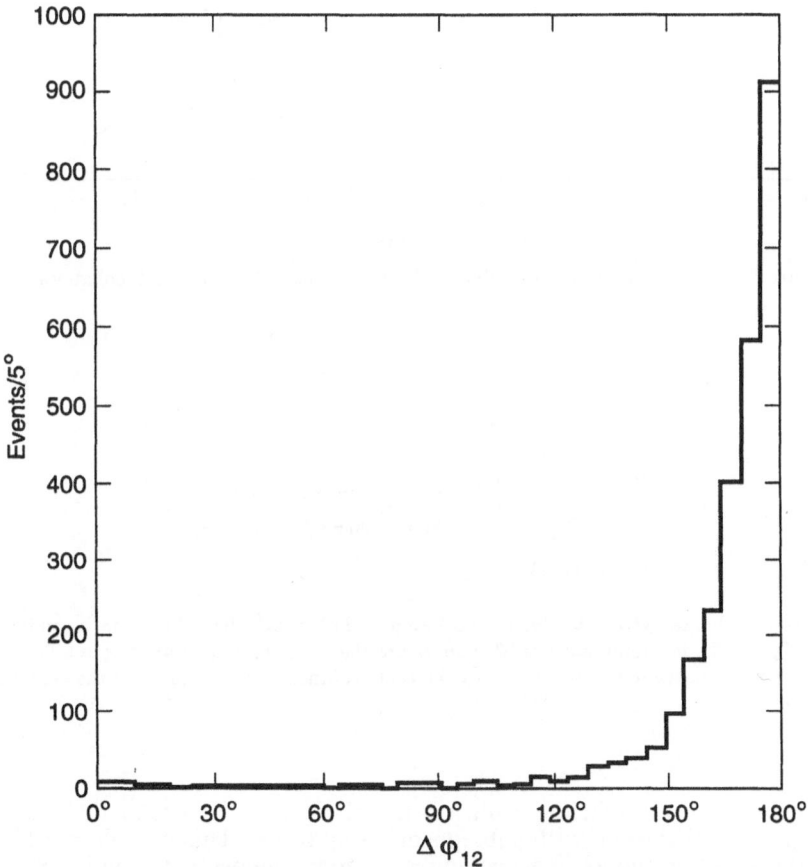

Δ φ between 2 highest P_t clusters

Figure 13. The distribution of difference in azimuth between the two clusters of highest transverse energy in a sample of events containing summed transverse energy > 60 GeV from the UA2 experiment.

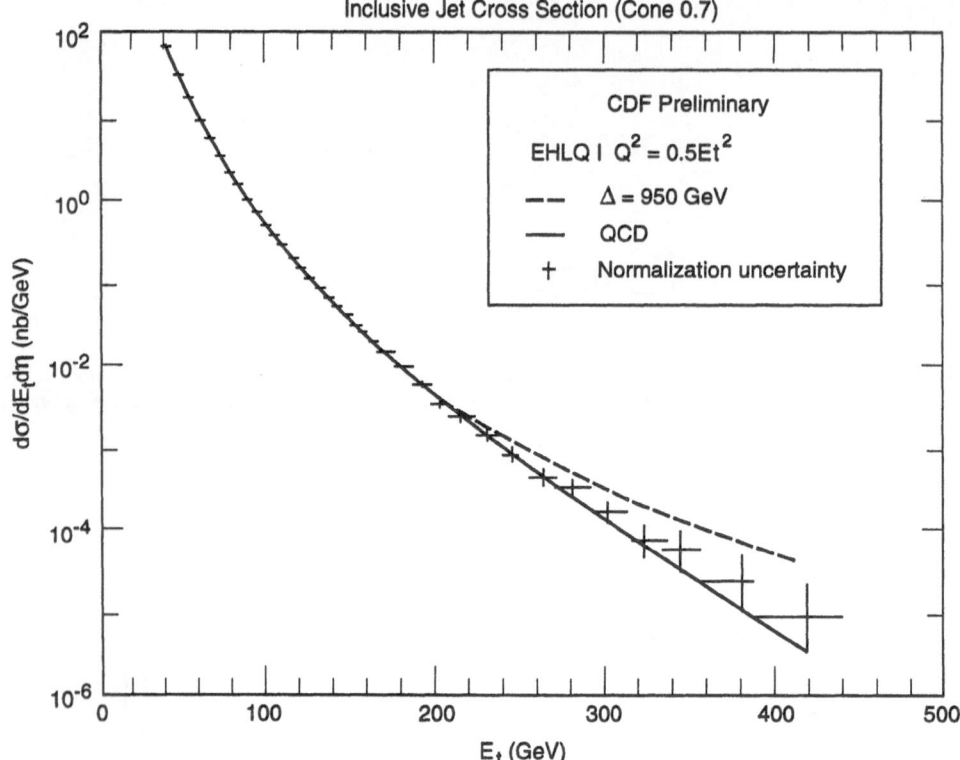

Figure 14. Transverse energy distribution for jets measured by the CDF collaboration.

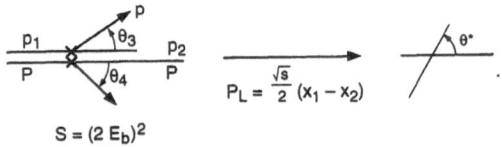

Figure 15. Coordinate system to describe collisions. The z axis lies along the incoming proton beam. P_1, P_2 are the incoming partons, θ_3 and θ_4 are the outgoing scattering angles. The boost by the longitudinal momentum P_L transforms to the center-of-mass system, showing the center-of-mass scattering angle θ^*.

Figure 16 shows a schematic illustration of electroweak production via proton constituents of a W boson, with subsequent decay to $e\nu$. This process can be calculated in the standard model in a manner completely analogous to the jet production cross section discussed earlier. In the standard model, $u+\bar{d}$ quarks couple to the W^+, $d+\bar{u}$ to W^-, and quark-antiquark pairs $u\bar{u}$, $d\bar{d}$, to the Z^0. The branching ratio of the W to $e\nu$ final states is about 8%. The characteristic of events resulting from W production is the presence of high-p_t leptons (e, μ, τ) in the final state, accompanied by large missing p_t from the (unmeasured) neutrinos.

Heavy flavor (top) quark pair production is also calculated in an analogous fashion. In order to set limits on top quark production, semileptonic decay modes are used to reduce background from standard model processes.

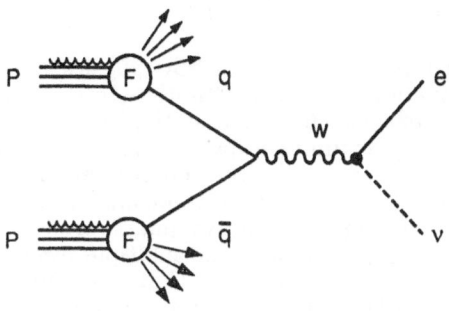

Figure 16. Schematic illustration of W production in pp collisions. F represents the proton structure function; subsequent decay of the W to an $e\nu$ final state is shown.

Standard Model Measurements with the Fermilab Main Injector Upgrade[1]

The principal focus of the main injector upgrade for the Tevatron is the accurate measurement of the W mass and the top quark mass. The W mass is measured by forming the "transverse" mass from the lepton transverse momentum and the missing transverse momentum from the unseen neutrino. The current uncertainty on the W mass is ~ 350 MeV/c^2, limited by statistics. This uncertainty is expected to be improved to about 50 MeV after the main injector is completed. The ultimate precision that can be achieved is limited by knowledge of the proton structure function at that point. Figure 17 shows the size of the errors expected in the main injector era in M_W vs. $M_{\rm top}$. The curves are standard model predictions for various Higgs masses. These precision measurements of M_W, $M_{\rm top}$ will extend our knowledge of the standard model parameters.

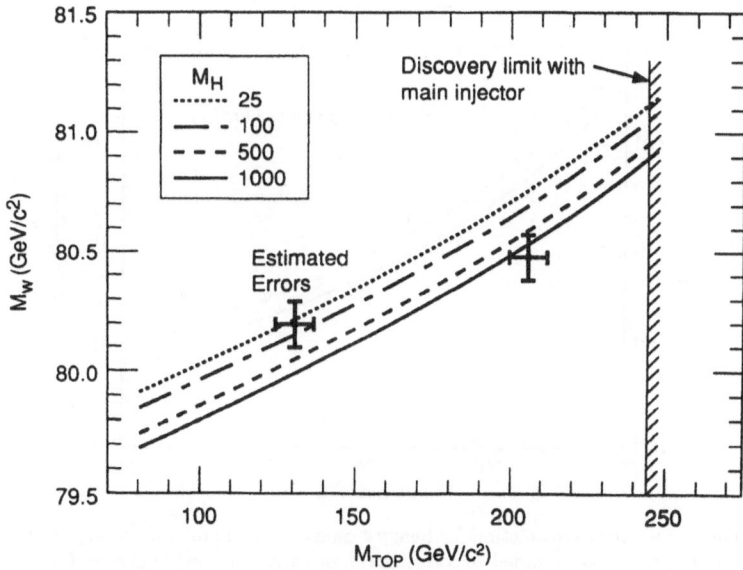

Figure 17. Expected precision on M_W and M_t with 1 pb^{-1} of data collected at the Tevatron collider. Also shown are the discovery reach on M_t and Standard Model predictions for various Higgs masses.

Standard Model Physics at SSC (SDC)[11]

Several of the predictions for important SSC physics are affected by theoretical uncertainty (structure functions, behavior of initial-state radiation, higher-order corrections underlying event effects, *etc.*). The initial lower-luminosity running period of SSC will offer a unique opportunity to determine some of the unknowns involved in the theoretical calculations and to understand critical components of the detector in a rather clean environment. Typical measurements will cover inclusive and multi-jet rates and fragmentation properties, heavy quark production cross sections and properties, electroweak boson production, and multiple production of gauge bosons. Since the top quark is the most obvious system that would benefit from high-statistics studies possible at SSC, the SDC has checked the capability of the detector design for this system rather extensively. The next section details the results of some of those studies.

Top Quark Physics at SSC

The top quark is one of the most important missing pieces in the standard model. While it may be discovered at the Tevatron before SSC begins operations, the production cross section (Figure 18) at the SSC is huge. The top will provide a window into physics beyond the standard model via its decays (*e.g.*, $t \rightarrow H^+ b$). The top is also a source of Higgs bosons via the associated production mode $pp \rightarrow t\bar{t}H$. The top is an important background to rare and exotic phenomena such as WW scattering.

The main identification tools to be used in SDC to identify top are semileptonic decays and b-tagging. Studies have been made using the SDC detector performance described previously. Three techniques for top detection will be discussed below. The first is isolated di-leptons, *e.g.*,

$$t\bar{t} \rightarrow (t \rightarrow e\nu b) + (\bar{t} \rightarrow \mu\nu\bar{b}) \,.$$

The second is isolated lepton and non-isolated muon, *e.g.*,

$$t\bar{t} \rightarrow (t \rightarrow X) + (\bar{t} \rightarrow l\,\nu_l\,\mu\nu_\mu\bar{c}) \,.$$

The third technique is to employ lepton plus jets events, *e.g.*,

$$t\bar{t} \rightarrow (t \rightarrow e\nu b) + (\bar{t} \rightarrow jets) \,.$$

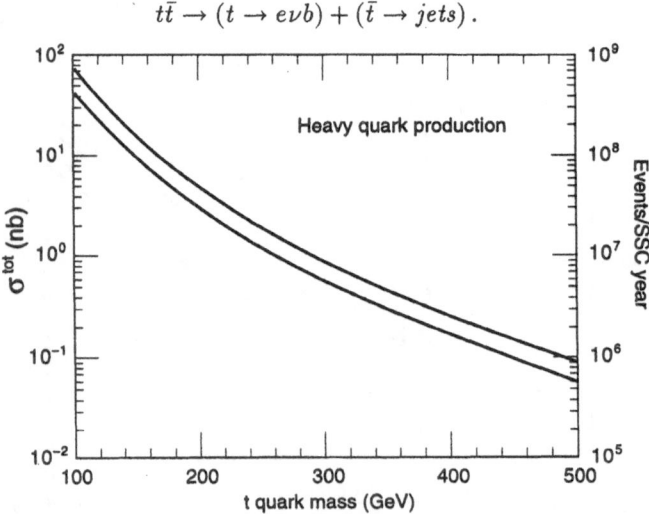

Figure 18. The production cross section for heavy t quarks. The band corresponds to the envelope of the smallest and largest rates obtained by varying the factorization scale between 0.5 and 2 times the quark mass, and by using the following sets of structure functions: DFLM160, DFLM360, HMRSB, and MTB.

Isolated Di-leptons

In the isolated di-lepton search, a pair of isolated e and μ, with $p_t > 20$ GeV and $|\eta| < 2.5$ are required. This sample is found to be background-free at the percent level, yielding 10^6 events per SSC year ($= 10^{40}$ cm^{-2}) for $m_t = 150$ GeV. A comparison of the measured and predicted cross sections yields a top quark mass measurement with a mass error in the 10–15 GeV range. After subtraction of the Z and Drell-Yan background, comparison of the $e\mu$ rate with the lepton pair rate can establish e/μ universality in the top decays. The $e\mu$ sample provides a clean sample of tagged b jets on which to develop b-tagging algorithms and to verify their performance.

Isolated Lepton and Non-Isolated Muon

Comparison of the rate for this process with the $e\mu$ rate measures the $t \rightarrow Wb$ branching ratio. This decay mode provides a better measurement of m_t. The selection cuts require that the electron have $p_t > 40$ GeV, with less than 4 GeV within an isolation cone of $\Delta R = 0.2$. The muon is required to have $p_t > 20$ GeV, with > 20 GeV within an isolation cone of $\Delta R = 0.4$. The p_t of the $e\mu$ system is required to be above 100 GeV, and the azimuthal angle between the e and μ must be less than 80°, to reduce opposite-charm background. The fit to the $e\mu$ mass distribution is shown in

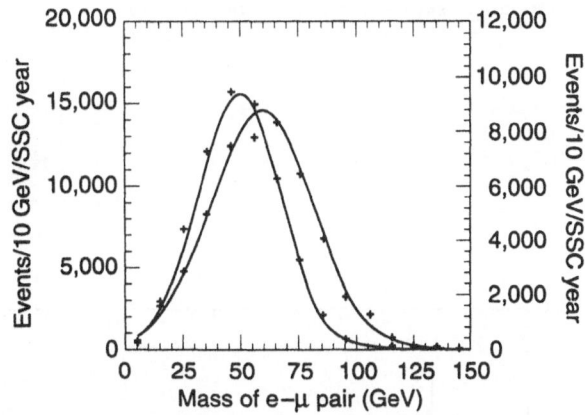

Figure 19. The invariant mass distribution for the $e\mu$ pair, $M(e\mu)$, for two different t-Quark masses. The lefthand scale and the leftmost curve are for $M_{\text{top}} = 150$ GeV, while the righthand scale and curve are for $M_{\text{top}} = 180$ GeV. The cut $p_t(e\mu) > 100$ GeV has been used. The superimposed curves represent Gaussian fits.

Figure 19. The left curve is for $m_t = 150$ GeV; the right is for $m_t = 180$ GeV. To check consistency, the results for different $p_t(e\mu)$ cuts ($> 60, 100, 140$ GeV) are shown in Figure 20. Using this technique, statistical errors are found to be 0.5 GeV ($m_t = 150$ GeV) or 0.8 GeV ($m_t = 250$ GeV), and backgrounds are small. Systematic errors on the t mass arise from the b quark fragmentation uncertainties, and the top quark p_t distribution uncertainties (Figure 21). The final result is that the top quark mass uncertainty for 150 GeV mass is ± 0.5 GeV (statistical) and ± 2.4 GeV (systematic); for 250 GeV mass, ± 0.8 GeV (statistical) and ± 3.9 GeV (systematic). These results are a factor 3–4 more accurate than for the e-μ cross section comparison.

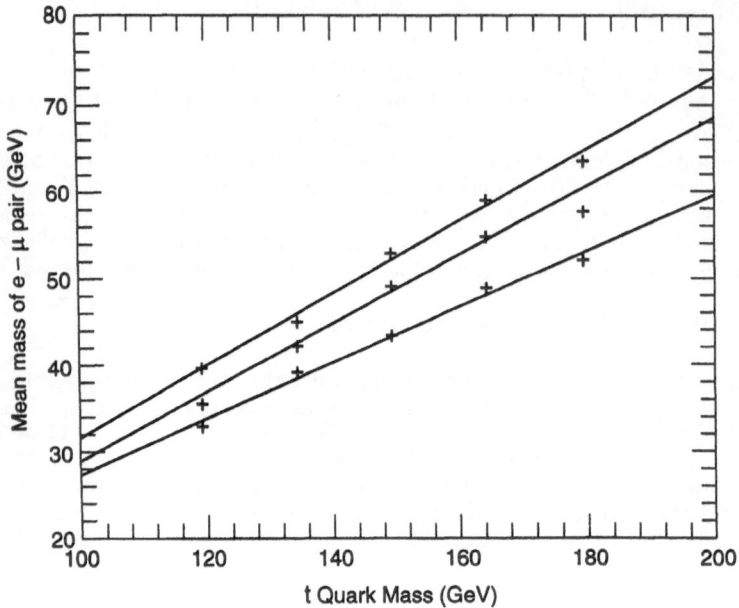

Figure 20. The mean invariant mass of the $e\mu$ pair as a function of the t-quark mass for $p_t(e\mu) >$ 60 GeV (lower curve), $p_t(e\mu) > 100$ GeV, and $p_t(e\mu) > 140$ GeV (upper curve). The data are derived from Gaussian fits to distributions such as those shown in Figure 19.

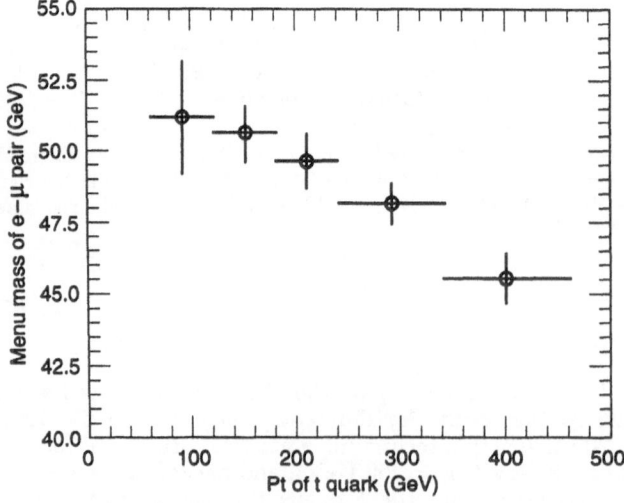

Figure 21. The invariant mass of the $e\mu$ pair as a function of the t-quark p_t for $p_t(e\mu) > 100$ GeV. The data are derived from Gaussian fits to distributions such as those shown in Figure 19, for different t-quark p_t bins.

Lepton and Jets

In the lepton and jets mode, the invariant mass of the $t \to 3$ jets system can be measured, using the large p_t of the top to reduce combinatorial background. For this analysis, jets in the region $|\eta| < 2.5$ are found using a cone size of 0.4. Leptons are required to have $|\eta| < 2.5$ and $p_t > 40$ GeV, with isolation cuts in a cone of radius 0.4 about the lepton. Energy corrections are applied to each individual jet to account for the calorimeter response discussed previously and for out-of-cone energy losses. The $t \to 3$ jets decay mode is searched for by requiring ≥ 3 jets with $p_t > 30$ GeV and $\Delta\phi$ from each jet to the lepton $> 90°$. A b tag (3 or more tracks $> 3\sigma$ from the primary vertex) is required among the jets, and the p_t of the 3-jet system is required to be > 200 (300) GeV for $m_t = 150$ (250) GeV. Figure 22 shows the 3-jet invariant mass for events where a pair of jets has a mass within 15 GeV of the W mass. Figure 23 shows the 2-jet mass distribution for events with 3-jet masses within 15 GeV of the top mass. If the 2-jet mass distribution is fixed to the W mass, the resulting 3-jet systematic mass error on the top mass is estimated to be ~ 3 GeV.

These studies show that detection and study of the top quark at SSC will allow detailed determination of its properties. After a look at standard model extensions and physics beyond the standard model, we will return to the issue of electroweak symmetry breaking.

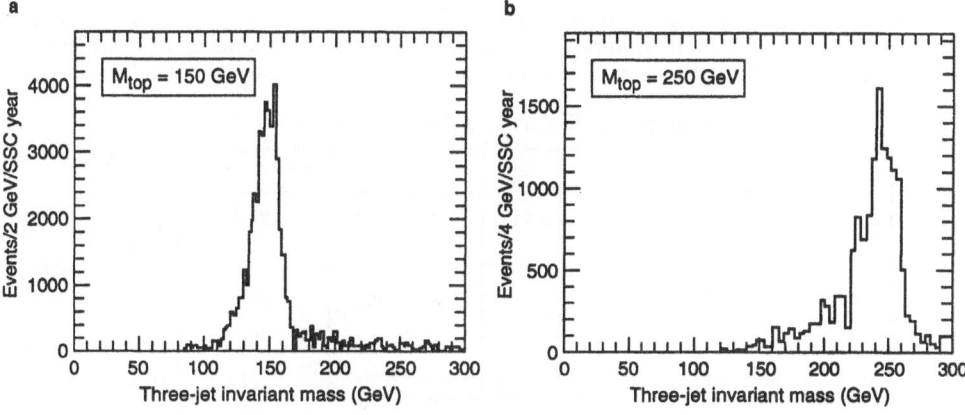

Figure 22. The corrected three-jet invariant mass distribution, for (a) $M_{\text{top}} = 150$ GeV and (b) $M_{\text{top}} = 250$ GeV. In this plot the two-jet invariant mass is required to be consistent with the W mass $65 < M(2\text{-}jet) < 95$ GeV.

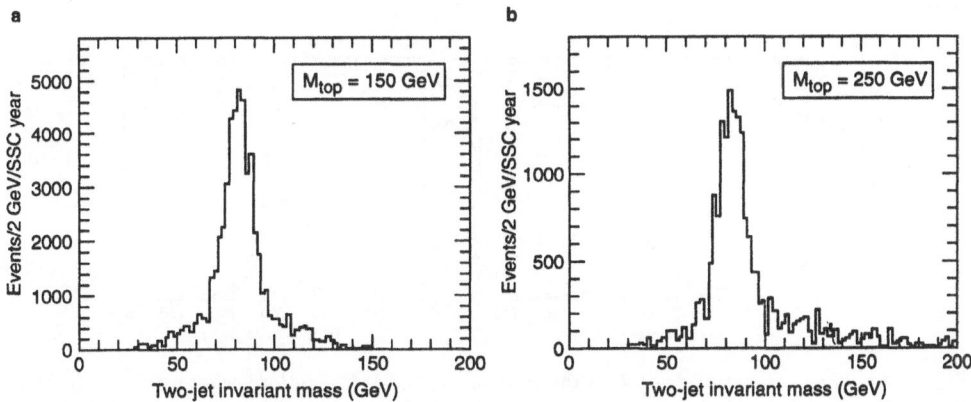

Figure 23. The corrected two-jet invariant mass distributions after requiring that the three-jet invariant mass be consistent with the t mass. (a) $M_{\text{top}} = 150$ GeV, requiring $135 < M(3\text{-}jet) < 165$ GeV. (b) $M_{\text{top}} = 250$ GeV, requiring $225 < M(3\text{-}jet) < 275$ GeV.

STANDARD MODEL EXTENSIONS

The FNAL Main Injector will extend searches for new physics in several areas. Quark or lepton substructure up to mass scales of order 3 TeV, supersymmetric particles (gluinos) up to 250-GeV mass, Z' bosons up to 750-GeV mass, and surprises in W^+W^- production can be seen after the Main Injector upgrade. To look in more detail at SDC capability for searching for simple extensions to the standard model, new heavier gauge bosons, W' and Z', have been considered. For new W', discovery up to 5 TeV should be straightforward, since there are ~ 10 events for typical L-R symmetric models for this mass. For the Z' case, a class of Z' which arises in E_6 models[14] with one free parameter $\cos\alpha$ has been considered. To untangle the properties of these Z's, the mass and width, production cross section, and angular distributions need to be measured. The production cross section measurement determines couplings, but it depends upon assumptions about branching ratios to exotic channels. The angular distributions directly measure the couplings, with the forward/backward asymmetry being a particularly powerful measurement. Table 2 summarizes the predictions on properties of various E_6-inspired Z', a Z' with standard Z^0 couplings, and $\cos\alpha = -0.6$, a value that makes the forward-backward asymmetry particularly large. Figures 24 and 25 show the mass distributions for a perfect detector (dotted), $Z' \to ee$ (solid), and $Z' \to \mu\mu$ (dashed) for 800-GeV and 4-TeV Z' masses. Figure 26 and 27 show the angular distributions for 800 (4000) GeV Z' with the longitudinal momentum of the Z' required to be > 500 (1000) GeV for a perfect detector (dotted), electron decay modes (solid), and muon modes (dashed). With adequate statistics, it is possible to measure the asymmetry and to distinguish among the models.

Table 2. Properties of Several Z' Arising in E_6 Models.

Property	$\cos\alpha = -0.6$	Z_η	Z_Ψ	Z_χ	SM Couplings	
$\Gamma(M = 800 \text{ GeV})$	8.5	5.0	4.2	9.2	21.4	GeV
$\Gamma(M = 4000 \text{ GeV})$	42.3	25.2	21.0	46.2	106.9	
$\sigma(M = 800 \text{ GeV})$	2.1	1.2	1.1	2.4	4.3	pb
$\sigma(M = 4000 \text{ GeV})$	0.004	0.0032	0.0027	0.0051	0.010	(into e^+e^-)

Other standard model extensions include searches for quark or lepton substructure. Such searches are performed by looking for deviations in the jet production cross section from the standard model prediction. Studies for SDC show that the detector should be able to set a limit on such substructure at about 30 TeV after one SSC year.

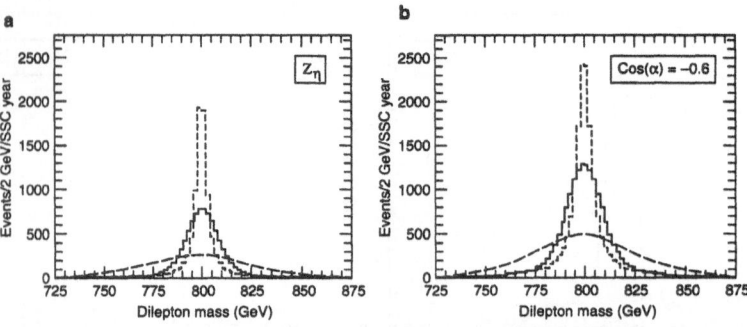

Figure 24. The cross section $d\sigma/dM$ for the production of an $\ell^+\ell^-$ pair for perfect resolution (dotted), the SDC resolution for electron pairs (solid), and the SDC resolution for muon pairs (dashed), as a function of the lepton pair invariant mass. The background is from the continuum production of lepton pairs (Drell-Yan). (a) The peak corresponds to a Z_η with a mass of 800 GeV. (b) The peak corresponds to a new Z with $\cos\alpha = -0.6$ and a mass of 800 GeV.

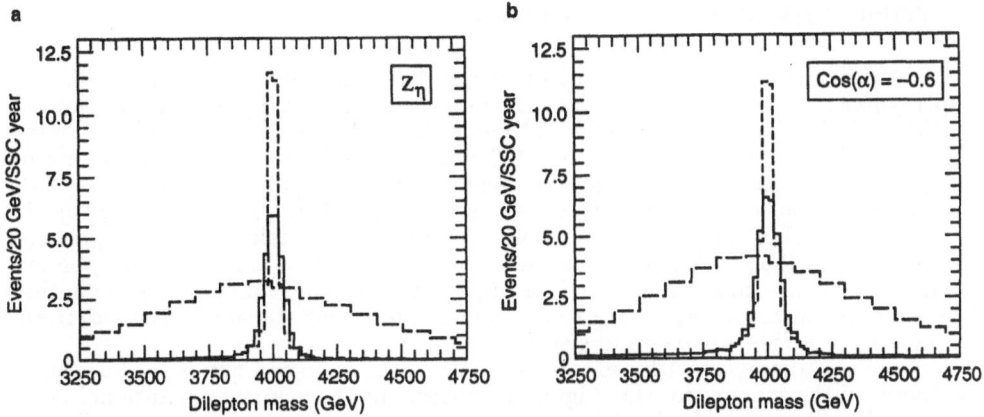

Figure 25. As Figure 24, except that the mass of the new Z is taken to be 4000 GeV.

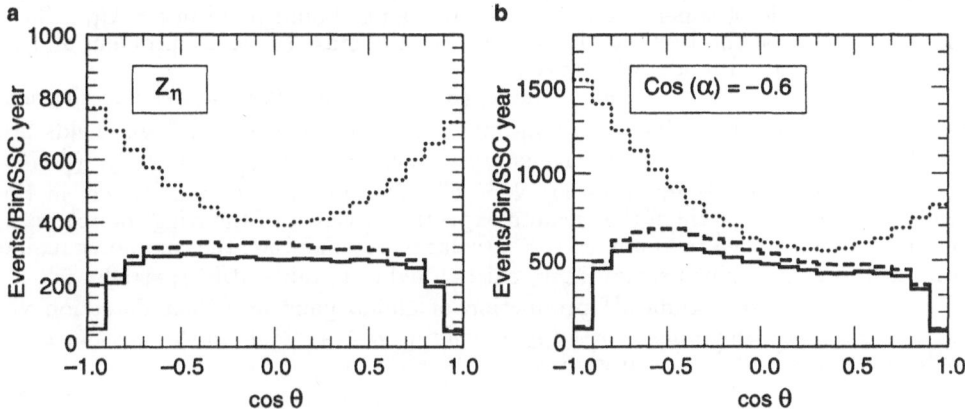

Figure 26. The cross section $d\sigma/d\cos\theta$ for the production of a lepton pair. The expected SDC result for e^+e^- (Solid Curve) and $\mu^+\mu^-$ (Dashed Curve) is Shown. A perfect detector, which neglects acceptance and resolution smearing, is shown for reference (dotted curve). The reconstructed dilepton invariant mass is required to be between 700 and 900 GeV. The longitudinal momentum of the dilepton pair is required to be greater than 500 GeV. Events appear in the plot if the total charge of the lepton pair as determined by the detector is zero. (a) for a Z_η with mass of 800 GeV. (b) for a new Z with $\cos\alpha = -0.6$ and a mass of 800 GeV.

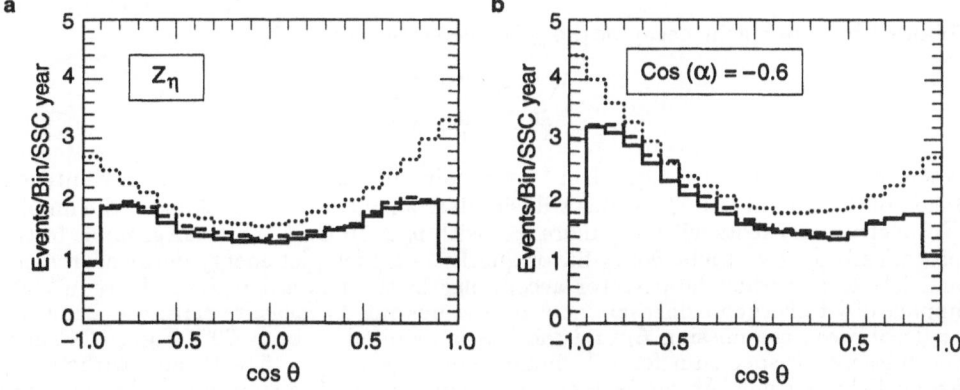

Figure 27. As Figure 26, except that a new Z with a mass of 4000 GeV has been assumed. The invariant mass of the dilepton pair was required to be between 3000 and 4000 GeV, and the longitudinal momentum was required to be greater than 1000 GeV.

BEYOND THE STANDARD MODEL

Many models propose frameworks that vastly extend the standard model picture. Supersymmetric models hypothesize an additional global symmetry that connects particles of differing spin. The minimal model ($N = 1$ supersymmetry) has a single generator Q_α that transforms as spin 1/2 under the Lorentz group. Q_α acting on an ordinary massless state of helicity h generates a superpartner of helicity $h - 1/2$. Because of various commutation rules satisfied by Q_α, when Q_α is applied again, the result vanishes. For this reason, the super multiplets are doublets, with two particles that differ by 1/2 unit of spin angular momentum. The super particles corresponding to the spin-1 gauge bosons are the spin-1/2 gauginos—gluino, wino, zino, and photino. The super particles corresponding to the spin-1/2 fermions are spin-0 partners to the quarks and leptons—squarks and sleptons.

Since Q_α commutes with the other operators, the quantum numbers of the super partners are the same as for the original particles, but they carry an additional quantum number R, that is absolutely conserved. Since supersymmetry is not readily seen in experiments, the symmetry must be broken. The conservation of R implies that the lightest super partners must be stable. To help solve the "naturalness" problem, the expected scale of supersymmetric scale breaking should be of order M_W. Such models seem to be the best candidates to solve the mass hierarchy problem and to connect the standard model with gravity.

In the minimal supersymmetric model (MSSM), the scalar squarks and sleptons are denoted by $\tilde{q}$ and $\tilde{l}$. The fermionic partners of the gauge and Higgs fields are usually represented by the mass eigenstates indicated as $\tilde{g}$, $\tilde{\chi}_i^\pm$ ($i = 1, 2$), $\tilde{\chi}_i^0$ ($i = 1, \ldots, 4$). The scalar Higgs fields are $H^\pm$, h^0, H^0, and A. The parameters in the model are $\tan\beta$, the ratio of the vacuum expectation values of the Higgs doublets; μ, the Higgs mass at the grand unified scale; scalar masses; and M, the common gaugino mass at the grand unified scale. There exists a lightest stable SUSY particle, $\tilde{\chi}_1^0$.

For SDC physics studies,[11] production of gluino pairs and their detection via decays to like-sign dileptons + jets ($\tilde{g}\tilde{g} \to l^\pm l^\pm + (n \geq 4)$ jets) and by missing E_t in multi-jet events have been considered.

Missing E_t

For the more difficult case of a light gluino, $M_{\tilde{g}} = 300$ GeV, with $m_{\tilde{q}} > m_{\tilde{g}}$, the gluino decays are dominated by the channels

$$\tilde{g} \to \chi_1^\pm q\bar{q}' \qquad (BR \sim 60\%)$$
$$\tilde{g} \to \tilde{\chi}_1^0 q\bar{q} \qquad (BR \sim 15\%)$$
$$\tilde{g} \to \tilde{\chi}_2^0 q\bar{q} \qquad (BR \sim 25\%),$$

followed by subsequent decay of the $\tilde{\chi}$ to the stable state $\tilde{\chi}_1^0$:

$$\tilde{\chi}_1^\pm \to \tilde{\chi}_1^0 + X$$
$$\tilde{\chi}_2^0 \to \tilde{\chi}_1^0 + X,$$

where X contains at least a quark or lepton pair. To search for the signal, advantage is taken of the stability of $\tilde{\chi}_1^0$—which will simulate a neutrino—by looking at the missing E_t spectrum in events with 3 or more jets with $p_t > 70$ GeV. The backgrounds to this process are $Z \to \nu\nu$+jets, heavy flavor quarks $\to \nu$+jets, jet energy mismeasurement, and jets lost outside the detector acceptance in the forward region. To reduce the impact of jet energy mismeasurement on the missing E_t measurement, the azimuthal angle between the missing E_t and the jets is required to be $> 20°$. Figure 28 shows the missing E_t spectrum for calorimeter coverage to $\eta = 5$ with and without the azimuthal angle cut. As can be seen, the ϕ cut is needed to suppress background and has little influence on the signal. Figure 29 shows the impact of forward calorimeter coverage to $|\eta| = 4$ or 5. As can be seen coverage for jets to $|\eta| = 5$ is required. Figure 30 shows the result for the SDC design for signal and backgrounds.

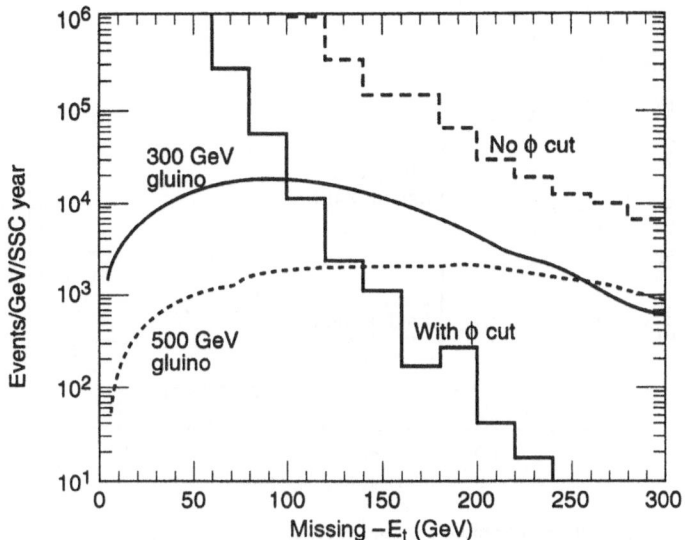

Figure 28. Comparison of the missing-E_t distributions for the background (to light gluino pair production) due to multijet events with mismeasurement of a jet. The two histograms are with (solid) and without (dashed) a cut on events containing a jet with $E_t > 70$ GeV within an azimuthal angle of $40°$ of the missing-E_t. The events are required to have three or more jets each with $E_t > 70$ GeV and $|\eta| < 3$ separated by ΔR of 0.7. The solid (dotted) curve is for pair production of 300 GeV (500 GeV) gluinos decaying as described in the text.

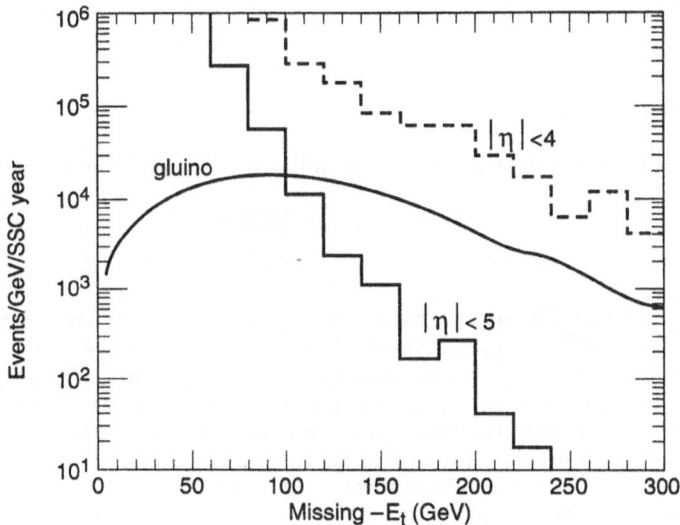

Figure 29. Comparison of the missing-E_t distributions for the background (to light gluino pair production) due to multijet events with energy loss out of the end of the detector, $|\eta| > 4$ (dashed histogram) or 5 (solid histogram). The events are required to have three or more jets each with $E_t > 70$ GeV and $|\eta| < 3$ separated by $\Delta R = 0.7$. Events are rejected if they contain a jet with $E_t > 70$ GeV within an azimuthal angle of $40°$ of the missing-E_t. The solid curve is for pair production of 300 GeV gluinos decaying as described in the text.

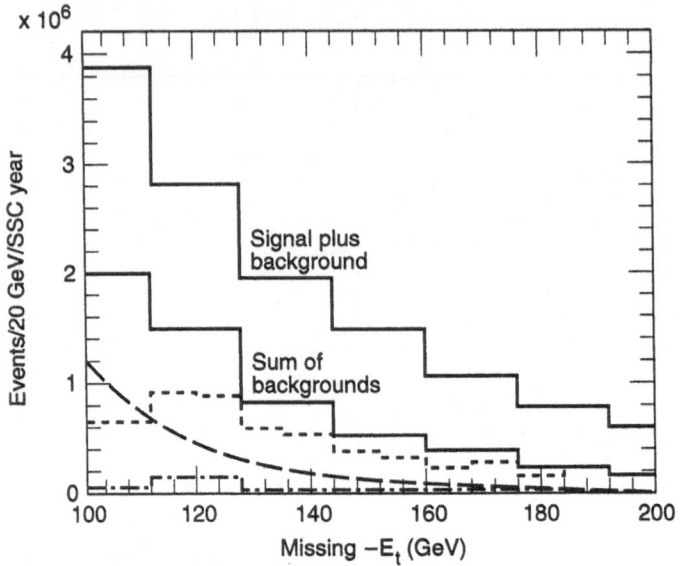

Figure 30. Search for evidence of gluino pair production in the distribution of missing-E_t for the final state of three or more jets each with $E_t > 70$ GeV separated by $\Delta R = 0.7$. The jets are required to have $|\eta| < 5$. Events are rejected if they contain a jet with $E_t > 70$ GeV within an azimuthal angle of 20° of the missing-E_t vector or if they have circularity $C < 0.2$. The signal is for pair productions of 300 GeV gluinos decaying as described in the text. The lower solid histogram is the sum of the individual background contributions, the upper solid histogram is the sum of the background and the gluino signal. The detector-dependent background due to multijet events with missing-E_t generated by calorimeter resolution or by energy loss out of the end of the detector ($|\eta| > 5$) is shown as a dashed curve. The dotted background arises from the final states $t\bar{t}$ and $b\bar{b}$, where the missing-E_t is due to neutrinos. The dash-dotted background is due to $Z + \text{multijet}$ events.

Like-Sign Di-Leptons

In this mode, a search is made for the following decay chain for each $\tilde{g}$:

$$\tilde{g} \to \tilde{\chi}_1^{\pm} q\bar{q}' \qquad (BR \sim 60\%)$$
$$\tilde{\chi}_1^{\pm} \to l^{\pm} \nu \tilde{\chi}_1^0 \qquad (BR \sim 20\%).$$

The total BR is about 2%, half of which corresponds to same-sign lepton pairs. This corresponds to 2×10^6 events (25 events) for $m_{\tilde{g}} = 180$ GeV (2 TeV). For the case $m_{\tilde{g}} = 300$ GeV, we require ≥ 4 jets with $|\eta| < 3$ and $p_t > 50$ GeV, $p_t > 20$ GeV for both leptons, and ΔR between the lepton and jet > 0.5 to reduce the $t\bar{t}$ background. Figure 31 shows the mass of the lower-p_t lepton and the two nearest jets for the signal and the top background *without* the separation cut. The top background is eliminated by the cut. Figure 32 shows the mass distributions for the high-p_t lepton plus the two nearest jets for two different gluino masses. The mass resolution using this method is about 10%.

These and other studies[11] show that the SDC design will be capable of detecting gluinos in the range $180 < m_{\tilde{g}} < 0$ (TeV) using two independent decay channels. Supersymmetry will manifest itself in several different phenomena. The coincidence and consistency of the various signals will help select the particular model chosen by nature.

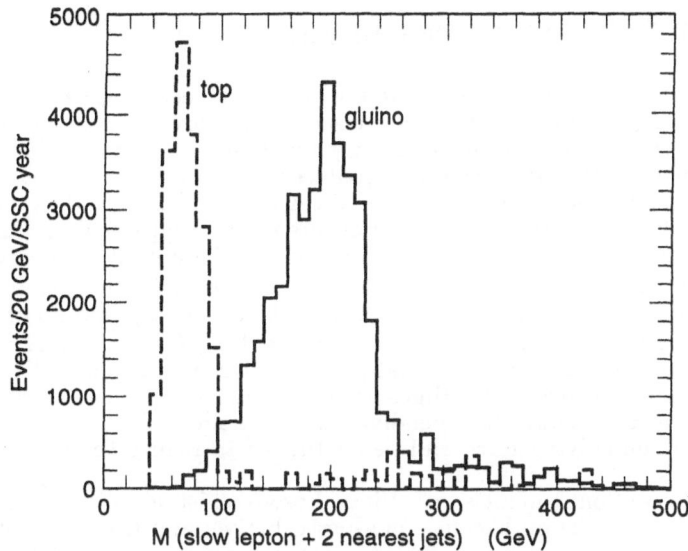

Figure 31. For events with two isolated same-sign leptons, the distribution of the invariant mass of the lower p_t lepton together with the two nearest jets chosen from the four jets with the highest p_t. Each lepton is required to have $p_t > 20$ GeV or alternatively one lepton must have $p_t > 40$ GeV and the other $p_t > 15$ GeV. The leptons must lie within $|\eta| < 2.5$. Events are required to have at least four jets with $p_t > 50$ GeV. The gluino mass was taken to be 300 GeV. The solid histogram is from gluino pair production and decay ($\tilde{g}\tilde{g} \to q\bar{q}\tilde{\chi}_1^+ q\bar{q}\tilde{\chi}_1^+ \to q\bar{q}\ell^+\nu\tilde{\chi}_1^0 q\bar{q}\ell^+\nu\tilde{\chi}_1^0$), whereas the dashed curve is due to $t\bar{t}$ production and decay ($t\bar{t} \to b\ell^+\nu\bar{c}\ell^+\nu$ jets, with $M_{\rm top} = 150$ GeV). Unlike the signal, the dashed curve has not had any isolation cut applied. Such a cut would have eliminated it entirely.

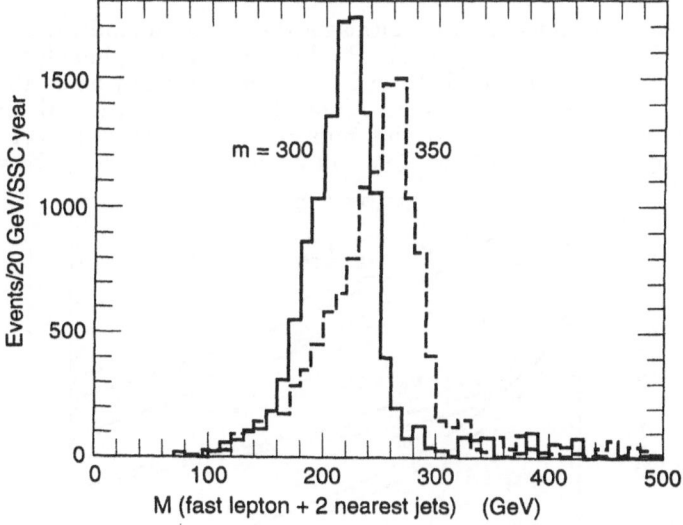

Figure 32. For events with two isolated same-sign leptons, the distribution of the invariant mass of the higher p_t lepton together with the two nearest jets chosen from the four jets with the highest p_t. Events are required to have one lepton with $p_t > 15$ GeV, one with $p_t > 65$ GeV, and at least four jets with $p_t > 50$ GeV. The leptons must lie within $|\eta| < 2.5$. The solid (dashed) curve is from $M = 300$ GeV ($M = 350$ GeV) gluino pair production and decay. The 300 GeV (350 GeV) gluino yields 6000 events (12,000 events) per year. (The cross section rises quickly with mass because of our cuts). The dashed curve has been divided by two for display purposes. This calculation was done using a parton-level Monte Carlo including resolution; a more realistic calculation will add tails to the peaks due to radiated jets that accidentally are near the lepton.

ELECTROWEAK SYMMETRY BREAKING

The theoretical notions concerning the nature of electroweak symmetry breaking reflect a wide range of ideas. The "conventional" picture involves one scalar particle that results when "spontaneous symmetry breaking"—a non-vanishing vacuum expectation value for the Higgs complex doublet field that selects a preferred direction in weak isospin and hypercharge space—breaks the $SU(2) \times U(1)$ symmetry of the standard model. The other three components of the Higgs field are combined with the transverse components of the gauge bosons in the unbroken theory to form the massive $W^{\pm}$ and Z^0 bosons. In minimal supersymmetric models, the situation is slightly more complicated, as two complex doublets are required for the Higgs field, yielding five remaining degrees of freedom denoted h^0, H^0, A^0 and $H^{\pm}$.

Deviations from this conventional picture, with fundamental Higgs scalars, are found in models in which scalar Higgs bosons exist, but they are composite particles, several neutral and charged, depending on the model. Such models (technicolor) require a new underlying force and new fermions to realize the symmetry-breaking sector.

In unconventional models, no Higgs bosons exist at all. Instead, the high-energy $W_L W_L$ interaction becomes non-perturbative, and the weak interactions become "strong" at high energy. In this case, resonances in gauge-boson scattering occur, and their effects can be seen even at SSC energies.

All these possibilities must be distinguished experimentally. To elucidate the nature of the electroweak symmetry-breaking sector is the main goal of the SSC experimental program. Much work has been done on the minimal standard model Higgs and its decay modes in various mass ranges. Details of the performance of the SDC design for this case and some results on supersymmetric models and $W_L W_L$ scattering follow.

Minimal Standard Model Higgs

The dominant production mechanisms for the standard model Higgs versus Higgs mass at the SSC are shown in Figure 33. Production by gluon-gluon fusion via a heavy quark loop is shown in solid, by WW or ZZ fusion in dots, by associated production with a $t\bar{t}$ pair in dot-dash, and by associated production with a W or Z boson in dashes in Figure 33. The curves are shown for top-quark mass values of 100, 150, and 200 GeV. A top quark mass of 150 GeV will be used in all subsequent work.

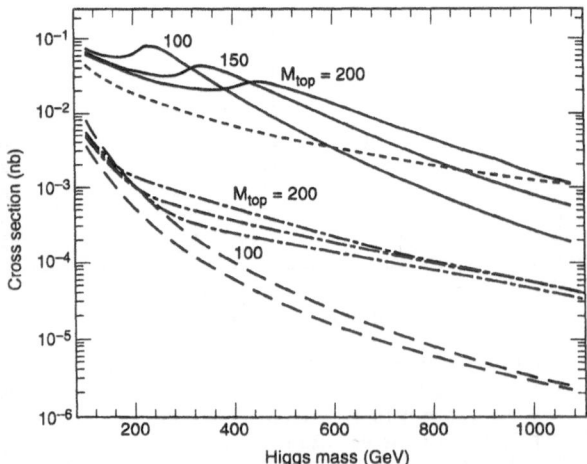

Figure 33. The cross section for the production of a Higgs boson in pp collisions at $\sqrt{s} = 40$ TeV as a function of the Higgs boson mass for several different production mechanisms: gg fusion (solid), WW/ZZ fusion (dotted), $t\bar{t} + H$ production (dot-dashed), $W + H$ production (upper dashed), and $Z + H$ production (lower dashed). When the cross section depends on the t-quark mass, several curves have been included for different values of M_{top}.

The decay modes of the standard Higgs are shown in Figure 34. For masses below the WW and ZZ threshold, the Higgs boson is very narrow (see Table 3), while at the highest masses it becomes quite broad. For good behavior of the lowest-order diagrams involving the Higgs, the Higgs is expected to be have mass less than ~ 1 TeV. Note from Figure 34 that decays $H \to ZZ^*$ or ZZ are significant for Higgs masses above approximately 140 GeV. Note also the small but significant $H \to \gamma\gamma$ branching ratio below 160 GeV. For the purpose of analysis, we divide the mass ranges into low mass ($80 < m < 130$ GeV), intermediate mass ($130 < m < 180$ GeV), and high mass (> 180 GeV).

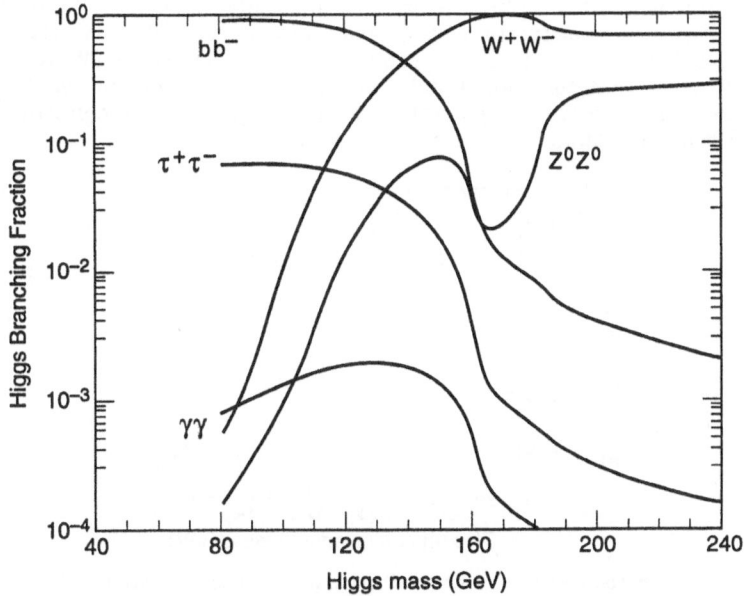

Figure 34. The branching ratio for a Higgs boson into various channels as a function of its mass.

Table 3. Decays of Standard Model Higgs.

Higgs Mass (GeV)	Higgs Width (GeV)
140	0.01
160	0.1
200	1.4
400	30
800	270

Low Mass Higgs

Searches at LEP-II are unlikely to extend much above 80 GeV, due to statistics and backgrounds from ZZ production. Direct production in pp collisions of $H \to \gamma\gamma$ suffers from a large continuum background from QCD production of photon pairs and misidentified 2-jet events. In addition, good mass resolution requires accurate knowledge of photon directions. For the SDC design, the signals for $M_H = 80$, 100, 120, 140, and 160 GeV are shown along with the background in Figure 35. (Only backgrounds from QCD photon pair production are included.) Figure 36 shows the result when a smooth two-photon background is subtracted in order to study the significance of signals. The significance of this marginal signal would increase slowly with improved calorimeter resolution, but the prohibitive cost of achieving the needed resolution suggests that other modes suffering less background would be more suitable for the SDC design.

Associated production of Higgs with a W or t quark gives an additional lepton tag that assists with triggering and significantly reduces the background. Figure 37 shows the efficiency for detection of $t\hat{t} + H \rightarrow l\gamma\gamma + X$ vs. p_t and η. The analysis requires $p_t > 20$ and $|\eta| < 2.5$ for the lepton and both photons. The dotted (solid) curves are for $M = 80$ (160) GeV, with four curves for η coverage of 1.5, 2.0, 2.5, 3.0. Several classes of background have been considered in this process. These include two-photon backgrounds $W + \gamma\gamma$, $b\bar{b} + \gamma\gamma$, $t\bar{t} + \gamma\gamma$; QED radiative decays W, $Z + \gamma\gamma$, and leptons faking γ in $Z + \gamma$; backgrounds with one real γ and one misidentified jet ($t\bar{t} + \gamma$); and backgrounds where both photons arise from misidentified jets. Figure 38 shows the resulting estimated background vs. Higgs mass. Figure 39 shows the signal and background for associated light Higgs production with signals at 80, 100, 120, 140, and 160 GeV. Note that the right-hand scale shows the number of events expected; the curves are generated with large statistics. This signal can be confirmed by SDC after several years' running.

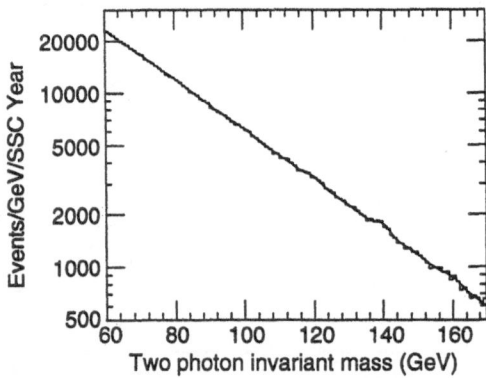

Figure 35. The two-photon invariant mass distribution including the signals from Higgs bosons with masses of 80, 100, 120, 140, and 160 GeV. The background includes only the irreducible backgrounds arising from the $q\bar{q} \rightarrow \gamma\gamma$ and $gg \rightarrow \gamma\gamma$ processes.

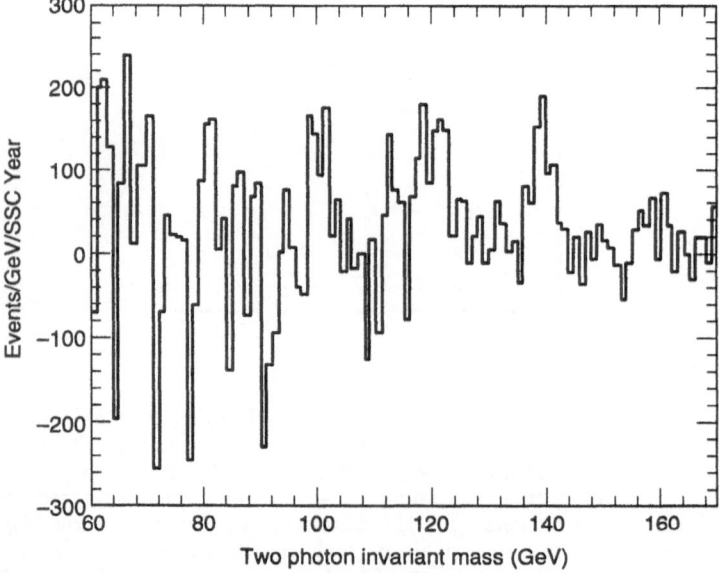

Figure 36. The expected signals from Higgs bosons with masses of 80, 100, 120, 140, and 160 GeV. The irreducible backgrounds displayed in Figure 35 have been statistically subtracted using an exponential fit to the background shape. The baseline calorimeter performance has been assumed.

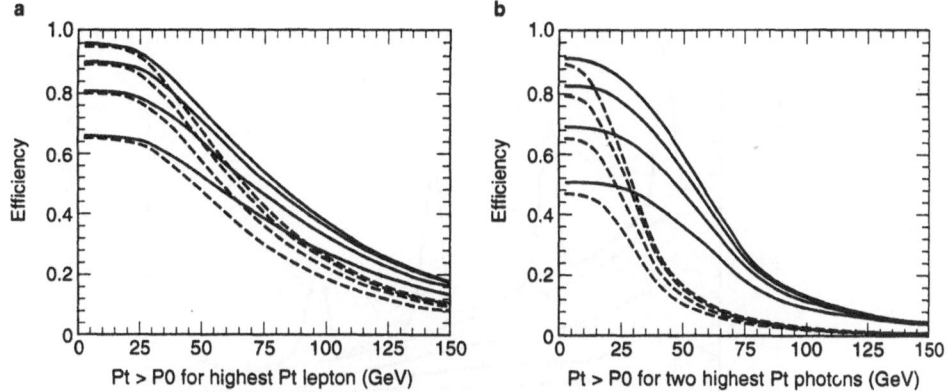

Figure 37. Families of acceptance curves for $t\bar{t} + H$, with $M_{\text{Higgs}} = 80$ GeV (dashed) and $M_{\text{Higgs}} = 160$ GeV (solid). (a) The fraction of events with at least one lepton with $p_t > p_0$ as a function of p_0. The lepton has $|\eta| < 1.5$ (lowest curve), 2.0 (lower middle curve), 2.5 (upper middle curve), or 3.0 (upper curve). (b) The fraction of events with at least two photons with $p_t > p_0$ as a function of p_0. The photons have $|\eta| < 1.5$ (lowest curve), 2.0 (lower middle curve), 2.5 (upper middle curve), or 3.0 (upper curve).

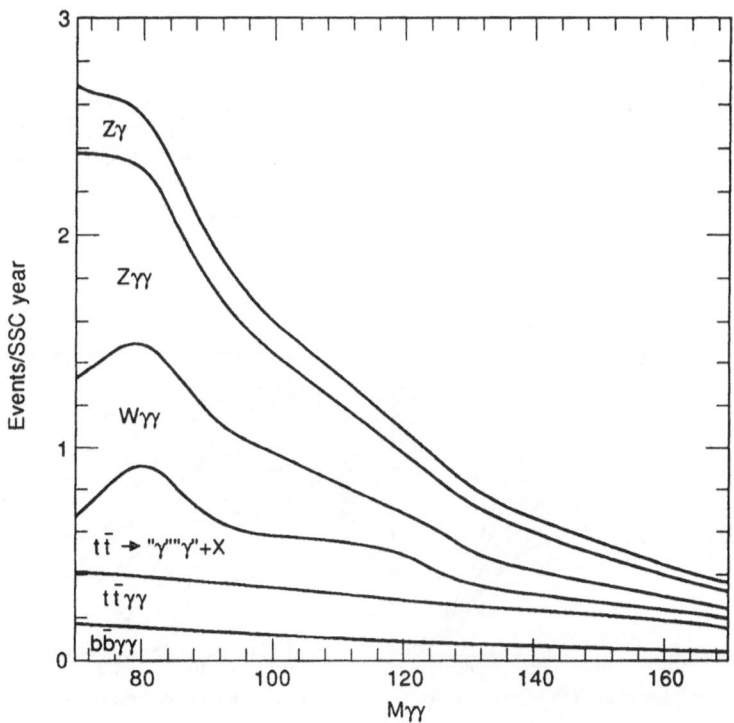

Figure 38. Backgrounds to $t\bar{t}+$ Higgs and $W+$ Higgs final state.

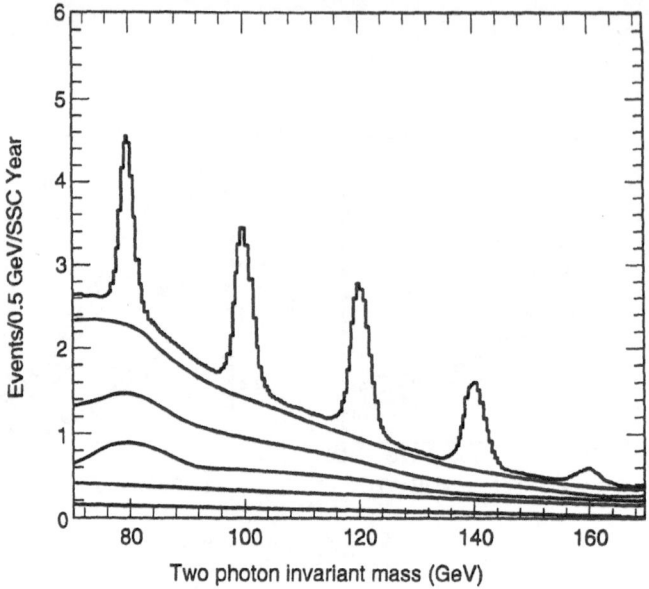

Figure 39. The two-photon invariant mass distribution for the expected signals from Higgs bosons of mass 80, 100, 120, 140, and 160 GeV. The background curves are cumulative, and are as shown in Figure 38. The baseline detector resolution has been assumed.

Intermediate Mass Higgs

In this mass region, the decay $H \to ZZ^* \to 4l$ provides a distinctive signature. The efficiency for $H \to ZZ^*$ detection versus p_t and η is shown in Figure 40. The curves are for $M_H = 120$ GeV (solid), 140 GeV (dotted) and 160 GeV (dashed). The four curves are for η coverages of 1.5, 2.0, 2.5, and 3.0. The p_t for two leptons is > 20 GeV, and all four must be > 10 GeV with $|\eta| < 2.5$.

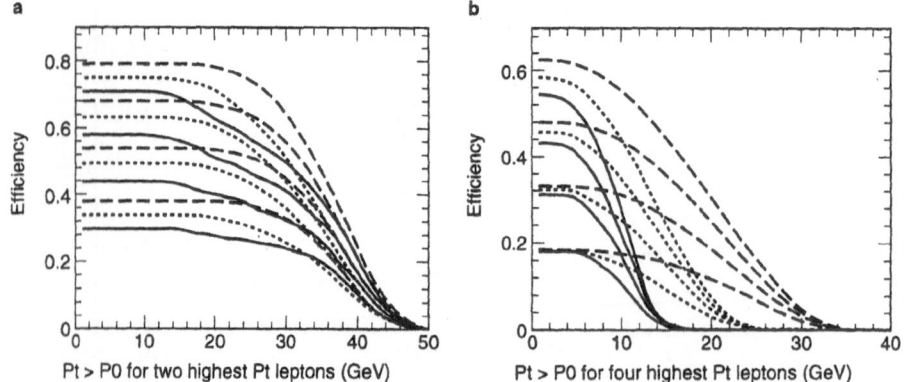

Figure 40. Families of acceptance curves for $H \to ZZ^*$, for $M_{\text{Higgs}} = 120$ GeV (solid), 140 GeV (dotted), and 160 GeV (dashed). (a) The fraction of events with at least two leptons with $p_t > p_0$ as a function of p_0. Both leptons have $|\eta| < 1.5$ (lower curve), 2.0 (lower middle curve), 2.5 (upper middle curve), or 3.0 (upper curve). (b) The fraction of events containing two leptons with $p_t > 20$ GeV and $|\eta| < 2.5$ plus two others with $p_t > p_0$ and $|\eta| < 1.5$ (lower curve), 2.0 (lower middle curve), 2.5 (upper middle curve), or 3.0 (upper curve).

Again, several classes of background have been considered. These include continuum $\bar{Z}Z$ production from $q\bar{q}$ and gg initial states where the latter has been approximated by scaling the former by 1.65, production of $Z + b\bar{b}$ and $Z + t\bar{t}$ with the heavy flavor decay providing two additional leptons, and production of $t\bar{t}$ with subsequent decay to four leptons. Topological requirements are effective at reducing the $t\bar{t}$ backgrounds. Figure 41 shows the distribution of excess E_t in a cone of radius 0.3 around the leptons for $H \rightarrow ZZ^*$, and the $t\bar{t}$ backgrounds. Leptons from b and lighter quark decays can be strongly suppressed while retaining high signal efficiency by requiring excess $E_t < 5$ GeV. This requirement is 94% efficient for signal leptons. Figure 42 shows the signal for Higgs masses of 130, 140, 150, 160, and 170 GeV, along with the background. For masses above 130 GeV, the signal will be visible in one SSC year.

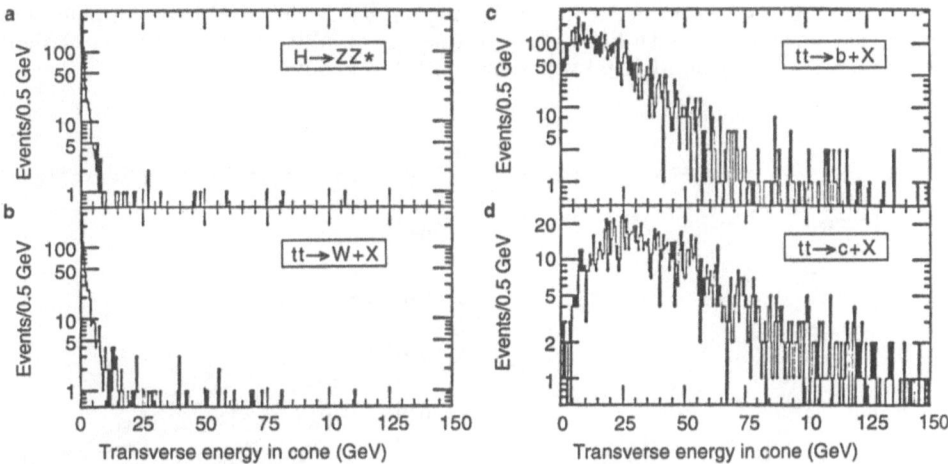

Figure 41. The distribution of excess E_t in a cone of radius $R = 0.3$ for different classes of electrons. The electrons are all in the range $10 < p_t < 20$ GeV. (a) Electrons from $H \rightarrow ZZ^*$ for $M_{\text{Higgs}} = 140$ GeV. (b) W electrons coming from t quark decays ($M_{\text{top}} = 150$ GeV). (c) b electrons coming from t quark decays. (d) c (or u, d) electrons coming from t-quark decays.

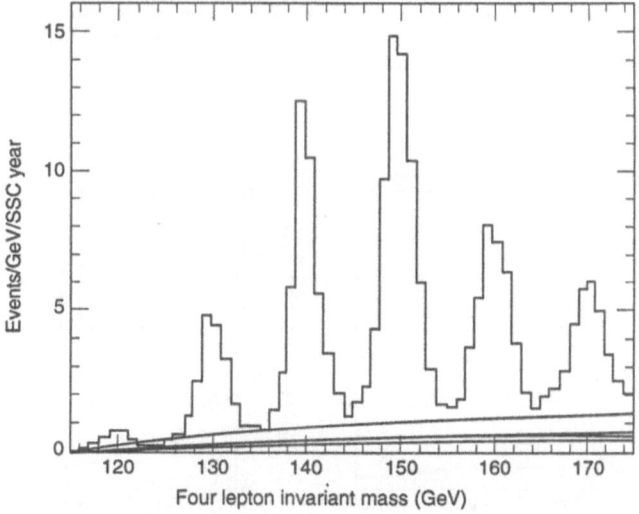

Figure 42. The reconstructed Higgs mass for ZZ^* decaying to $4e$, 4μ, and $2e2\mu$ with $M_{\text{Higgs}} = 120$, 130, 140, 150, 160, and 170 GeV, including the expected backgrounds. The backgrounds curves are cumulative, and are (from lowest to highest): $q\bar{q} \rightarrow ZZ^*$, multiplied by 1.65 to account for $gg \rightarrow ZZ^*$, $Z + b\bar{b}$, $Z + t\bar{t}$, and $t\bar{t}$. The invariant mass has been calculated using calorimeter measurements for the electrons.

Heavy Higgs

Clean but statistically limited modes $H \to ZZ \to 4l^{\pm}$ and $H \to ZZ \to 2l^{\pm}\nu\bar{\nu}$ have been considered in detail by SDC. Other modes with larger branching fractions are accompanied by much larger backgrounds. Figure 43 shows the efficiencies for $H \to ZZ \to 4l^{\pm}$ detection vs. p_t and η. Results are for $M_H = 200$ (solid), 400 (dotted) and 800 (dashed) GeV. The four curves are for η coverage of 1.5, 2.0, 2.5 and 3.0. The analysis requires $p_t > 20$ GeV for two leptons, and $p_t > 10$ GeV and $|\eta| < 2.5$ for all four leptons. The backgrounds are the same as those discussed previously for the ZZ^* mode, but the second Z mass constraint eliminates all but the ZZ continuum background. Figure 44 (45) shows the expected signal along with the continuum background for Higgs of mass 200 and 400 (800) GeV. Since the statistics are not high in the 800-GeV sample, a cut requiring $p_t(Z) > 200$ GeV was applied that reduces the background with little effect on the signal. With this cut, about 14 signal events above a background of 6 events are expected for one SSC year.

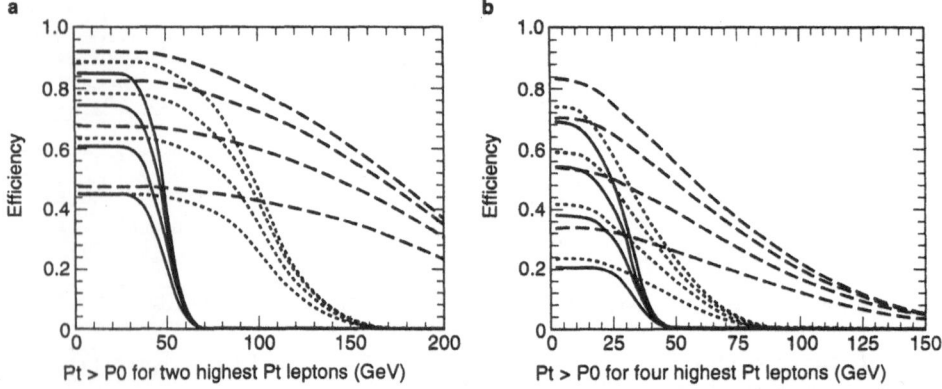

Figure 43. Families of acceptance curves for $H \to ZZ$, for $M_{\text{Higgs}} = 200$ GeV (solid), 400 GeV (dotted), and 800 GeV (dashed). (a) The fraction of events with at least two leptons with $p_t > p_0$ as a function of p_0. Both leptons have $|\eta| < 1.5$ (lower curve), 2.0 (lower middle curve), 2.5 (upper middle curve), or 3.0 (upper curve). (b) The fraction of events containing two leptons with $p_t > 20$ GeV and $|\eta| < 2.5$ plus two others with $p_t > p_0$ and $|\eta| < 1.5$ (lower curve), 2.0 (lower middle curve), 2.5 (upper middle curve), or 3.0 (upper curve).

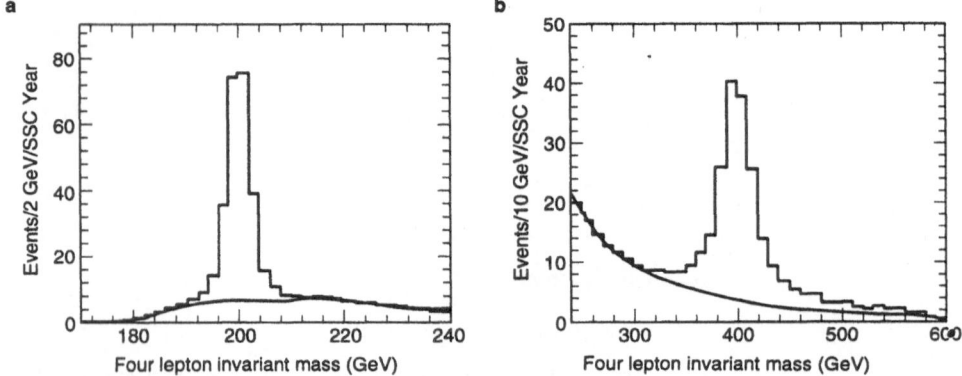

Figure 44. (a) The ZZ invariant mass distribution showing a peak due to a Higgs of mass 200 GeV. The two lepton pairs were both required to have $M_{\ell\ell} = M_Z \pm 10$ GeV. The background curves have the same significance as those of Figure 42, but the ZZ continuum background gives the only visible contribution. (b) Same as (a), except that the Higgs mass is 400 GeV.

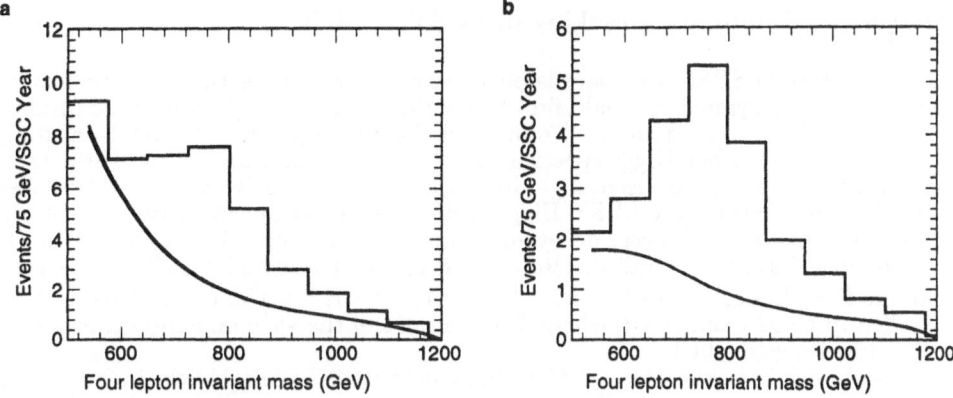

Figure 45. (a) Same as Figure 44(a), except that the Higgs mass is 800 GeV. (b) Same as (a), except that both Z's were required to satisfy $p_t(Z) > 200$ GeV.

To improve on the rate at high mass, the mode $H \to ZZ \to 2l^\pm \nu\bar{\nu}$ was considered. The branching ratio is a factor six larger than the previous case, but the 2ν mode is more sensitive to background. The major additional background is $Z+$ jets, where the $Z \to l^\pm l^\mp$, and one of the jets is mismeasured or lost to simulate the neutrinos. Continuum ZZ production and production of $Z + b\bar{b}$ and $Z + t\bar{t}$ also contribute to the background. Figure 46 shows the expected number of events as a function of the missing E_t in the events for the signal and background. The dashed line is the continuum background, the dot-dashed line is $Z+$ jets, and the dotted line is $Z + t\bar{t}$. This signal, in combination with the $4l^\pm$ final state, would provide strong evidence for an 800-GeV Higgs in one SSC year.

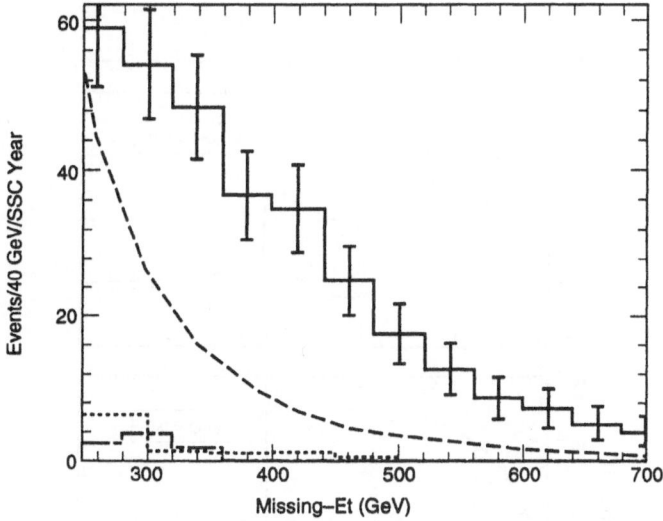

Figure 46. The distribution (solid histogram) in missing-E_t for the final state $Z(\to \mu^+\mu^-, e^+e^-)+$ missing-E_t including the effect of a Higgs boson of mass 800 GeV and the various backgrounds. The reconstructed Z is required to have $p_t > 250$ GeV and the events are rejected if they contain a jet with $E_t > 300$ GeV. The background shown as a dashed curve arises from $q\bar{q} \to ZZ$ (multiplied by 1.65 to account for the $gg \to ZZ$ process). The dot-dashed background arises from the final state $Z + jets$ where the missing-E_t is generated by calorimeter resolution or by losing energy out of the end of the detector. The dotted background arises from the final state $t\bar{t}$ where there is an e^+e^- (or $\mu^+\mu^-$) pair of mass $M_Z \pm 20$ GeV and the missing-E_t is due to neutrinos; the $Z + Q\bar{Q}$ background is negligible in this figure.

Electroweak Symmetry Breaking in SUSY Models

The nature of SUSY models was outlined in a previous section. As was shown, evidence for supersymmetry would first be seen in relatively high-rate gluino production. Once SUSY is seen there, a more complicated Higgs sector would naturally be expected. The richer Higgs spectrum predicted in such models is a considerable experimental challenge to disentangle from the data. Extensive work with detailed detector simulations on the SUSY Higgs sector has not yet been carried out by the community. In the SDC case, decay modes of $\gamma\gamma$, ZZ^*, or ZZ, and top decay to charged Higgs have been studied. (Recall that for the two-doublet SUSY Higgs case, there are five physical Higgs bosons, h^0, H^0, A^0, and $H^\pm$.) The two free parameters are taken to be M_A and $\tan\beta$, where β is the ratio of the vacuum expectation values for the two Higgs doublets.

The branching ratios for neutral Higgs decays to $\gamma\gamma$ are shown in Figure 47, where the solid curve is for h^0 and the dotted (dashed) curve is for H^0 (A^0). For large values of M_h^0, $h^0 \rightarrow \gamma\gamma$ is observable. For small $\tan\beta$, $A^0 \rightarrow \gamma\gamma$ should be observable. Figure 48 shows the branching ratios for neutral Higgs to ZZ. Again, the solid curve is for h^0, and the dotted (dashed) curve is for H^0 (A^0). For large values of $\tan\beta$, $H^0 \rightarrow ZZ$ is observable. For small values of $\tan\beta$, $H^0 \rightarrow ZZ$ is observable. Studies of top $\rightarrow H^+b$ have shown that the charged Higgs can be seen for all values of $\tan\beta > 0.1$ via the $c\bar{s}$ and $\tau\nu_\tau$ decay modes.

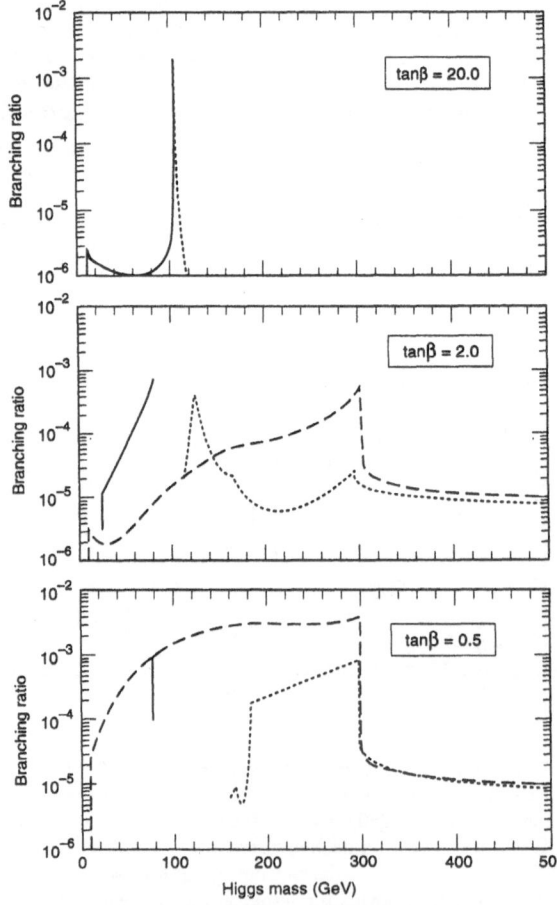

Figure 47. The branching ratios for the decay of the neutral Higgs bosons in the MSSM to the $\gamma\gamma$ final state, assuming $M_{\text{top}} = 150$ GeV. The solid curve is for the h^0, the dotted curve is for the H^0, and the dashed curve is for the A^0. The three different plots are for $\tan\beta = 0.5$, 2, and 20.

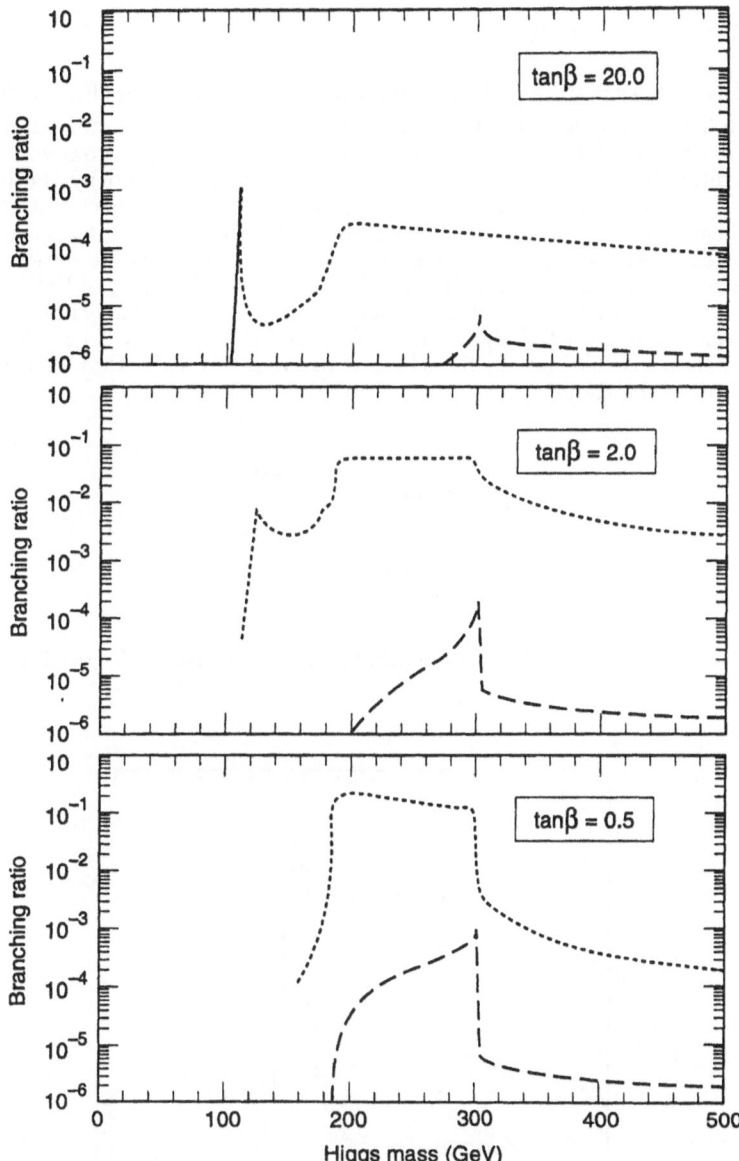

Figure 48. The branching ratios for the decay of the neutral Higgs bosons in the MSSM to the ZZ or ZZ^* final state, assuming $M_{\text{top}} = 150$ GeV. The solid curve is for the h^0, the dotted curve is for the H^0, and the dashed curve is for the A^0. The three different plots are for $\tan\beta = 0.5$, 2, and 20.

Assuming 3–5 years at SSC design luminosity, the experimental situation can be summarized for SDC by the following comments. For small M_A, h^0 is observable at LEP-II, and $t \rightarrow H^\pm b$ is observable at SSC. For moderate M_A and small $\tan\beta$, h^0 is observable at LEP-II, and $H^0 \rightarrow ZZ$, $t \rightarrow H^\pm b$ are observable at SSC. For moderate M_A and large $\tan\beta$, none of the Higgs bosons may be observable in ZZ or $\gamma\gamma$. For large M_A, $h^0 \rightarrow \gamma\gamma$ is observable at SSC.

Clearly, the experimental situation needs further work and clarification within the context of realistic detector models. Tau decays and b tagging may provide some of the tools necessary to further extend the parameter range over which a viable search may be carried out.

Strongly Coupled Models

If the Higgs mass is $\lesssim 1.8$ TeV, the longitudinal W scattering (Figure 49) process has good high-energy behavior—the cross section growth from lower energies is cut off by the exchange of narrow spin zero bosons (the Higgs). However, if $M_H \geq 1$ TeV, the interaction of the longitudinal W bosons (represented by the interaction lagrangian $\mathcal{L}_H$ in Figure 49) becomes strong, and a complex spectrum of particles and resonances is likely to appear. Such resonances are expected to be in the mass range 1–3 TeV. Even though the resonances may be too heavy to be directly detected at SSC, the interaction lagrangian $\mathcal{L}_H$ can still be probed via study of WW scattering.

Resonant production also occurs in specific models such as technicolor models. Figure 50 shows the expected signal for a Techni-ω decaying to $Z\gamma$ in SDC. As can be seen, the signal is distinctive, but the low event rate means that high luminosity is required to see the signal.

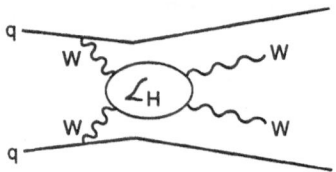

Figure 49. Schematic of WW scattering, showing the interaction Lagrangian $\mathcal{L}_H$.

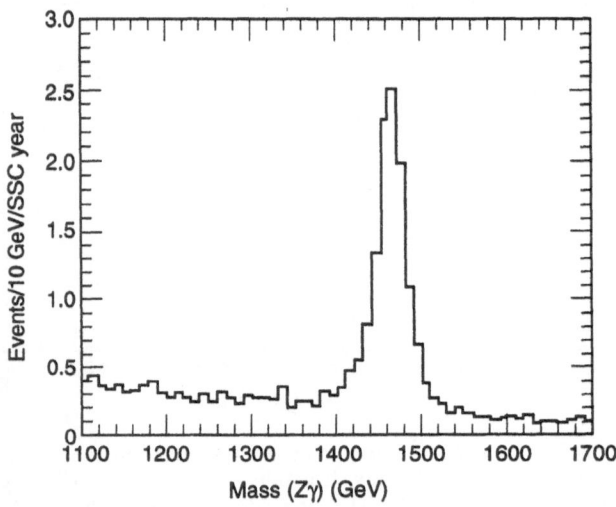

Figure 50. The distribution of invariant mass for the final state system of a photon and an e^+e^-. The lepton pair is required to have mass $M_Z \pm 10$ GeV. The leptons have $|\eta| < 2.5$ and the photon has $|\eta| < 3.0$. The peak corresponds to the production and decay of the techni-omega particle of mass 1.46 TeV discussed in the text.

Non-resonant WW scattering has also been studied by SDC. The solid curve of Figure 51 shows the expected signal in a strong-coupling model due to Chanowitz.[15] The major experimental backgrounds, $q\bar{q} \rightarrow W^+W^-$ (dotted) and $gg \rightarrow t\bar{t} \rightarrow W^+W^- + X$ (dashed), are also shown. To suppress these backgrounds, same-sign W pairs may be used; the like-sign backgrounds from $gg \rightarrow t\bar{t} \rightarrow W + b\bar{b} + X \rightarrow l^+l^+ + X$ are also shown (dot-dashed) in Figure 51.

Clearly, accurate charge measurement is critical to carry out this measurement. For p_t of the lepton of 100 (500) GeV, rejection of the opposite sign background of 10^{-5} (10^{-3}) is needed. The like-sign $t\bar{t}$ events are relatively easily rejected by means of isolation cuts on the leptons.

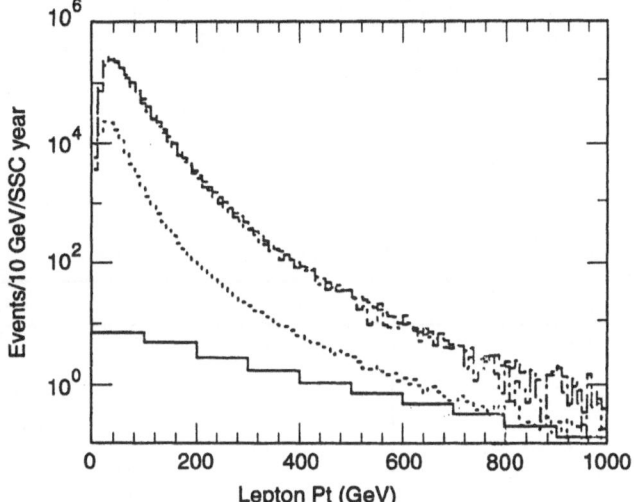

Figure 51. The lepton transverse momentum distribution expected for the W^+W^+ signal and backgrounds (there are two entries per event, one for each lepton). The signal is shown as the solid histogram, the dashed histogram is for $t\bar{t} \to W^+W^- + X$, and the dot-dashed histogram is for $t\bar{t} \to W^+ b\bar{b} + X \to \ell^+\ell^+ + X$. Note that the curves for these two backgrounds sources lie on top of each other. The dotted histogram is for the direct production of opposite-sign $q\bar{q} \to W^+W^-$ background.

Summary

The SDC design is capable of finding a standard model Higgs over the full mass range to even the highest masses. SUSY models are more difficult to disentangle, but some signals will likely be visible. (Evidence for SUSY itself is no problem.) The strong WW scattering signal—difficult to extract even at SSC—appears possible albeit requiring a few years' running time. The theoretical uncertainty surrounding the nature of electroweak symmetry breaking implies that detection of a signal at, say, 200 GeV, decaying to ZZ will not be enough to be sure of the physics. Full exploration of the complete mass range and as many decay modes as possible are necessary to strongly limit the available theoretical phase space. Studies performed thus far with realistic detector simulations show that the SSC can cover the full mass range necessary. Detector designs for LHC experiments should be available within a year to properly assess the limits to the capability of the proposed LHC machine.

DETECTOR DESIGN IMPLICATIONS

The process of optimizing the SDC detector design to best match the physics goals of the collaboration requires identification of those physics processes that put the most stringent requirements on the detector performance. From the discussion above, one readily sees that detection of the Higgs sector, with its typical four-body decay modes, *requires* that the detector subsystems be designed to work together effectively.

Space does not permit a full discussion here of all the tradeoffs in the SDC design. A few comments concerning the impact of the physics goals on the calorimeter design are appropriate here. The angular coverage of the calorimeter is set by need to retain high efficiency for the Higgs signal (lepton ID for $|\eta| < 3$), and by the need to have clean signals for gluino production ($|\eta| < 6$). The energy resolution and linearity requirements become mass resolution requirements for width determination or background suppression capability. Position resolution and particle identification capability are determined by the need to provide adequate segmentation to reduce backgrounds. Timing information must be accurate enough to tag the beam crossing in which the event occurred to simplify the trigger and DAQ system design. Tables 4–6 summarize the SDC calorimeter requirements.

Table 4. Performance Requirements on the SDC Calorimeter.

Parameter	Requirement	Basis
η max for $e^{\pm}$ ID	2.5	$H \rightarrow 4e, 2e2\mu$
EM efficiency loss in $|\eta| < 2.5$	$< 5\%$	electron ID
η max for jets	5	SUSY searches
gaps in full jet coverage, $|\eta| < 5$	$\leq 1\%$	Missing-E_t
EM energy resolution		$H \rightarrow \gamma\gamma, Z' \rightarrow ee$
stochastic term	$\leq 15\%/\sqrt{E_t}$	
constant term	$\leq 1\%$	
EM transverse segmentation	0.05	$H \rightarrow 4e, H \rightarrow \gamma\gamma$
Hadronic energy resolution		dijet mass resolution
stochastic term	$\leq 70\%/\sqrt{E_t}$	
constant term (single $\pi^{\pm}$)	$\leq 6\%$	
Hadronic transverse segmentation	0.10	dijet mass resolution
EM residual nonlinearity	$\leq 1\%, E_t > 10$ GeV	$ee, \gamma\gamma$ mass resolution
Jet residual nonlinearity	$\leq 1\%/$ TeV, $E_t > 2$ TeV	compositeness search
Dynamic range (EM and HAC)	20 MeV–4 TeV	e ID, compositeness
EM depth	$22/25\ X_0$	$ee, \gamma\gamma$ mass resolution
Calorimeter depth ($\eta = 0$)	$\geq 10\ \lambda$	dijet mass resolution

Table 5. Performance Requirements on the Shower Maximum Detector.

Parameter	Requirement
Strip width	< 1.5 cm
Resolution $r - \phi$ and z	< 3 mm
Resolution on relative energy (strip-strip)	$< 10\%$
Strip length ($\Delta\eta$ or $\Delta\phi$)	≤ 0.2
Cross talk	$\leq 0.5\%$ after correction

Table 6. Performance Requirements on the Forward Calorimeter.

Parameter	Requirement
Energy resolution	$< 10\%$ at 1 TeV
Transverse segmentation, $\eta = 3$	0.2
$\eta = 5$	0.4
Time resolution	$\sigma_t < 5$ ns, $E_t > 50$ GeV
Noise	$\sigma_{E_t} < 30$ GeV
η coverage, jet axis	$3 \leq |\eta| \leq 5$
(physical)	$2.8 \leq |\eta| \leq 6$

CONCLUSIONS

The SSC is the tool of choice for future progress in particle physics. The SDC design is well advanced and appears capable of exploiting many of the physics opportunities made available by the SSC. We will see within a year designs for the SSC GEM experiment, as well as other high-p_t experiments at LHC. Studies of those designs with detailed detector models will allow proper assessment of the capabilities and limitations of the SSC and LHC experimental programs. The proposed upgrade program at the FNAL Tevatron will provide experience relevant to the eventual SSC program, as the experimental community continues to push there the limits of understanding of standard model phenomenology in the high-rate, high-p_t environment. However, the risk of discovery is much higher for any of a number of possible standard model extensions at the SSC. The ability of the proposed detector designs to cover a wide range of *expected* new physics gives confidence that the designs will be capable of discovering *unexpected* new phenomena hoped for at the SSC.

ACKNOWLEDGEMENTS

The author thanks K. Einsweiler and M. Mangano for their valuable help in preparing these notes. Discussions with M. Chanowitz, G. Kane, and F. Paige helped clarify numerous points. Thanks also to my SDC collaborators, with whom I have had many useful discussions. Lastly, thanks to T. Ferbel, who did a fine job of organizing a very pleasant environment for the school.

REFERENCES

1. J. Peoples *et al.*, Fermilab HEPAP subpanel presentations, 1992.
2. SSC Site-Specific Conceptual Design Report, July 1990.
3. SDC Technical Design Report, April 1992.
4. GEM Expression of Interest, July 1991.
5. W. Burgett *et al.*, "Full Power Test of a String of Magnets Comprising a Half-Cell of the Superconducting Super Collider," SSCL-Preprint-162, October 1992.
6. SSC-250, Dec. 1989.
7. CDF, PRD 45, 3921 (1992).
8. CDF, PRL 68, 1463 (1992).
9. CDF, PRL 68, 1104 (1992).
10. A. Beretvas, *Proc. of the 1990 Summer Study on High Energy Physics*, Snowmass (1990).
11. SDC Technical Design Report, April 1992.

12. For a review, see Eichten, Hincliffe, Lane, and Quigg (EHLQ), *Rev. Mod. Phys.* **56**, 579 (1984).

13. For a useful review of the kinematics, see the Appendix in Owens, *Rev. Mod. Phys.* **59**, 465 (1987).

14. Barger, Snowmass 1986, p. 224.

15. Berger and Chanowitz, *Phys. Lett.* **B263**, 509 (1991).

TECHNICAL CHALLENGES OF THE LHC/SSC COLLIDERS

D.A. Edwards

DESY/FNAL

Basic Parameters and Their Implications

The LHC (Large Hadron Collider) proposed at CERN and the SSC (Superconducting Super Collider) under construction in the United States have much in common from an accelerator physics point of view. I have worked on the SSC in one way or another since 1983, and so I am much more familiar with the history and plans for execution of that project. As a result, more of my examples and illustrations will come from the SSC.

Which project will be completed first, even if both efforts survive to completion, is of course not clear. Both are subject to as yet unknown funding prospects and profiles. Although the LHC is still in its R&D phase, it has the considerable advantage that an enclosure - the LEP tunnel - already exists for it, the major injector components have operated for years, and CERN provides a well-established infrastructure.

The fundamental parameters of the LHC and the SSC are set down in Table 1. These facilities are intended to explore physics at the 1 TeV mass scale through the use of proton-proton collisions. In order to make use of the LEP tunnel, the LHC designers trade-off energy against luminosity to achieve a comparable discovery potential to that of the SSC.

The LHC parameters are taken from the "Pink Book"[1]; the SSC parameters have remained unchanged since the Reference Designs Study in 1984.

These are "many bunch" colliders, in order that the number of interactions per bunch-bunch hit remain within reason. At 10^{33}cm^{-2}sec^{-1} the reaction rate is 100 MHz. If half of these reactions result in a track in a general purpose detector, then the detector must cope with event rates at the 50 MHz level. For the SSC, the intent is that there be approximately one event per bunch-bunch hit. Then the bunch-bunch hit frequency of 50 MHz implies a bunch spacing of about 6 m. The actual figure in the SSC design is 5 m. Thus the bunch spacing is typical of that used in fixed target physics rather than the larger spacing found in past electron-positron or proton-antiproton rings.

The number of particles per bunch is consistent with present practice. The luminosity can be written in the form

$$\mathcal{L} = \frac{c}{S_B} \frac{1}{\beta^*} \frac{\gamma}{4\epsilon_N} N_B^2. \tag{1}$$

Here, c is the speed of light, S_B is the bunch spacing, β^* is the value of the ampli-

---

[1] *Design Study of the Large Hadron Collider* CERN 91-03, 2 May 1991

Techniques and Concepts of High Energy Physics VII
Edited by T. Ferbel, Plenum Press, New York, 1994

Table 1. Basic Parameters of the LHC/SSC

	LHC	SSC
Energy(TeV)	8×8	20×20
Luminosity (cm^{-2}sec^{-1})	2×10^{34}	10^{33}
Events per hit		≈ 1
IR's	up to 6	up to 6

tude function at the interaction point, γ is the Lorentz factor, ϵ_N is the normalized emittance with definition $\epsilon_N \equiv \gamma\pi\sigma\sigma'$, and N_B is the number of protons per bunch. The beams are assumed to be "round" Gaussians at the interaction point.

For the SSC, we take $\beta^* = 0.5$ m, $\epsilon_N = \pi$ mm mrad and find $N_B = 0.7 \times 10^{10}$ particles per bunch. The normalized emittance of the beam from the SPS is about a factor of 4 larger than that assumed for the yet to be constructed SSC injector; then with the same β^* and bunch spacing, the bunch population for the LHC will 10^{11}.

So it is readily apparent that, in either project, each beam contains a large number of particles. We know the circumference of the LEP tunnel, and with the assumption of a 5 m bunch spacing, each beam will contain about 5×10^{14} protons. Interestingly enough, this figure is approximately the same as the number of protons in one of the ISR rings at 70 amperes years ago. Therefore it is not surprising that relatively little consideration was given to the proton-antiproton approach. A dozen or so hours are required to collect 10^{12} antiprotons at Fermilab. A two orders-of-magnitude extrapolation is not attractive, when cost and reliability issues are taken into account.

The Tevatron was commissioned in 1983 and HERA in 1992. The main magnets of the LHC and SSC will represent a "third generation" of these components, and drawing upon the technological progress of the past twenty years, will continue to use NbTi as the superconducting material.

Now, for the first time, synchrotron radiation emerges as an important design consideration for proton rings. The energy radiated per turn by a proton circulating in a synchrotron is

$$W = 78 \left[\frac{E_{TeV}^4}{10}\right] \frac{1}{\rho_{km}} \frac{\text{keV}}{\text{turn}}. \tag{2}$$

where E_{TeV} is the proton energy in TeV and ρ_{km} is the radius of curvature in kilometers. Table 2 shows the loss per turn and power dissipated due to synchrotron radiation for the LHC and SSC as well as the Tevatron. The last column indicates the operating circumstance.

In the Tevatron, synchrotron radiation is just not a consideration. It is amusing that at 900 GeV particles lost from a bucket take over an hour to spiral inward and strike the walls of the vacuum chamber. For the SSC, a particle lost from a bucket would strike the vacuum chamber in less than a minute, but more to the point, the synchrotron radiation load must be removed be the refrigeration system. We will return to the consequences of this situation in the section on design and construction issues.

The Main Challenge - Integrated Luminosity

It is, of course, the integrated, in contrast to the instantaneous, luminosity that really matters. For the SSC, the specification states that the annual integrated luminosity

Table 2. Synchrotron radiation parameters from selected hadron colliders.

Device	U (keV/turn)	P (kW)	$P/2\pi\rho$ (W/m)	
Tevatron	0.006	1.6×10^{-5}	3.5×10^{-6}	$p\bar{p}$ Collider, 88-89
SSC	125	8.8	0.14	@ 10^{33}
LHC	2×12	2×10	2×0.63	@ $4 \cdot 10^{34}$

will be 10^4 pb^{-1}. That is to say, the nominal luminosity will in effect be delivered at each of the two high luminosity interaction regions for 10^7 seconds out of the year.

The luminosity is a function of time during a period of storage. To emphasize that, let us modify Equation 1 to read

$$\mathcal{L} = \frac{c}{S_B} \frac{1}{\beta^*} \frac{\gamma}{4\epsilon_N(t)} N_B(t)^2 [1 - H(t_k)]. \qquad (3)$$

The time dependence of the emittance and the number of particles per bunch is shown explicitly, and the fact that the store is terminated abruptly is indicated by the factor that includes the Heaviside function, H. The argument of the Heaviside function, t_k, is the duration of the kth store.

Ideally, the t_k would be chosen to optimize the integrated luminosity taking into account the initial luminosity of a store, the other time dependencies in Equation 3, and the turn around time from the end of one store to resumption of high luminosity collisions in the next. However, it is in the t_k that a major impact of reliability is to be found.

It is instructive to look at the reasons for store termination in the Tevatron. The list below is a summary of the reasons for store termination in the 1988-1989 run.

REASON	NUMBER of CASES
Intentional, low luminosity	99
Intentional, M&D or studies	21
Abort kicker pre-fire	31
Quench protection system	17
Tevatron power system	16
Lightning and power glitches	15
Detector access	13
Correction magnet system	12
Tevatron radiofrequency system	11
Human error	11
Cryogenics	9
Controls	8
Experimental area operations	7
Vacuum circuit breakers opening	5
Tevatron vacuum system	4
Low beta quadrupoles	4
Water system pumps	3
Other - 7 sources	9

In this 52-week collider run, there were 295 stores, 133 of which were ended intentionally. For the intentionally terminated stores, the mean store length was 20.5 hours. For the stores that ended in some failure mode, the mean store length was 9.5 hours.

In the first four months of the 1992-93 run, there have been 112 stores thus far, 75 of which were terminated intentionally. Three of the first four of the main sources of failure in the list above have played no role in the present run. A greater proportion of the stores are now terminated intentionally. So there is a clear improvement in reliability, but it is a slow and painful process. Each corrective action is a story in itself, and it would take us too far afield to go into detail. But, for instance, the isolation of the reason for the abort kicker pre-fires in the 1988-89 run was a process lasting some six months or so (as I recall) involving the pursuit of many false trails.

It was realized at the outset of the LHC/SSC activity that reliability would likely be the major challenge. Since then, the Tevatron has been commissioned and operated in both fixed target and collider modes; that experience has reinforced the conviction that reliability is a major issue. There is nothing new about this; each major step in size and complexity of accelerator facilities has required greater attention to quality assurance and control. When HERA nears its design luminosity, it will be very interesting to see the corresponding list to the one for the Tevatron above. The DESY electron–proton rings were designed as a colliding beams system from the outset, and that emphasis, coupled with the second generation superconducting synchrotron magnets, may show a step forward in performance.

At the risk of oversimplification, the three major ingredients to achieving the integrated luminosity goal are (1) execution of a design that reaches the initial luminosity specification, (2) production and assembly of systems with reliability sufficient to permit long term storage, and (3) management of the turn-around time from store termination to resumption of colliding-beam physics so that this time is a reasonably small fraction of store time. The first could be characterized as the traditional concern of accelerator physics. Can one get the right initial luminosity with the expectation that, provided the systems continue to work as specified, the time dependence of luminosity will be as calculated? Some comment on accelerator physics input will be found in the next section. The second, reliability, has been the main topic of this section.

Let me close this section with remarks on turn-around time. The word "management" is used in the preceding paragraph in connection with this subject. The typical fast turn-around for the Tevatron is two hours, even though the ramp times of the Tevatron and its injectors are measured in minutes. There are a number of check-out procedures that are regarded as essential, and a number that are discretionary. In both categories, the time devoted to each is also a matter of judgement. After 7 years of collider physics at the Tevatron, the value of these time allocations is rather well understood. In earlier years, particularly when a store was terminated with a system failure, 4 hours or so was typical.

What are the implications for the next generation of hadron colliders for turn-around time? The ramp times are no longer negligible. For the SSC, the ramp-down, inject, and ramp-up times alone consume 1.5 hours! As noted above, these times are negligible for the Tevatron. So the pressure on check-out times should be stronger, and that flows back to systems design. It looks like that under the best conditions it will be difficult to turn around in under four hours.

Input from Accelerator Physics

Much of the time in the lectures at St. Croix was devoted to basic accelerator physics. That material was taken directly from a textbook of which I am coauthor[2] and can be found in many other references as well. There is no point in reproducing those fundamentals in this paper. Here, I limit myself to two topics discussed in the lectures that do not appear in the text referenced above.

Most of the periphery of the LHC and SSC will be occupied by many near-identical alternating gradient cells; a series of quadrupoles separated by bending magnets and other accelerator components. To get as high an energy as possible for a given bend field, one would like the quadrupole to occupy a small fraction of the quadrupole to quadrupole spacing. Treated as a thin lens, for focal length, f, of a quadrupole is

$$\frac{1}{f} = \frac{e}{p} G\ell \tag{4}$$

where G is the quadrupole gradient and ℓ its length. One of the standard results of accelerator physics mentioned above is that the quadrupole focal length is approximately equal to the quadrupole spacing, L. If we say that we will allow the quadrupole to occupy only 5% of quadrupole to quadrupole distance, then

$$L^2 = 20\frac{p}{e}\frac{1}{G} \tag{5}$$

Superconducting quadrupoles are capable of gradients in excess of 200 Tesla per meter – the design values are 210 T/m and 250 T/m for the SSC and LHC respectively. For the SSC, the estimate above yields $L = 80$m, to be compared with the 90 m spacing in the current design. For the LHC, the estimate yields 46 m, while the design value is 48 m. It's interesting that this quick way of making a first cut at the number of alternating gradient cells has survived the change in technology from conventional to superconducting magnets.

Given the higher design luminosity of the LHC, the question naturally was asked "Why not raise the SSC luminosity to match?" The answer is that luminosity costs money, and some comments will be made in the next section is association with cryogenics. From a purely accelerator physics point of view, even larger numbers are conceivable. The maximum achievable luminosity in a hadron collider was the subject of a study at a Snowmass workshop.[3] If the only limiting effect is the beam-beam interaction the luminosity expression, since it contains the ratio of particles per bunch to normalized emittance, can be recast in terms of the head-on beam-beam tune shift parameter. The total tune spread is the sum of the head-on and long-range contributions. If it is assumed that it is the total tune spread that is the limiting parameter, then the maximum luminosity at a single interaction point is given by

$$\mathcal{L} \leq 5.3 \times 10^{34} cm^{-2} sec^{-1} \; \frac{\gamma^{7/2}\alpha^4}{\epsilon_N L^*}. \tag{6}$$

In this expression, $L*$ is the distance from the interaction point to the first quadrupole

---

[2]Edwards, D. A., and M. J. Syphers, *An Introduction to the Physics of High Energy Accelerators*, Wiley, New York (1993)

[3]*Proceedings of the 1990 Workshop on High Energy Physics*, World Scientific, Singapore, 1992

and the maximum tune spread is taken to be 0.024. This latter figure is inferred from experience at the Spp̄S and the Tevatron.

This formula sets a challenging luminosity scale indeed. For typical SSC parameters, the maximum achievable luminosity under these ground rules is $\sim 7 \times 10^{35}$. Why do both the SSC and LHC fall short of this ultimate goal? The technological constraints of the next section play a role, as does the budget.

There are standard accelerator physics concerns that demand attention. Preservation of the normalized emittance in the injection process is once such; this subject has been discussed frequently elsewhere. Also, the mechanisms that lead to emittance dilution during storage have deservedly been placed at a high priority for exammination. Again, these concerns are not unique to the LHC and SSC, and are covered elsewhere.

Design and Construction Issues

Whatever long range promise high temperature superconductivity may hold, the magnets for the next generation of hadron colliders will be based on NbTi or Nb_3Sn materials. Magnetic fields in the 10 Tesla range imply stresses due to magnetic pressure in the neighborhood of 400 atmospheres; significantly higher fields are unlikely due to strength limits of materials.

Coil cooling will be provided by liquid helium at a temperature at or near 4 K. If the refrigeration system operated at the ideal Carnot efficiency, 75 W input at room temperature would be required to remove 1 W at 4 K. In reality, the best one can do is worse by a factor of about four.

To set a scale for refrigeration load, suppose one could construct a cryostat with 0.1 W/m static heat loss. Then, for instance, two synchrotrons each of circumference 87 km would represent a total refrigeration load of 17.4 kW, or 5 MW at room temperature. Already we see that 0.1 W/m is a significant amount. For comparison, the static heat load of the Tevatron is somewhat in excess of 1 W/m. So, as we turn to look at processes that add to the static heat load, we will look cautiously at levels much in excess of 0.1 W/m.

Energy Deposition from Beam Loss at the IP

Most any process which leads to beam loss will deposit energy in the superconducting magnets and add to the static heat load. Rather than attempt to catalogue all beam loss processes, we comment only on the one process which is inescapable in a collider — the energy streaming out in reaction products from the interaction point. For example, for SSC parameters the power generated at the interaction point is

$$
\begin{aligned}
P &= \mathcal{L} \cdot \sigma_{int} \cdot E \\
&= \left(\frac{10^{33}}{cm^2 sec} \right) \left(0.1 \times 10^{-24} cm^2 \right) \left(20 \times 10^{12} eV \right) \\
&= 320 \ Watts
\end{aligned}
\tag{7}
$$

in each direction. Despite all protective measures, much of this energy will be deposited in the focussing elements near the interaction point. These elements are apt to have very high gradients and are among the most technologically challenging pieces of hardware in the accelerator.

The energy deposition is calculated with the aid of a Monte Carlo program which contains an accurate model of the superconducting magnet geometry and fields as

Table 3. Maximum energy deposition dose rate D' and annual dose D in the super-conducting coils of the SSC low-β IR beam elements. Interaction rate is 10^8/sec at $\mathcal{L} = 10^{33}$ cm^{-2} sec^{-1}. Here, the operational year is taken to be 10^7 sec. From Baishev, et al.

Name	Distance from IP (m)	D' (mW/gm)	D (MGy/year)
IP	0		
QL1	35	0.32	3.20
QL2a	47	0.19	1.92
QL2b	59	0.22	2.22
QL3	73	0.10	0.96
BV1	85	0.01	0.064
BV1	91	0.01	0.11
BV1	97	0.02	0.21

well as the shower process. An example of the results of such calculations is shown in Table 3 for the nominal SSC luminosity and the optics associated with a high luminosity interaction region.[4] The energy deposition rates are listed in the third column. In the Tevatron, measurements have shown that an energy deposition rate of 8 mW/gram could be tolerated. Though it is not known what dose rate can be tolerated in SSC magnets, it is likely that the situation is reasonably safe at the design luminosity. Upgrades to higher luminosity could pose more severe problems. The annual radiation dose is shown in the next column. Here, the situation is perhaps less comfortable. Table 4 indicates that a careful selection of materials will be necessary to achieve a ten year lifetime of the cold mass of the interaction region quadrupoles. At a luminosity of 10^{34}, the lifetime would shrink to one year, while at the challenging number set forth by the dynamicists the lifetime would shrivel to less than a week.

Synchrotron Radiation

A selection of synchrotron radiation parameters were listed in Table 2. Note that the synchrotron radiation load for the SSC at its design luminosity is the same as our static heat load goal of 0.1 W/m and so is significant. An increase in luminosity to 10^{34} by increasing the beam current by a factor of three would require a doubling of the cryogenic plant capacity. Further increase by this route is not attractive, but the SSC would still lag the LHC in luminosity!

An attractive proposal is to remove the synchrotron radiation load at a higher temperature. Because of the increased Carnot efficiency, the refrigeration power would be lowered. The magnet design is, of course, somewhat more complicated due to the introduction of the higher temperature intercept. The superconducting coils are maintained at temperature T_1 outside the bore tube of the magnet. The liner is held at a temperature $T_2 > T_1$ by coolant flowing through the passage attached to the bottom of the liner. While the coils are at $T_1 = 2\text{-}4$ K, the temperature T_2 could be in the range of 20-80 K implying an improvement in Carnot efficiency of an order of magnitude or so. Slots or holes provide an escape route for hydrogen which is desorbed from the inner surface of the liner so that it can be trapped at the lower

---

[4]Baishev, I. S., A. I. Drozhdin, and N. V. Mokhov, *Beam Loss and Radiation Effects in the SSC Lattice Elements*, SSC Report SSCL-306, 1990

Table 4. Radiation resistance of selected materials. From Baishev, et al.

Material	Tolerable Dose (MGy)
Kapton, polyimide Kapton film Carbon-fiber reinforced tube Carbon-fiber-filled epoxy rods	50
G11 CR tube	20
PK102 (epoxy) Crest 7450 epoxy Fiberglass (epoxy impregnated)	10
Fiberglass rein. polyester resin Aluminum mylar Superinsulation	5 2 2
Electrical insulation	0.1-10
Tefzel adhesive Cerex spunbonded polyester Teflon	0.5 0.06 0.01

temperature of the bore tube. The LHC has incorporated a bore tube liner into its magnet design and a similar step is under consideration for the SSC.[5]

Vacuum Stability

Even at 10^{33} luminosity there is a potentially worrisome vacuum situation in the SSC. To achieve the desired 150 hr partial lifetime of the beam due to interactions with the residual gas implies a gas density of $6 \times 10^8 H_2$ molecules/cm^3 or less. The saturated vapor pressure of hydrogen gas at 4.3 K corresponds to a gas density of 6×10^{11} molecules/cm^3. Therefore, a significant fraction of a monolayer of hydrogen must not be allowed to condense on the inner surface of the beam tube.

But we know that hydrogen is released even from clean baked surfaces that are exposed to synchrotron radiation. Experiments indicate that the time to build up the first monolayer at a luminosity of 10^{33} will be approximately 50 hours.[6] Then, a warm-up to 20 K would be necessary to pump out the hydrogen following which the system would be cooled down and operation resumed. After a number of such cycles this situation would eventually improve.

This is a fine picture if the entire ring cleaned up simultaneously. Unfortunately, the ring is made up of many isolated pieces. Each one is a ticking clock and clean-ups would need to be synchronized.

Further, at the saturated vapor pressure at 4.3 K, the energy deposition due to the beam-gas interaction is about 8 W/m — unsustainable by the cryogenic system. So, a conservative approach would be to have a liner as sketched above with T_1 at about 3 K consistent with a saturated vapor pressure appropriate to a beam lifetime of 150 hours. Obviously, further study is needed on these matters.

---

[5] Edwards, H. T., *Study on Beam Tube Vacuum with Consideration of Synchrotron Light, Potential Liner Intercept, and Collider Quad/Spool Coil Diameter*, SSC Report SSCL-N-771, 1991.

[6] SSC Site Specific Design Report, June 1990

Concluding Remarks

A few years ago, some of us thought that reliability was the only major issue to be addressed in the next generation of hadron colliders. I still believe that it is the most significant single problem.

The magnet technology problem has been solved. The string test at the SSC last Summer demonstrated that. But the question that remains open is whether or not these magnets must be cold-tested before installation. That is a step beyond the Tevatron or HERA, and has cost implications.

There are other considerations of an engineering nature that we have not mentioned. The stored energy in the beam of these devices can be considerable; for the SSC, the 400 MJ in the beam of one ring represents a potential threat to the hardware. The nominal parameters of these collider designs involve many closely spaced bunches; the suppresion of multi-bunch instabilities will surely be a problem. Only passing mention has been made of the interaction region optics; the numbers advanced for β^* at a high luminosity interaction point may be most difficult to realize.

But the upgrade potential cannot be ignored. Some upgrade capability has to be addressed at the outset, within cost and schedule bounds. As a result, these synchrotrons are becoming a good deal more interesting as design challenges as they move out of the paper stages.

And the motion out of the paper stage is rapid for the SSC. As I write this in late 1992, over half of the tunnel contracts for the SSC arcs have been let, and by February 1, 3 tunnel boreing devices will be underground. The Linac enclosure will be finished by the end of June of this year. The SSC is churning ahead. It seems to me that the LHC and the SSC have traded leads over the years – it will be very interesting to see which project ends up in front.

From left to right (as would uncover pulling paper across the photo) —
M. Spiro, G. Jarlskog, C. Talamonti, M. Ciara, C. Jarlskog, E. Hyatt, P. Schenk, S. Maselli, P. Maas, K. Gounder,
M. Stavrianakou, C. DaVia, L. Demortier, J. Fast, M. Shaw, J. Steuerer, J. Schwindling, M. Spurio, R. Poggiani,
M. Fabbri, M. Jimack, C. Avila, M. Cobal, P. Luffman, D. Sperka, T. LeCompte, A. Montanari, A. Erdogan, M. Punturo,
I. Albuquerque, R. Ferreira, S. Rudaz, D. Borden, L. Keeble, L. Taylor, R. Forty, I. Lopes, R. Graessler, R. Roser,
J.deCarvalho, G. Bruckner, T. Ferbel, L. del Pozo, M. Vagins, K. Tzamariudaki, C. Zhang, D. Skow, K. Strahl, A. Gougas,
F. Zarnecki, A. Perrotta, S. Takach, P. Gu, S. Gruenendahl, P. Privitera, A. Dyring, F. Dydak, D. Saxon, A. Witzmann,
G. Maelhum, M. Danilov, P. Tipton, J. Boudreau, M. Paterno, G. Raselli, M. Frank, E. Machado. Missing: P. Bjorkholm,
D. Edwards, and J. Siegrist.

INDEX